Steel Design
for
Engineers and Architects

Rene Amon
Bruce Knobloch
Atanu Mazumder

VNR VAN NOSTRAND REINHOLD COMPANY

Copyright © 1982 by Van Nostrand Reinhold Company Inc.

Library of Congress Catalog Card Number: 81-16139
ISBN: 0-442-20297-0

Manufactured in the United States of America

Published by Van Nostrand Reinhold Company Inc.
135 West 50th Street
New York, New York 10020

Van Nostrand Reinhold Company Limited
Molly Millars Lane
Wokingham, Berkshire RG11 2PY, England

Van Nostrand Reinhold
480 Latrobe Street
Melbourne, Victoria 3000, Australia

Macmillan of Canada
Division of Gage Publishing Limited
164 Commander Boulevard
Agincourt, Ontario M1S 3C7, Canada

15 14 13 12 11 10 9 8 7 6 5 4 3

Library of Congress Cataloging in Publication Data

Amon, Rene.
  Steel design for engineers and architects.

  Includes index.
  1. Building, Iron and steel. 2. Steel,
Structural. I. Knobloch, Bruce. II. Mazumder,
Atanu. III. Title.
TA684.A5   624.1'821   81-16139
ISBN 0-442-20297-0   AACR2

TO
Our Families

Preface

This is an introductory book on the design of steel structures. Its main objective is to set forth steel design procedures in a simple and straightforward manner. We chose a format such that very little text is necessary to explain the various points and criteria used in steel design, and we have limited theory to that necessary for understanding and applying code provisions. This book has a twofold aim: it is directed to the practicing steel designer, whether architect or engineer, and to the college student studying steel design.

The practicing structural engineer or architect who designs steel structures will find this book valuable for its format of centralized design requirements. It is also useful for the veteran engineer who desires to easily note all the changes from the seventh edition of the *AISC Steel Design Manual* and the logic behind the revisions. Yet, the usefulness as a textbook is proven by field-testing. This was done by using the appropriate chapters as texts in the following courses offered by the College of Architecture of the University of Illinois at Chicago Circle: Steel Design; Additional Topics in Structures; and Intermediate Structural Design (first-year graduate). The text was also field-tested with professionals by using the entire book, less Chapter 8, as text for the steel part of the Review for the State of Illinois Structural Engineers' License Examination, offered by the University of Illinois at Chicago Circle. The text, containing pertinent discussion of the numerous examples, was effectively combined with supplemental theory in classroom lectures to convey steel design requirements. Unsolved problems follow each chapter to strengthen the skill of the student.

The American Institute of Steel Construction (AISC) is the authority that codifies steel construction as applied to buildings. Its specifications are used by most governing bodies, and, therefore, the material presented here has been written with the AISC specifications in mind. We assumed—indeed, it is necessary—that the user of this book have an eighth edition of the *AISC Manual of Steel Construction*. Our book complements, but does not replace, the ideas presented in the *Manual*. Various tables from the *Manual* have been reprinted here, with permission, for the convenience of the user.

We assume that the user is familiar with methods of structural analysis. It is the responsibility of the designer to assure that proper loadings are used and proper details employed to implement the assumed behavior of the structure.

In this book, the chapters follow a sequence best suited for a person familiar with steel design. For classroom use, the book has been arranged for a two-quarter curriculum as follows:

First quarter—Chapters 1 through 6. Section 3.2 and Examples 3.7 through 3.11 should be omitted and left for discussion in the second quarter. Furthermore, examples in Section 3.1 could be merely introduced initially and discussed fully after Chapters 5 and 6 have been covered. This amount of material, if properly covered, will enable the student to understand the fundamentals of steel design and to comprehensively design simple steel structures.

Second quarter—Chapters 7 through 10 and Section 3.2. This material is complementary to steel design. An instructor can follow any desired order because these chapters are totally independent of each other.

This book can also be used for a one-semester course in steel design by covering Chapters 1 through 6, as discussed above for First quarter, and Chapters 9 and 10. An instructor may also want to include some material from Chapter 7.

In conclusion, we would like to thank all our friends, families, and colleagues who provided help and understanding during the writing of this manuscript. Without their help and cooperation, we might never have made this book a reality.

RENE AMON
BRUCE KNOBLOCH
ATANU MAZUMDER

Chicago, Illinois

Symbols and Abbreviations

A — Cross-sectional area (in.^{2})

Gross area of an axially loaded compression member (in.^{2})

A_b — Nominal body area of a fastener (in.^{2})

A_c — Actual area of effective concrete flange in composite design (in.^{2})

A_e — Effective net area of an axially loaded tension member (in.^{2})

A_f — Area of compression flange (in.^{2})

A_n — Net area of an axially loaded tension member (in.^{2})

A_s — Area of steel beam in composite design (in.^{2})

A_s' — Area of compressive reinforcing steel (in.^{2})

A_{st} — Cross-sectional area of stiffener or pair of stiffeners (in.^{2})

A_w — Area of girder web (in.^{2})

A_1 — Area of steel bearing on a concrete support (in.^{2})

A_2 — Total cross-sectional area of a concrete support (in.^{2})

AISCC — American Institute of Steel Construction specifications commentary

AISCM — American Institute of Steel Construction *Manual*

AISCS — American Institute of Steel Construction specifications

B_x, B_y — Area of column divided by its appropriate section modulus

B_c — Load per bolt, including prying action (kips)

C — Coefficient for determining permissible loads in kips for eccentrically loaded connections

C_b — Bending coefficient dependent upon moment gradient

$$= 1.75 + 1.05 \left(\frac{M_1}{M_2}\right) + 0.3 \left(\frac{M_1}{M_2}\right)^2$$

C_c — Column slenderness ratio separating elastic and inelastic buckling

C_m — Coefficient applied to bending term in interaction formula for prismatic members and dependent upon column curvature caused by applied moments

C_t Reduction coefficient in computing effective net area of an axially loaded tension member

C_v Ratio of "critical" web stress, according to the linear buckling theory, to the shear yield stress of web material

C_1 Coefficient for web tear-out (block shear)
Increment used in computing minimum spacing of oversized and slotted holes

C_2 Coefficient for web tear-out (block shear)
Increment used in computing minimum edge distance for oversized and slotted holes

D Factor depending upon type of transverse stiffeners
Outside diameter of tubular member (in.)
Number of $\frac{1}{16}$-in. in weld size

D_c Uniform load deflection constant (in./ft^2)

E Modulus of elasticity of steel (29,000 ksi)

E_c Modulus of elasticity of concrete (ksi)

F_a Axial compressive stress permitted in a prismatic member in the absence of bending moment (ksi)

F_{as} Axial compressive stress permitted in the absence of bending moment, for bracing and other secondary members (ksi)

F_b Bending stress permitted in a prismatic member in the absence of axial force (ksi)

F_b' Allowable bending stress in compression flange of plate girders as reduced for hybrid girders or because of large web depth-to-thickness ratio (ksi)

F_e' Euler stress for a prismatic member divided by factor of safety (ksi)

F_p Allowable bearing stress (ksi)

F_t Allowable axial tensile stress (ksi)

F_u Specified minimum tensile strength of the type of steel or fastener being used (ksi)

F_v Allowable shear stress (ksi)

F_y Specified minimum yield stress of the type of steel being used (ksi). "Yield stress" denotes either the specified minimum yield point (for those steels that have a yield point) or specified minimum yield strength (for those steels that do not have a yield point)

F_y' The theoretical maximum yield stress (ksi) based on the width-thickness ratio of one-half the unstiffened compression flange, beyond which a particular shape is not "compact." See AISCS Section 1.5.1.4.1.2.
$$= \left[\frac{65}{b_f/2t_f} \right]^2$$

F_y''' The theoretical maximum yield stress (ksi) based on the depth-thickness ratio of the web below which a particular shape may be considered "compact" for any condition of combined bending and axial stresses. See AISCS Section 1.5.1.4.1.4.

$$= \left[\frac{257}{d/t_w}\right]^2$$

F_{yc}	Column yield stress (ksi)
F_{yst}	Stiffener yield stress (ksi)
G	Shear modulus of elasticity of steel (11,200 ksi)
H_s	Length of a stud shear connector after welding (in.)
I	Moment of inertia of a section (in.^{4})
I_d	Moment of inertia of steel deck supported on secondary members (in.^{4})
I_{eff}	Effective moment of inertia of composite sections for deflection computations (in.^{4})
I_s	Moment of inertia of steel beam in composite construction (in.^{4})
I_{tr}	Moment of inertia of transformed composite section (in.^{4})
I_x	Moment of inertia of a section about the X - X axis (in.^{4})
I_y	Moment of inertia of a section about the Y - Y axis (in.^{4})
J	Torsional constant of a cross section (in.^{4})
K	Effective length factor for a prismatic member
L	Span length (ft)
	Length of connection angles (in.)
L_c	Maximum unbraced length of the compression flange at which the allowable bending stress may be taken at $0.66\,F_y$ or as determined by AISCS Formula (1.5-5a) or Formula (1.5-5b), when applicable (ft)
	Unsupported length of a column section (ft)
L_u	Maximum unbraced length of the compression flange at which the allowable bending stress may be taken at $0.6\,F_y$ (ft)
L_v	Span for maximum allowable web shear of uniformly loaded beam (ft)
M	Moment (kip-ft)
	Factored bending moment (kip-ft)
M_1	Smaller moment at end of unbraced length of beam-column
M_2	Larger moment at end of unbraced length of beam-column
M_D	Moment produced by dead load
M_L	Moment produced by live load
M_p	Plastic moment (kip-ft)
M_R	Beam resisting moment (kip-ft)
N	Length of base plate (in.)
	Length of bearing of applied load (in.)
N_e	Length at end bearing to develop maximum web shear (in.)
N_r	Number of stud shear connectors on a beam in one rib of a metal deck, not to exceed 3 in calculations
N_1	Number of shear connectors required between point of maximum moment and point of zero moment

N_2 Number of shear connectors required between concentrated load and point of zero moment

P Applied load (kips)
Force transmitted by a fastener (kips)
Factored axial load (kips)

P_{bf} Factored beam flange or connection plate force in a restrained connection (kips)

P_{cr} Maximum strength of an axially loaded compression member or beam (kips)

P_e Euler buckling load (kips)

P_{fb} Force, from a beam flange or moment connection plate, that a column will resist without stiffeners, as determined using Formula (1.15-3) (kips)

P_R Beam reaction divided by the number of bolts in high-strength bolted connection (kips)

P_{wb} Force, from a beam flange or moment connection plate, that a column will resist without stiffeners, as determined using Formula (1.15-2) (kips)

P_{wi} Force, in addition to P_{wo}, that a column will resist without stiffeners, from a beam flange or moment connection plate of 1-in. thickness, as derived from Formula (1.15.1) (kips)

P_{wo} Force, from a beam flange or moment connection plate of zero thickness, that a column will resist without stiffeners, as derived from Formula (1.15.1) (kips)

P_y Plastic axial load, equal to profile area times specified minimum yield stress (kips)

Q Prying force per fastener (kips)

R Maximum end reaction for $3\frac{1}{2}$ in. of bearing (kips)
Reaction or concentrated load applied to beam or girder (kips)
Radius (in.)

R_{BS} Resistance to web tear-out (block shear) (kips)

R_i Increase in reaction (R) in kips for each additional inch of bearing

R_v Shear capacity of the net section of connection angles

S Elastic section modulus (in.^{3})

S' Additional section modulus corresponding to $\frac{1}{16}$-in. increase in web thickness for welded plate girders (in.^{3})

S_{eff} Effective section modulus corresponding to partial composite action (in.^{3})

S_s Section modulus of steel beam used in composite design, referred to the bottom flange (in.^{3})

S_t Section modulus of transformed composite cross section, referred to the top of concrete (in.^{3})

S_{tr} Section modulus of transformed composite cross section, referred to the bottom flange; based upon maximum permitted effective width of concrete flange (in.^{3})

S_x Elastic section modulus about the $X - X$ axis (in.^{3})

T Horizontal force in flanges of a beam to form a couple equal to beam end moment (kips)

V Maximum permissible web shear (kips)
Statical shear on beam (kips)

V_h Total horizontal shear to be resisted by connectors under full composite action (kips)

V_h' Total horizontal shear provided by the connectors in providing partial composite action (kips)

W_c Uniform load constant (kip-ft)
Weld capacity (k/in.)

Z Plastic section modulus (in.^{3})

Z_x Plastic section modulus with respect to the major $(X - X)$ axis (in.^{3})

Z_y Plastic section modulus with respect to the minor $(Y - Y)$ axis (in.^{3})

a Distance from bolt line to application of prying force Q (in.)
Clear distance between transverse stiffeners (in.)
Dimension parallel to the direction of stress (in.)

a' Distance required at ends of welded partial length cover plate to develop stress (in.)

b Actual width of stiffened and unstiffened compression elements (in.)
Dimension normal to the direction of stress (in.)
Fastener spacing vertically (in.)
Distance from the bolt center line to the face of tee stem or angle leg in determining prying action (in.)

b_e Effective width of stiffened compression element (in.)

b_f Flange width of rolled beam or plate girder (in.)

d Depth of column, beam or girder (in.)
Nominal diameter of a fastener (in.)

d_c Column web depth clear of fillets (in.)

d_h Diameter of hole (in.)

e_o Distance from outside face of web to the shear center of a channel section (in.)

f Axial compression stress on member based on effective area (ksi)

f_a Computed axial stress (ksi)

f_b Computed bending stress (ksi)

f_c Concrete working stress (ksi)

f_c' Specified compression strength of concrete at 28 days (ksi)

f_p Actual bearing pressure on support (ksi)

f_t Computed tensile stress (ksi)

f_v Computed shear stress (ksi)

f_{vs} Shear between girder web and transverse stiffeners (kips per linear inch of single stiffener or pair of stiffeners)

g Transverse spacing locating fastener gage lines (in.)

h Clear distance between flanges of a beam or girder at the section under investigation (in.)

k Coefficient relating linear buckling strength of a plate to its dimensions and condition of edge support

Distance from outer face of flange to web toe of fillet of rolled shape or equivalent distance on welded section (in.)

l For beams, distance between cross sections braced against twist or lateral displacement of the compression flange (in.)

For columns, actual unbraced length of member (in.)

Unsupported length of a lacing bar (in.)

l_b Actual unbraced length in plane of bending (in.)

l_v Distance from center line of fastener hole to free edge of part in the direction of the force (in.)

l_h Distance from center line of fastener hole to end of beam web (in.)

l_w Length of weld (in.)

m Factor for converting bending to an approximate equivalent axial load in columns subjected to combined loading conditions

Cantilever dimensions of base plate (in.)

n Number of fasteners in one vertical row

Cantilever dimension of base plate (in.)

Modular ratio (E/E_c)

n' An equivalent cantilever dimension of a base plate (in.)

q Allowable horizontal shear to be resisted by a shear connector (kips)

r Governing radius of gyration (in.)

r_b Radius of gyration about axis of concurrent bending (in.)

r_T Radius of gyration of a section constituting the compression flange plus $\frac{1}{3}$ of the compression web area, taken about an axis in the plane of the web (in.)

r_v Allowable shear or bearing value for one fastener (kips)

r_x Radius of gyration with respect to the X - X axis (in.)

r_y Radius of gyration with respect to the Y - Y axis (in.)

s Longitudinal center-to-center spacing (pitch) of any two consecutive holes (in.)

t Girder, beam, or column web thickness (in.)

Thickness of a connected part (in.)

Wall thickness of a tubular member (in.)

Angle thickness (in.)

t_b Thickness of beam flange or moment connection plate at rigid beam-to-column connection (in.)

t_f	Flange thickness (in.)
t_s	Stiffener plate thickness (in.)
t_w	Web thickness (in.)
w	Length of channel shear connectors (in.)
x	Subscript relating symbol to strong axis bending
y	Subscript relating symbol to weak axis bending
α	Moment ratio used in prying action formula
β	Ratio S_{tr}/S_s or S_{eff}/S_s
Δ	Beam deflection (in.)
	Displacement of the neutral axis of a loaded member from its position when the member is not loaded (in.)
δ	Ratio of net area (at bolt line) to the gross area (at the face of the stem on angle leg)
ν	Poisson's ratio, may be taken as 0.3 for steel
kip	1000 pounds
ksi	Expression of stress in kips per square inch

Contents

Introduction

0.1 STEEL AS A BUILDING MATERIAL

From the dawn of history, man has been searching for the perfect building material to construct his dwellings, bridge across rivers and make his tools. Not until the discovery of iron and its manufacture into steel did he find the needed material to largely fulfill his dreams. All other building materials discovered and used in construction until then proved to be either too weak (wood), too bulky (stone), too temporary (mud and twigs), or too deficient in resisting tension and fracture under bending (stone and concrete). Other than its somewhat unusual ability to resist compression and tension without being overly bulky, steel has many other properties that have made it one of the most common building materials today. Brief descriptions of some of these properties follow. Some of the other concepts not discussed here have been introduced in treatments of the strength of materials and similar courses and will not be repeated.

High strength. Today steel comes in various strengths, designated by its yield stress F_y or by its ultimate tensile stress F_u. Even steel of the lowest strengths can claim higher ratios of strength to unit weight or volume than any of the other common building materials in current use. This allows steel structures to be designed for smaller dead loads and larger spans, leaving more room (and volume) for use due to slenderer members.

Ease of erection. Steel construction allows virtually all members of a structure to be prefabricated in the mills, leaving only erection to be completed in the field. As most structural components are standard rolled shapes, readily available from suppliers, the time required to produce all the members for a structure can often be shortened. Because steel members are generally standard shapes having known properties and are available throughout the country,

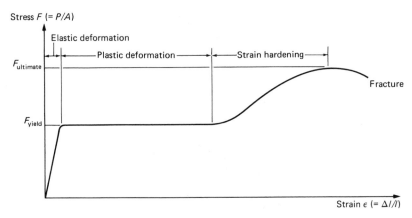

Fig. 0.1. Stress-strain diagram of mild structural steel.

analyzing, remodeling, and adding to an existing structure are very easily accomplished.

Uniformity. The properties of steel as a material and as structural shapes are so rigidly controlled that engineers can expect the members to behave reasonably as expected, thus reducing overdesign due to uncertainties.

Ductility. The property of steel that enables it to withstand extensive deformations under high tensile stresses without failure, called *ductility*, gives steel structures the ability to resist sudden collapse. This property is extremely valuable when one considers the safety of the occupants of a building subject to, for instance, a sudden shock, such as an earthquake.

Some of the other advantages of structural steel are: (1) speed of erection, (2) weldability, (3) possible reuse of structural components, (4) scrap value of unreusable components, and (5) permanence of structure with proper maintenance. Steel also has several disadvantages, among which are: (1) the need to fireproof structural components to meet local fire codes, (2) the maintenance costs to protect the steel from excessive corrosion, and (3) its susceptibility to buckling of slender members capable of carrying its axial loads, but unable to prevent lateral displacements. Engineers should note that under high temperatures, such as those reached during building fires, the strength of structural steel is severely reduced, and only fireproofing or similar protection can prevent the structural members from sudden collapse in case of a fire. Heavy timber structural members usually resist collapse much longer than unprotected structural steel. The most common methods of protecting steel members against fire are a sprayed-on

Fig. 0.2. Fabrication of special trusses at a job site to be used for the wall and roof structural system for the atrium lobby of an office building. Note that the trusses, made from tubular sections, are partially prefabricated in the shop to transportable lengths and assembled at the site to form each member.

coating (about 2 in.) of a cementitious mixture, full concrete embedment, or encasement by fire-resistant materials, such as gypsum board.

0.2 LOADS AND SAFETY FACTORS

Components of a structure must be designed to resist applied loads without excessive deformations or stresses. These loads are due to the dead weight of

Fig. 0.3. A typical, steel-frame high-rise building under construction. Note how, in steel construction, the entire structure can be framed and erected before other trades commence work, thus reducing conflict and interference among various trades.

the structure and its components, such as walls and floors; snow; wind; earthquakes; and people and objects supported by the structure. These loads can be applied to a member along its longitudinal axis (axially), causing it to elongate or shorten depending on the load; perpendicular to its axis (transversely), causing it to flex in a bending mode; by a moment about its axis (torsionally), causing the member to twist about that axis; or by a combination of any two or all three. It is very important for the engineer to recognize all the loads acting on each and every element of a structure and on the entire structure as a whole and to determine which mode they are applied in and the combinations of loads that critically affect the individual components and the entire structure. The study of these loads and their effects is primarily the domain of structural analysis.

Loads are generally categorized into two types, dead and live. Loads that are permanent, steady, and due to gravity forces on the structural elements (dead weight) are called *dead loads*. Estimating the magnitudes of dead loads is usually quite accurate, and Table 0.1 can be used for that purpose. *Live loads*, however, are not necessarily permanent or steady and are due to forces acting on a struc-

Fig. 0.4. Wide-flange beam and girder floor system with sprayed-on cementitious fireproofing. Note that openings in the webs are reinforced because they are large and they occur at locations of large shear.

ture's superimposed elements, such as people and furniture, or due to wind, snow, earthquakes, etc. Unlike dead loads, live loads cannot be accurately predicted, but can only be estimated. To relieve the engineer of the burden of estimating live loads, building codes often dictate the magnitude of the loads, based on structure type and occupancy. National research and standardization organizations, such as the American National Standards Institute (ANSI), and other national and city building codes dictate the magnitude of wind, snow, and earthquake loads, based on extensive research and data. The engineer must determine which codes apply to his structure and find the applicable loads in them. Some of the dead and live loads for buildings recommended by ANSI, for preliminary design, are shown in Table 0.1.

Nevertheless, a structure cannot be designed just to resist estimated dead loads and estimated or code-specified live loads. If that were allowed, the slightest variation of loads toward the high side would cause the structure or member to

Table 0.1. Some Common Loads Recommended by ANSI.

Material	Weight	Unit
Dead Loads (Weights) of Some Common Building Materials[a]		
Concrete	150	pcf
Masonry (brick, concrete block)	120–145	pcf
Earth (soil, clay, sand)	100–120	pcf
Steel	490	pcf
Stone (limestone, marble)	165	pcf
4″ Brick wall	40	psf
8″ Brick wall	80	psf
12″ Brick wall	120	psf
4″ Hollow concrete slab	30	psf
6″ Hollow concrete slab	43	psf
8″ Hollow concrete slab	55	psf
12″ Hollow concrete slab	80	psf
4″ Lightweight concrete blocks	25	psf
6″ Lightweight concrete blocks	30	psf
8″ Lightweight concrete blocks	33	psf
12″ Lightweight concrete blocks	45	psf
Live Loads (psf)[b]		
Rooms (residences, hotels, etc.)	40	—
Offices	50	—
Corridors	80–100	—
Assembly rooms, lobbies, theaters, etc.	100	—
Wind (depends on location and terrain)	15–50	—
Snow (depends on location and roof type)	10–80	

[a] Also see pp. 6-8, 6-9, and 6-10 of the AISCM for other weights.
[b] Also see AISC Specification 1.3 for specific rules about loading and Specification 1.5.6 for allowable stress increases when design takes wind into account along with dead and live loads, provided the required section is satisfactory for only dead and live loads without the increase.

deform unacceptably (considered failure). To avoid this, the stresses in the members are knowingly kept to a safe level below the ultimate limit. This safe level is usually specified to be between one-half and two-thirds of the yield stress level, which means that from a half to a third of a member's capacity is kept on reserve for uncertainties in loading, material properties, and workmanship. This reserve capacity is the *safety factor*.

In the United States, the American Institute of Steel Construction (AISC) recommends what safety factors should be used for every type of structural steel component for buildings. These safety factors are usually determined from experiments conducted or authorized by the AISC. Most municipalities in the United States have local building codes that require that the AISC specifications be met. Structures other than buildings are designed according to other specifications, such as the American Association of State Highway and Transportation Officials (AASHTO) for highway bridges and the American Railway Engineering Association (AREA) for railway bridges. Because this book is directed toward architects and engineers, only the AISC specifications are mentioned, hereafter referred to as AISCS. These specifications can be found in the AISC *Manual*, from here on called AISCM.

1
Tension Members

1.1 TENSION MEMBERS

A tension member is defined as an element capable of resisting tensile loads along its longitudinal axis. Classic examples are bottom chords of trusses and sag rods used for alignment guides (Fig. 1.1). For the most part, the shape of the cross section has little effect on the tensile capacity of a member. The net cross-sectional area will be uniformly stressed except at points of load applications and their vicinity (St. Venant's principle). If fasteners (bolts or rivets) are used, it may become necessary to design for stress concentrations near the fasteners, referred to as *shear lag*. Other stress buildups, in the form of bending stresses, will develop if the centers of gravity of the connected members do not line up. This effect is usually neglected, however, in statically loaded members (AISCS 1.15.3).

The design concept of tension members has been altered by the latest AISC specifications. Allowable stresses are now computed for both gross member area and effective net area. The gross area stress is designed to remain below yield stress, at which point excessive deformations will occur, and the effective net area is designed to prevent local fracture.

To account for the effective net area, it is now necessary to use reduction coefficients for tension members that are not connected through all elements of the cross section. This provision is intended to account for the phenomenon of shear lag. For example, the angle in Fig. 1.3 is connected through one leg. The shear stress being transferred through its bolts will concentrate at the connection. The effect of shear lag will diminish, however, as the number of fasteners increases.

1.2 GROSS, NET, AND EFFECTIVE NET SECTIONS

The gross section of a member, A_g, is defined as the products of the thickness and the gross width of each element as measured normal to the axis of the mem-

A_g = GROSS SECTION = THICKNESS × WIDTH

1

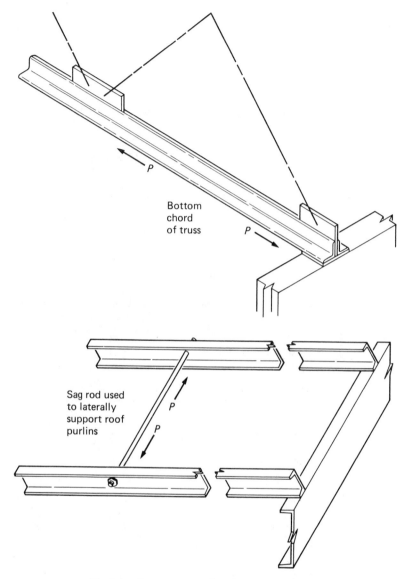

Bottom chord of truss

P

P

Sag rod used to laterally support roof purlins

P

P

Fig. 1.1. Common examples of tension members.

Fig. 1.2. John Hancock Building, Chicago. (*Courtesy of U.S. Steel*)

ber (AISCS 1.14.1). This is the cross-sectional area of the member with no parts removed.

The net width multiplied by the member thickness is the net section. Net width is determined by deducting from the gross width the sum of all holes in the section cut. In AISCS 1.14.4, the code states: "In computing net area, the width of a rivet or bolt shall be taken as $\frac{1}{16}$ in. larger than the nominal dimension of the hole normal to the direction of applied stress." AISC Table 1.23.4 lists the diameter of holes as a function of fastener size. For standard holes, the hole diameter is $\frac{1}{16}$ in. larger than the nominal fastener size. Thus, a value of fastener diameter plus $\frac{1}{8}$ in. $(d + \frac{1}{16} + \frac{1}{16})$ must be used in computing net sections.

A_N = NET SECTION = NET WIDTH × THICKNESS

(GROSS WIDTH − (HOLES)(THICKNESS))
($d + \frac{1}{8}$)

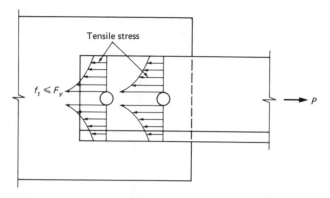

Fig. 1.3. Shear lag concept. Because the stress is transferred through the bolts, a concentration of tensile stress occurs at the bolt holes. As the number of bolts increases, the magnitude of shear lag decreases.

Example 1.1. Determine the net area of a $4 \times 4 \times \frac{1}{2}$ angle with one line of $3\text{-}\frac{3}{4}$-in. bolts as shown.[1]

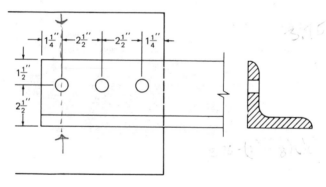

Solution. The net area is equal to the gross area less the sum of the nominal hole dimension plus $\frac{1}{16}$ in. (AISCS 1.14.4).

$$A_{\text{gross}} = 3.75 \text{ in.}^2$$

$$A_{\text{net}} = 3.75 \text{ in.}^2 - [\tfrac{3}{4} \text{ in.} + (\tfrac{1}{16} \text{ in.} + \tfrac{1}{16} \text{ in.})] \times \tfrac{1}{2} \text{ in.} = 3.31 \text{ in.}^2$$

If there is a chain of holes on a diagonal or forming a zigzag pattern, as in Fig. 1.4, the net width is taken as the gross width minus the diameter of all the holes

---

[1]As A36 steel and E70 weld are commonly used, unless specifically stated otherwise, the reader should assume A36 steel (F_y = 36 ksi) and E70 welds for all examples and problems throughout this book.

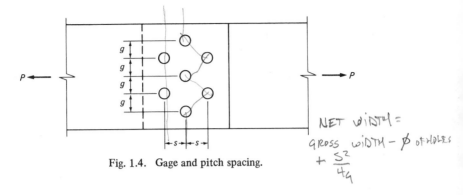

Fig. 1.4. Gage and pitch spacing.

[handwritten: NET WIDTH = GROSS WIDTH − ∅ of HOLES + $\frac{S^2}{4g}$]

in the chain and then adding for each gage space in the chain the quantity

$$\frac{s^2}{4g} \tag{1.1}$$

where s is the longitudinal spacing (pitch), in inches, of two consecutive holes, and g is the transverse spacing (gage), in inches, of the same two holes (AISCS 1.14.2.1).

The critical net section is taken at the chain that yields the least net width. In no case, however, shall the net section, when taken through riveted or bolted splices, gusset plates, or other connection fittings subject to a tensile force, exceed 85% of the gross section (AISCS 1.14.2.3). This 85% rule, previously applicable to all tension members, now applies only to relatively short fittings.

For determining the areas of angles, the gross width is the sum of the widths of the legs less the angle thickness. The gage for holes in opposite legs, as

[handwritten: L → GROSS WIDTH = Σ WIDTHS of LEGS − ANGLE THICKNESS]

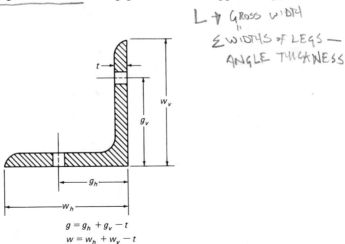

$$g = g_h + g_v - t$$
$$w = w_h + w_v - t$$

Fig. 1.5. Measurement of gage dimension and gross width for angles.

shown in Fig. 1.5, is the sum of the gages from the back of the angle less the angle thickness (AISCS 1.14.3).

Example 1.2. Determine the net area of the plate below if the holes are for $\frac{7}{8}$-in. bolts.

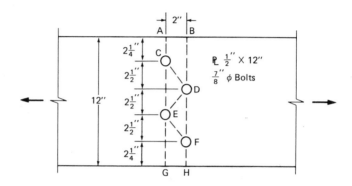

Solution. To find the net section, consult the AISC Code Section 1.14.2.1. The net width must be calculated by considering all possible lines of failure and deducting the diameters of the holes in the chain. Then for each diagonal path, the quantity $(s^2/4\,g)$ is added, where s = longitudinal spacing (pitch) of any two consecutive holes, and g = transverse spacing (gage) of the same two holes. The critical net section is the chain that gives the least net width. The net critical width is then multiplied by the thickness to obtain the net area (AISCS 1.14.2.1).

Chain	Width – Holes + $\dfrac{s^2}{4\,g}$ for each diagonal path
ACEG	$12 - 2 \times (\frac{7}{8} + \frac{1}{8}) = 10.0$
BDFH	$12 - 2 \times (\frac{7}{8} + \frac{1}{8}) = 10.0$
ACDEG	$12 - 3 \times (\frac{7}{8} + \frac{1}{8}) + (2^2/4(2.5)) + (2^2/4(2.5)) = 9.8$
ACDFH	$12 - 3 \times (\frac{7}{8} + \frac{1}{8}) + (2^2/4(2.5)) = 9.4$
ACDEFH	$12 - 4 \times (\frac{7}{8} + \frac{1}{8}) + 3 \times (2^2/4(2.5)) = 9.2$

$$\text{Critical section} = 9.2 \text{ in.} \times \tfrac{3}{4} \text{ in.} = 6.90 \text{ in.}^2$$

$$\text{Net section} = 6.90 \text{ in.}^2$$

Example 1.3. For the two lines of bolt holes shown, determine the pitch (s) that will give a net section along ABCDEF equal to a net section through two holes. Holes are for $\frac{3}{4}$-in. bolts.

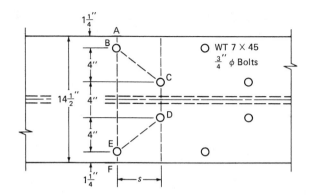

Solution. The section through four holes plus the quantity $2 \times (s^2/4\,g)$ must equal the gross section minus two holes.

$$14.5 - 4 \times \left(\frac{3}{4} + \frac{1}{8}\right) + 2 \times \frac{s^2}{4 \times 4} = 14.5 - 2 \times \left(\frac{3}{4} + \frac{1}{8}\right)$$

$$11.0 + \frac{s^2}{8} = 12.75$$

$$s^2 = 14 \qquad s = 3.74 \text{ in.}$$

Use pitch of $3\frac{3}{4}$ in.

Example 1.4. A single angle tension member $6 \times 4 \times \frac{3}{4}$ has gage lines in the legs as shown. Determine the pitch (s) for $\frac{3}{4}$-in. rivets, so that the reduction in area is equivalent to two holes in line.

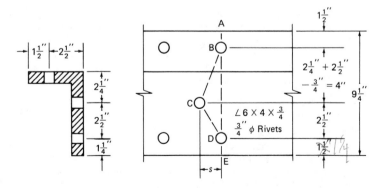

Solution. Because of the angle thickness, the gross width of the angle must be $(6 + 4 - \frac{3}{4}) = 9.25$ in.

Path ABDE = path ABCDE

$$\text{Path ABDE} = 9.25 - 2 \times \left(\frac{3}{4} + \frac{1}{8} \right) = 7.5$$

$$\text{Path ABCDE} = 9.25 - 3 \times \left(\frac{3}{4} + \frac{1}{8} \right) + \frac{s^2}{4 \times 4} + \frac{s^2}{4 \times 2.5} = 6.625 + \frac{s^2}{16} + \frac{s^2}{10}$$

$$7.5 = 6.625 + \frac{s^2}{16} + \frac{s^2}{10}$$

$$0.875 = \frac{(5 + 8)\, s^2}{80}$$

$$s^2 = 5.385 \qquad s = 2.32 \text{ in.}$$

Use pitch of $2\frac{3}{8}$ in.

The effective net area, A_e, must be calculated if the tensile force is transmitted by bolts or rivets through some, but not all, of the cross-sectional elements of the member (AISCS 1.14.2.2). This new provision in the AISC specifications is to account for the effect of shear lag, which is the concentration of shear stress in the vicinity of connections. The effective net area is, therefore, computed from the formula   EFFECTIVE NET AREA= CT X NET AREA

$$A_e = C_t A_n \tag{1.2}$$

where A_n is the net area of the member, and C_t is a reduction factor. The value of C_t is assumed to be 1.0 unless determined otherwise. Values for C_t based on AISCS 1.14.2.2 can be found in Table 1.1.

Table 1.1. Values for the Reduction Coefficient C_t Based on AISCS 1.14.2.2.

Shape[a]	No. of Fasteners	C_t
1) W, M, or S shapes where $b_f \geqslant \frac{2}{3} d$. Tee sections from above shapes	3 or more	0.90
2) W, M, or S shapes not meeting above conditions and all other shapes including built-up sections	3 or more	0.85
3) All members	2	0.75

[a] b_f = flange width; d = member depth.

Example 1.5. What is the effective net area of the angle in Example 1.1?

Solution. The effective net area is the product of the net area and the required reduction coefficient (AISCS 1.14.2.2).

$$A_e = C_t \times A_{net}$$

$$C_t = 0.85 \text{ (from Table 1.1)}$$

$$A_n = 3.31 \text{ in.}^2 \text{ (from Example 1.1)}$$

$$A_e = 0.85 \times 3.31 \text{ in.}^2 = 2.81 \text{ in.}^2$$

1.3 ALLOWABLE TENSILE STRESSES

A tension member can fail in either of two modes: excessive elongation of the gross section or localized fracture of the net section.

As an applied tensile force increases, the strain will increase linearly until the stress reaches its yield stress F_y (Fig. 1.6). At this point, inelastic strain will develop and continue in the ultimate stress (F_u) region, where additional stress capacity is realized. Once the yield stress has been reached and inelastic elongation occurs, the member's usefulness is diminished. Furthermore, the failure of other members in the structural system may result. To safeguard against yield failure, the AISC states that stress on the gross member section (except pin-connected members) shall not exceed

$$F_t = 0.60\,F_y \tag{1.3}$$

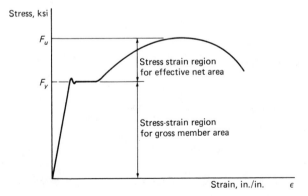

Fig. 1.6. Stress-strain diagram of mild steel. The design of tension members is based on strain (rate of elongation) on the gross member area and stress on the effective net area.

Localized fracture will occur at the net section of least resistance. The value of the load may be less than that required to yield the gross area. Therefore, the stress on the effective net section, as defined in Section 1.3, shall not be greater than

$$F_t = 0.50 \, F_u \tag{1.4}$$

For pin-connected members the allowable stress on its net section is

$$F_t = 0.45 \, F_y \tag{1.5}$$

It is known that the type of failure depends on the ratio of the effective net area (or net area if $C_t = 1.0$) to gross area and the stress limits of the steel. When

$$\frac{A_e}{A_g} \geq \frac{0.60 \, F_y}{0.50 \, F_u} = \frac{F_y}{0.833 \, F_u} \tag{1.6}$$

the member failure will be due to general yielding. Should the reverse be true, the member will fail by fracture at the weakest net section (AISCS 1.5.1.1).

The tensile stress produced by an axial tensile load P can be determined by

$$f_t = \frac{P}{A} \qquad f_t \leq F_t \tag{1.7}$$

where A is the cross-sectional area of the member at the point under investigation, and f_t must be less than or equal to F_t. Considering the gross member section,

$$A_g \geq \frac{P}{0.60 \, F_y} \tag{1.8}$$

and the minimum effective net section is

$$A_e \geq \frac{P}{0.50 \, F_u} \tag{1.9}$$

When the failure mode is known, the member can be designed by the use of one of the above equations.

The design of threaded rods was revised in the 1978 AISC specifications to incorporate the use of the nominal area of the rod, that is, the area corresponding to its gross diameter. To allow for the reduced area through the threaded

Table 1.2. Allowable Stresses for Tension Members.

F_y	On Net Area $0.45\,F_y$	On Gross Area $0.60\,F_y$	F_u	On Nominal Rod Area $0.33\,F_u$	On Eff. Net Area $0.50\,F_u$
36	16.2	22.0	58.0	19.1	29.0
42	18.9	25.2	60.0	19.8	30.0
45	20.3	27.0	60.0	19.8	30.0
50	22.5	30.0	65.0	21.5	32.5
55	24.8	33.0	70.0	23.1	35.0
60	27.0	36.0	75.0	24.8	37.5
65	29.3	39.0	80.0	26.4	40.0
90	40.5	54.0	100.0	33.0	50.0
100	45.0	60.0	110.0	36.3	55.0

part, the allowable stress for threaded bars is now limited to

$$F_t = 0.33\,F_u \qquad (1.10)$$

Computed values of allowable yield stress and allowable ultimate stress are provided in Table 1.2.

Example 1.6. At what effective net area of a $4 \times 4 \times \frac{1}{2}$ angle will the design be to prevent fracture failure? Determine the member capacity for the section shown in Example 1.1.

Solution. (AISCM, Table 1, p. 1–5; Table 1.2)

$$F_y = 36.0 \text{ ksi} \quad \text{(for A36 steel)}$$
$$F_u = 58.0 \text{ ksi}$$
$$A_g = 3.75 \text{ in.}^2$$
$$A_e \geqslant \frac{A_g\,F_y}{0.833\,F_u}$$
$$A_e \geqslant 0.745\,A_g$$
$$A_e \geqslant 2.79 \text{ in.}^2$$

A $4 \times 4 \times \frac{1}{2}$ angle in tension will be designed to prevent fracture failure at the weakest effective net section when the section is less than 2.79 in.^{2}. The use of Eq. 1.9 is required for sections less than 2.79 in.^{2}. Effective net sections equal to or greater than 2.79 in.^{2} will be designed to resist excessive deformation by the use of Eq. 1.8.

The effective net area of the member in Example 1.1 is

$$A_e = 2.81 \text{ in.}^2 \quad \text{(from Example 1.5)}$$

$$A_e > 2.79 \text{ in.}^2$$

The use of Eq. 1.8 is required

$$P = 0.60 F_y \times A_g = 0.60 \times 36.0 \text{ ksi} \times 3.75 \text{ in.}^2$$

$$P = 81.0 \text{ k}$$

Example 1.7. What tensile load can a $4 \times 4 \times \frac{3}{8}$ carry with the connections shown.

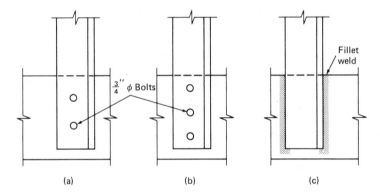

(a)   (b)   (c)

Solution.

For $A_e \geqslant 0.745 A_g$ (Example 1.6)

$$F_t = 0.60 F_y$$

For $A_e < 0.745 A_g$

$$F_t = 0.50 F_u$$

For a $4 \times 4 \times \frac{3}{8}$ angle

$$A_g = 2.86 \text{ in.}^2 ; \quad 0.745 A_g = 2.13 \text{ in.}^2$$

$$A_n = 2.86 \text{ in.}^2 - [\tfrac{3}{8} \text{ in.} \times (\tfrac{3}{4} \text{ in.} + \tfrac{1}{16} \text{ in.} + \tfrac{1}{16} \text{ in.})] = 2.53 \text{ in.}^2$$

$$A_e = C_t \times A_n$$

a) $\qquad C_t = 0.75 \qquad$ (from Table 1.1)

$$A_e = 0.75 \times 2.53 \text{ in.}^2 = 1.90 \text{ in.}^2$$

$$A_e < 2.13 \text{ in.}^2$$

Use Eq. 1.7

$$P = 0.50 \, F_e \times A_e = 0.50 \times 58.0 \text{ ksi} \times 1.90 \text{ in.}^2 = 55.1 \text{ k}$$

F_u

Note: If member capacity with respect to general yielding is checked

$$P = 0.60 \, F_y \times A_g = 0.60 \times 36.0 \text{ ksi} \times 2.86 \text{ in.}^2 = 61.8 \text{ k}$$

Use least value.

b) $\qquad C_t = 0.85$

$$A_e = 0.85 \times 2.53 \text{ in.}^2 = 2.15 \text{ in.}^2$$

$$A_e > 2.13 \text{ in.}^2$$

Use Eq. 1.6

$$P = F_t \times A_g$$

$$F_t = 0.60 \times 36.0 \text{ ksi} = 21.6 \text{ ksi}$$

The value can be rounded to 22.0 ksi (AISC, Appendix A)

$$P = 22.0 \text{ ksi} \times 2.86 \text{ in.}^2 = 62.9 \text{ k}$$

c) $\qquad C_t = 1.0$ (no reduction factor applies)

$$P = F_t \times A_g$$

$$F_t = 0.60 \, F_y = 22.0 \text{ ksi}$$

$$P = 22.0 \text{ ksi} \times 2.86 \text{ in.}^2 = 62.9 \text{ k}$$

Example 1.8. Determine the maximum tensile load a $\frac{3}{4} \times 7$-in. plate connection fitting can carry if it has welded connections and punched holes as shown. Use A572 Gr 50 steel.

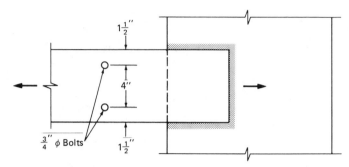

Solution.

$$\frac{F_y}{0.833\,F_u} = \frac{50.0\text{ ksi}}{0.833 \times 65.0\text{ ksi}} = 0.923$$

$$A_g = \frac{3}{4}\text{ in.} \times 7\text{ in.} = 5.25\text{ in.}^2$$

$$A_n = 5.25\text{ in.}^2 - 2 \times \left[\frac{3}{4}\text{ in.} \times \left(\frac{3}{4}\text{ in.} + \frac{1}{8}\text{ in.}\right)\right] = 3.94\text{ in.}^2$$

$$A_e = 1.0 \times 3.94\text{ in.}^2 = 3.94\text{ in.}^2$$

$$A_{\max} = 0.85 \times \left(\frac{3}{4}\text{ in.} \times 7\text{ in.}\right) = 4.46\text{ in.}^2 \qquad \text{AISCS 1.14.2.3}$$

Use $A_e = 3.94\text{ in.}^2$

$$A_e < 0.923\,A_g = 4.85\text{ in.}^2$$

$$P = 0.50\,F_u \times A_e = 0.50 \times 65.0\text{ ksi} \times 3.94\text{ in.}^2 = 128.0\text{ k}$$

Example 1.9. Calculate the allowable tensile load in the $\frac{1}{2}$-in. $\times$ 14-in. plate. The holes are for $\frac{3}{4}$-in. bolts; use A36 steel.

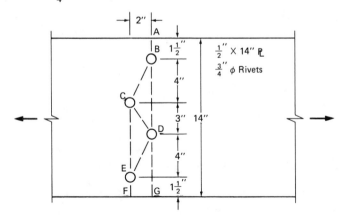

Solution.

Chain	Width – Holes + $\dfrac{s^2}{4g}$ for each diagonal path	(AISCS 1.14.2.1)

ABDG $14 - 2 \times (\frac{3}{4} + \frac{1}{8}) = 12.25$ in.

ABDEF $14 - 3 \times (\frac{3}{4} + \frac{1}{8}) + 2^2/4(4) = 11.625$ in.

ABCEF $14 - 3 \times (\frac{3}{4} + \frac{1}{8}) + 2^2/4(4) = 11.625$ in.

ABCDG $14 - 3 \times (\frac{3}{4} + \frac{1}{8}) + 2^2/4(4) + 2^2/4(3) = 11.96$ in.

ABCDEF $14 - 4 \times (\frac{3}{4} + \frac{1}{8}) + 2 \times (2^2/4(4)) + 2^2/4(3) = 11.33$ in.

Critical section = 11.33 in. $\times \frac{1}{2}$ in. = 5.67 in.^{2}

A_N (C_T)

$$A_e = 1.0\, A_n = 5.67 \text{ in.}^2$$

$$A_e > 0.745\, A_g = 5.22 \text{ in.}^2$$

$$P = 0.60\, F_y \times A_g = 22.0 \text{ ksi} \times 7.0 \text{ in.}^2$$

$$P = 154.0 \text{ k}$$

Example 1.10. A single angle tension member $6 \times 6 \times \frac{1}{2}$ has two gage lines in both legs as shown. Determine the allowable tension load that can be carried. Holes are for $\frac{3}{4}$-in. bolts.

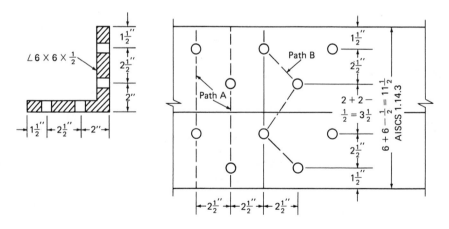

Solution.

$$\text{Path A} \quad \left(6 + 6 - \frac{1}{2}\right) - 2 \times \left(\frac{3}{4} + \frac{1}{8}\right) = 9.75 \text{ in.}$$

$$\text{Path B} \quad \left(6 + 6 - \frac{1}{2}\right) - 4 \times \left(\frac{3}{4} + \frac{1}{8}\right) + 2 \times \frac{(2.5)^2}{4 \times 2.5}$$

$$+ \frac{(2.5)^2}{4 \times \left(2 + 2 - \frac{1}{2}\right)} = 9.70 \text{ in.}$$

$$A_n = 9.70 \text{ in.} \times \frac{1}{2} \text{ in.} = 4.85 \text{ in.}^2$$

$$A_g = 5.75 \text{ in.}^2$$

$$A_e = 1.0 \, A_n = 4.85 \text{ in.}^2$$

$$A_e > 0.745 \, A_g = 4.28 \text{ in.}^2$$

$$P = 0.60 \, F_y \times A_g = 22.0 \text{ ksi} \times 5.75 \text{ in.}^2 = 126.5 \text{ k}$$

Example 1.11. A threaded rod cut from A36 stock is to be used as a tie bar carrying a 10-kip tensile force. Determine the diameter required.

Solution. The allowable tensile force can be found in AISC, Appendix A Table 2, and is repeated in Table 1.2 for convenience.

$$F_t = 0.33 \, F_u = 19.1 \text{ ksi}$$

The stress is to be calculated on the nominal body area

$$A = \frac{P}{F_t} = \frac{10 \text{ k}}{19.1 \text{ ksi}} = 0.52 \text{ in.}^2$$

$$\frac{\pi d^2}{4} > 0.52 \text{ in.}^2$$

$$d^2 > 0.66 \text{ in.}^2$$

$$d > 0.81 \text{ in.}$$

Use minimum $\frac{7}{8}$-in. ϕ rod.

1.4 AISC DESIGN AIDS

The AISCM provides a chart to determine net areas for double angles in tension. Common double angles used as tension members are given on p. 4-94 for two, four, and six holes out. To use, simply find the angle designation in the left

Fig. 1.7. Wind bracing for an office building, each made from four angles. Note the stiffeners on the beam web to resist the large concentrated reactions from the braces.

column and find the number of holes out in the fastener size in the top row. To find A_{net}, determined in accordance with AISCS 1.14.3 and 1.14.4, carry the angle designation across and the fastener diameter down until the two lines intersect. The appropriate C_t value is then applied to determine the effective net section. Values for single angles can also be determined by assuming one hole out instead of two for double angles, two holes out instead of four, and three holes out instead of six, and noting that the area will be *one-half* of the value listed.

AISCM (p. 4–96) also provides a table for area reductions due to holes. The thickness of steel along the left margin is matched with the hole diameter, and the value determined is the area of the hole.

Values for $s^2/4g$ can be determined from the chart on p. 4–97, labeled "Net Section of Tension Members." Entering the chart on the side with the gage value g, carrying the line across until it intersects the curve for the appropriate pitch s. The value found vertically above or below that point is $s^2/4g$.

Example 1.12. Using the AISC Design Aids, find the net area for two $4 \times 4 \times \frac{3}{8}$ angles with two rows of $\frac{3}{4}$-in. bolts in each angle.

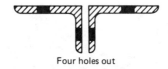

Four holes out

Solution. The $4 \times 4 \times \frac{3}{8}$-in. angle can be found in the "Connections" section of AISCM. With two $\frac{3}{4}$-in. bolts in each of the angles, there are a total of four holes out. Finding the correct column for $\frac{3}{4}$-in. fasteners with four holes out, follow down until it intersects with the row of L $4 \times 4 \times \frac{3}{8}$ in. The value for the net area is 4.41 in.^{2}

$$A_{net} = 4.41 \text{ in.}^2$$

Example 1.13. Using the AISC chart, find the net area for an angle $6 \times 4 \times \frac{3}{8}$ with two rows of $\frac{3}{4}$-in. bolts.

Solution. The angle $6 \times 4 \times \frac{3}{8}$ can be considered as one-half of a double angle. Therefore, use the column for two $\times$ two holes out with a double angle. Reading down the column for $\frac{3}{4}$-in. fasteners and across for the L $6 \times 4 \times \frac{3}{8}$ in. yields a value of 5.91 in.^{2}. Because this gives the net area for two angles, the value must be divided by 2, which gives a net area for a single angle as 2.95 in.^{2}

$$A_{net} = 2.95 \text{ in.}^2$$

Example 1.14. Two $6 \times 4 \times \frac{1}{2}$ angles are connected with long legs back to back. Assume two rows of $\frac{3}{4}$-in. bolts are used in the long legs and one row of $\frac{3}{4}$-in. bolts is used in the short legs. Determine the maximum tensile load that can be carried.

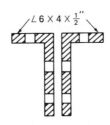

$L\ 6 \times 4 \times \frac{1}{2}''$

Solution. Two methods will be shown, one using the tables in the AISCM.

a) Using the AISCM, determine the number of holes out, and read down the column to the row corresponding to the angle designation.

From the AISCM

$$A_g = 2 \times 4.75 \text{ in.}^2 = 9.50 \text{ in.}^2$$
$$A_n = 6.88 \text{ in.}^2$$

$$C_t = 1.0$$

$$A_e = 1.0 \times A_n = 6.88 \text{ in.}^2$$

$$0.745 \, A_g = 7.08 \text{ in.}^2$$

$$A_e < 7.08 \text{ in.}^2$$

$$P = 0.5 \, F_u \times A_e = 0.5 \times 58.0 \text{ ksi} \times 6.88 \text{ in.}^2 = 199.5 \text{ k}$$

b) $A_n = 2 \times [4.75 \text{ in.}^2 - 3 \times \frac{1}{2} \text{ in.} \times (\frac{3}{4} \text{ in.} + \frac{1}{8} \text{ in.})] = 6.88 \text{ in.}^2$
The rest of the computation is the same as part a.

Example 1.15. Using the AISC charts for tension member net areas, determine the net areas and maximum tensile load for

a) $6 \times 6 \times \frac{3}{4}$ angle connected by two lines of $3\text{-}\frac{3}{4}$-in. bolts in one leg.

b) $6 \times 6 \times \frac{3}{4}$ angle connected by one line of $3\text{-}\frac{3}{4}$-in. bolts in each leg, and connected with lug angles.

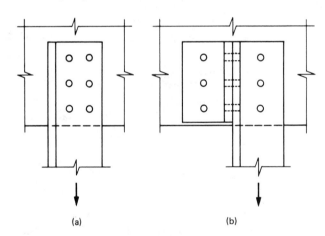

(a)   (b)

Solution.

$$A_g = 8.44 \text{ in.}^2$$

a) Using the column for $\frac{3}{4}$-in. fasteners with four holes out (two holes out in a single angle), and the row for L $6 \times 6 \times \frac{3}{4}$

$$A_n = 14.3 \text{ in.}^2 / 2 = 7.15 \text{ in.}^2$$

$$A_e = C_t \times A_n$$

$$C_t = 0.85$$

$$A_e = 0.85 \times 7.15 \text{ in.}^2 = 6.08 \text{ in.}^2$$

$A_e < 0.745\,A_g = 6.29$ in.^{2} Example 1.6

$P = 0.50\,F_u \times A_e = 0.50 \times 58.0$ ksi $\times 6.08$ in.$^2 = 176.3$ k

b) The load is transferred through both legs of the angle, and therefore no reduction need be taken

$A_n = 14.3$ in.$^2/2 = 7.15$ in.^{2}

$C_t = 1.0$

$A_e = 7.15$ in.^{2}

$A_e > 0.745\,A_g = 6.29$ in.^{2}

$P = 0.60\,F_y \times A_g = 22.0$ ksi $\times 8.44$ in.$^2 = 185.7$ k

1.5 SLENDERNESS AND ELONGATION

To prevent lateral movement or vibrations, the AISC recommends limits to the slenderness ratio l/r for tension members other than rods. Table 1.3 lists the limits of l/r, where l is the length of the member in inches, and r is its least radius of gyration, equal to $\sqrt{I/A}$.

Provided tension members are designed within stated allowable stresses, elongations of tension members should not be critical. Should the elongation of a member be desired, however, it can be calculated in the elastic range $(f_t \leqslant F_y)$ by

$$\Delta = \frac{Pl}{AE} \tag{1.11}$$

where l is the member length in inches, and E is the modulus of elasticity. For this calculation, the area should be taken as the gross area, though at net sections the strain value will locally be greater.

Table 1.3.   Maximum Allowable Slenderness Ratios
Based on AISCS 1.8.4.

Member Type	AISC Recommended Slenderness Ratio Limit
Main members	$l/r \leqslant 240$
Secondary members and bracing	$l/r \leqslant 300$

Example 1.16. A WT 8×13 structural tee is used as a main tension member with a length of 20 ft. Determine if the member is within recommended AISC limits to the slenderness ratio.

Solution. (AISCS 1.8.4)

$$\frac{l}{r} \leqslant 240$$

The member length in inches is 20 ft $\times$ 12 in./ft = 240 in.

Checking properties for designing, r_{xx} = 2.47 in., r_{yy} = 1.12 in. Use r_{yy} = 1.12 in. (least value)

$$\frac{l}{r} = \frac{240 \text{ in.}}{1.12 \text{ in.}} = 214$$

$$214 < 240 \quad \text{ok}$$

The member satisfies the AISC recommended slenderness limit.

Example 1.17. A $5 \times 3 \times \frac{5}{16}$-in. angle is used as a bracing member carrying tension. The member is to be within recommended AISC slenderness ratio limits. Determine the maximum length of the member to be within AISC limits.

Solution. (AISCS 1.8.4)

$$\frac{l}{r} \leqslant 300$$

$$l \leqslant 300 \times r$$

$$r_{xx} = 1.61 \text{ in.}; \quad r_{yy} = 0.853 \text{ in.}; \quad r_{zz} = 0.658 \text{ in.}$$

Use r_{zz} = 0.658 in.

$$l < 300 \times 0.658 \text{ in.} = 197.4 \text{ in.} = 16.45 \text{ ft}$$

To satisfy recommended AISC slenderness limits, the length of the member cannot exceed 16 ft, 5 in.

Example 1.18. Design the 12 ft WT 4 structural tee shown to carry 60 kips and satisfy the recommended slenderness ratio. Use $\frac{7}{8}$-in. bolts.

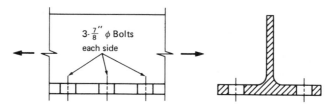

Solution.

$$\text{Gross area required} = \frac{60 \text{ k}}{22.0 \text{ ksi}} = 2.73 \text{ in.}^2$$

$$\text{Effective net area required} = \frac{60 \text{ k}}{29.0 \text{ ksi}} = 2.07 \text{ in.}^2$$

Using the tables for structural tees cut from W shapes, find the most economical section by choosing one with the required area and the least weight.

Try Wt 4 × 10.5

$A_g = 3.08 \text{ in.}^2$

$A_n = 3.08 \text{ in.}^2 - 2 \times [0.400 \text{ in.} \times (\frac{7}{8} \text{ in.} + \frac{1}{8} \text{ in.})] = 2.28 \text{ in.}^2$

$C_t = 0.90$

$A_e = C_t \times A_n = 0.90 \times 2.28 \text{ in.}^2 = 2.05 \text{ in.}^2 < 2.07 \text{ in.}^2$ NG

Try WT 4 × 12

$A_g = 3.54 \text{ in.}^2$

$A_n = 3.54 \text{ in.}^2 - 2 \times \left[0.400 \text{ in.} \times \left(\frac{7}{8} \text{ in.} + \frac{1}{8} \text{ in.} \right) \right] = 2.74 \text{ in.}^2$

$A_e = 0.90 \times 2.74 \text{ in.}^2 = 2.47 \text{ in.}^2 > 2.07 \text{ in.}^2$ ok

$r_{xx} = 0.999 \text{ in.}; \quad r_{yy} = 1.61 \text{ in.}$

$\dfrac{l}{r} = \dfrac{12 \text{ ft} \times 12 \text{ in/ft}}{0.999 \text{ in.}} = 144 < 240$ ok

Use WT 4 × 12.

1.6 PIN-CONNECTED MEMBERS

Eyebar members and pin-connected plates are designed to carry the tensile load through the bar and transfer the load through the pinhole to the pin. The allow-

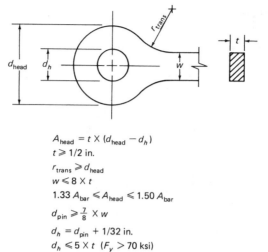

$A_{head} = t \times (d_{head} - d_h)$

$t \geqslant 1/2$ in.

$r_{trans} \geqslant d_{head}$

$w \leqslant 8 \times t$

$1.33 A_{bar} \leqslant A_{head} \leqslant 1.50 A_{bar}$

$d_{pin} \geqslant \frac{7}{8} \times w$

$d_h = d_{pin} + 1/32$ in.

$d_h \leqslant 5 \times t$ $(F_y > 70$ ksi)

Note: Eyebar shall be of uniform thickness, without reinforcement at the pinholes.

Fig. 1.8. Design requirements of eyebars according to AISCS.

able stress in the eyebar is $F_t = 0.45 F_y$ and is taken across the member net area. Figures 1.8 and 1.9 show the requirements for eyebars and pin-connected plates, as stated in AISCS 1.14.5.

Example 1.19. Design an eyebar to carry a tensile load of 150 kips.

Solution. (AISCS 1.14.5)

$$F_t = 0.45 F_y \quad \text{(AISCS 1.5.1.1)}$$

$$F_t = 16.2 \text{ ksi}$$

$$A_{bar} = \frac{150 \text{ k}}{16.2 \text{ ksi}} = 9.26 \text{ in.}^2$$

Try plate $1\frac{1}{4}$ in. $\times$ $7\frac{1}{2}$ in.; $A = 9.38$ in.^{2}

$$w \leqslant 8 t$$

$$7\frac{1}{2} \text{ in.} < 8 \times 1\frac{1}{4} \text{ in.} = 10 \text{ in.} \quad \text{ok}$$

$$d_p \geqslant \frac{7}{8} w$$

$$d_p \geqslant \frac{7}{8} \times 7\frac{1}{2} \text{ in.} = 6\frac{18}{32} \text{ in.}$$

Use $6\frac{5}{8}$ in. diameter pin

$$d_h = d_p + \frac{1}{32} \text{ in.} = 6\frac{21}{32} \text{ in. (6.66 in.)}$$

$$1.33\, A_{bar} \leq A_{net} \leq 1.50\, A_{bar}$$

$$1.33 \times 9.38 \text{ in.}^2 = 12.48 \text{ in.}^2$$

$$1.50 \times 9.38 \text{ in.}^2 = 14.07 \text{ in.}^2$$

Try $A_{net} = 13.0 \text{ in.}^2$

$$w_{net} = \frac{13.0 \text{ in.}^2}{1.25 \text{ in.}} = 10.4 \text{ in.}$$

$$w_{gross} = 10.4 \text{ in.} + 6.66 \text{ in.} = 17.06 \text{ in.}, \quad \text{say } 17\frac{1}{4} \text{ in.}$$

$$A_{net} = \left(17\frac{1}{4} \text{ in.} - 6\frac{21}{32} \text{ in.}\right) \times 1\frac{1}{4} \text{ in.} = 13.24 \text{ in.}^2 \quad \text{ok}$$

$$r_{trans} \geq d_{head} \quad \text{Use 7-in. radius}$$

Use eyebar as shown.

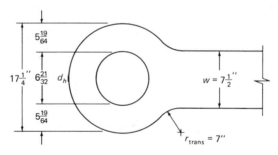

Example 1.20. Design a pin-connected plate to carry a tensile load of 150 kips. Assume the pin diameter to be $5\frac{1}{4}$ in.

Solution. (AISCS 1.14.5)

$$F_t = 0.45\, F_y = 16.2 \text{ ksi}$$

$$A_{req} = \frac{150 \text{ k}}{16.2 \text{ ksi}} = 9.26 \text{ in.}^2$$

$$b_1 \leq 4\, t$$

Try $t = 1\frac{1}{8}$ in.

$$b_1 \leqslant 4\, t = 4\frac{1}{2} \text{ in.}$$

Try $b_1 = 4\frac{1}{8}$ in.

$$A = (2 \times 4\frac{1}{8} \text{ in.}) \times 1\frac{1}{8} \text{ in.} = 9.28 \text{ in.}^2$$

$$d_h \geqslant 1.25 \times 4\frac{1}{8} \text{ in.} = 5.16 \text{ in.}$$

$$d_h \geqslant 5\frac{1}{4} \text{ in.} + \frac{1}{32} \text{ in.} = 5\frac{9}{32} \text{ in.} > 5.16 \text{ in.} \quad \text{ok}$$

Use pin-connected plate as shown.

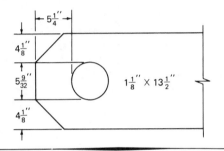

1.7 BUILT-UP MEMBERS

Requirements for built-up tension members are discussed in AISCS 1.18.3. For two plates or a plate and a rolled shape, the longitudinal spacing of rivets, bolts,

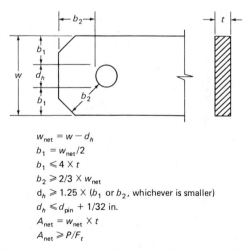

$$w_{net} = w - d_h$$
$$b_1 = w_{net}/2$$
$$b_1 \leqslant 4 \times t$$
$$b_2 \geqslant 2/3 \times w_{net}$$
$$d_h \geqslant 1.25 \times (b_1 \text{ or } b_2, \text{ whichever is smaller})$$
$$d_h \leqslant d_{pin} + 1/32 \text{ in.}$$
$$A_{net} = w_{net} \times t$$
$$A_{net} \geqslant P/F_t$$

Fig. 1.9. Design requirements of pin-connected plates according to AISCS.

or intermittent fillet welds shall not exceed 24 times the thickness of the thinner plate nor 12 in. The longitudinal spacing of rivets, bolts, or intermittent welds connecting two or more rolled shapes shall not exceed 24 in. For members separated by intermittent fillets, connections must be made at intervals such that the slenderness ratio of either component between the fasteners does not exceed 240.

Perforated cover plates or tie plates without lacing may be used on the open sides of the built-up tension members (Fig. 1.10). Such tie plates must be designed to satisfy the criteria below. The spacing shall be such that the slenderness ratio of any component in the length between tie plates will not exceed 240.

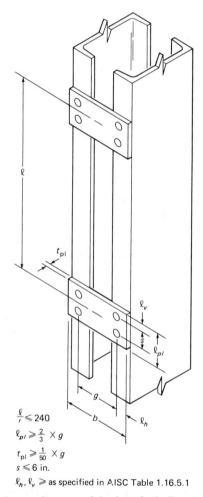

$$\frac{\ell}{r} \leqslant 240$$

$$\ell_{pl} \geqslant \tfrac{2}{3} \times g$$

$$t_{pl} \geqslant \tfrac{1}{50} \times g$$

$$s \leqslant 6 \text{ in.}$$

$\ell_h, \ell_v \geqslant$ as specified in AISC Table 1.16.5.1

Fig. 1.10. Spacing requirements of tie plates for built-up tension members.

Lengths of tie plates shall not be less than two-thirds the distance between the lines of rivets, bolts, or welds connecting them to the components of the member. The thickness is not to be less than $\frac{1}{50}$ of the distance between these lines. Longitudinal spacing of rivets, bolts, or intermittent welds at tie plates shall not exceed 6 in. The minimum width shall be determined by AISC Table 1.16.5, which lists the minimum distance from the center of a rivet or bolt hole to any edge.

Example 1.21. A 30-ft pinned member is to consist of four equal leg angles arranged as shown. The tensile load is to be 200 kips. Two $\frac{5}{8}$ in. bolts will be used in each angle. Include tie plates in the design.

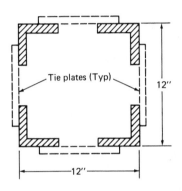

Solution.

$$\text{Gross area required} = \frac{200 \text{ k}}{22.0 \text{ ksi}} = 9.09 \text{ in.}^2$$

$$\text{Gross area for one angle} = \frac{9.09 \text{ in.}^2}{4} = 2.27 \text{ in.}^2$$

$$\text{Effectve net area required} = \frac{200 \text{ k}}{29.0 \text{ ksi}} = 6.90 \text{ in.}^2$$

$$\text{Effective net area for one angle} = \frac{6.90 \text{ in.}^2}{4} = 1.73 \text{ in.}^2$$

Try L $4 \times 4 \times \frac{5}{16}$

$A = 2.40 \text{ in.}^2$

$r_{zz} = 0.791 \text{ in.}$

$A_n = 2.40 \text{ in.}^2 - 2 \times [(\frac{5}{8} \text{ in.} + \frac{1}{8} \text{ in.}) \times \frac{3}{8} \text{ in.}] = 1.84 \text{ in.}^2 > 1.73 \text{ in.}^2$ ok

Because the member is symmetrical, $I_{xx} = I_{yy}$

$$I = I_0 + Ad^2$$

$$I = 4 \times [3.71 \text{ in.}^4 + 2.40 \text{ in.}^2 \times (6 \text{ in.} - 1.12 \text{ in.})^2] = 243.5 \text{ in.}^4$$

$$r = \sqrt{I/A} = \sqrt{\frac{243.5 \text{ in.}^4}{4 \times 2.40 \text{ in.}^2}} = 5.0 \text{ in.}$$

$$\frac{l}{r} = \frac{360 \text{ in.}}{5.0 \text{ in.}} = 72.0 < 240 \quad \text{ok}$$

Design of tie plates (AISCS 1.18.3.2)

Maximum spacing of tie plates, $r_{zz} = 0.791$ in.

$$l = \frac{240 \times 0.791 \text{ in.}}{12 \text{ in./ft}} = 15.82 \text{ ft} \quad \text{Use 15 ft-0 in.}$$

Plate length

$$l_{pl} \geqslant \tfrac{2}{3} g$$

Use $g = 12$ in. $- (2 \times 1\tfrac{1}{2}$ in.$) = 9$ in.

$$l_{pl} \geqslant \tfrac{2}{3} \times 9 \text{ in.} = 6 \text{ in.} \quad \text{Use 6 in.}$$

Plate thickness

$$t \geqslant \tfrac{1}{50} \times 9 \text{ in.} = 0.18 \text{ in.}, \quad \text{say } \tfrac{3}{16} \text{ in.}$$

Minimum width of tie plates

$$9 \text{ in.} + (2 \times 1\tfrac{1}{8} \text{ in.}) = 11\tfrac{1}{4} \text{ in.} \quad \text{Use } 11\tfrac{1}{2} \text{ in.}$$

Use tie plates $\tfrac{3}{16} \times 6 \times 11\tfrac{1}{2}$ in. at a maximum spacing of 15 ft-0 in.

Example 1.22. Design the most economical W 6 or W 8 shape to carry a 100-kip tensile load. The length is to be 20 ft, and two rows of 3-$\tfrac{3}{4}$-in. bolts will be used in each flange.

Solution.

$$\text{Gross area required} = \frac{100 \text{ k}}{22.0 \text{ ksi}} = 4.55 \text{ in.}^2$$

$$\text{Effective net area required} = \frac{100 \text{ k}}{29.0 \text{ ksi}} = 3.45 \text{ in.}^2$$

Try W 6 X 16

$$A = 4.74 \text{ in.}^2, \quad t_f = 0.405 \text{ in.}, \quad b_f = 4.03 \text{ in.}, \quad d = 6.28 \text{ in.}$$

$$A_n = 4.74 \text{ in.}^2 - 4 \times [(\tfrac{3}{4} \text{ in.} + \tfrac{1}{8} \text{ in.}) \times 0.405 \text{ in.}] = 3.32 \text{ in.}^2$$

$$C_t = 0.90 \, (b_f/d = \tfrac{2}{3})$$

$$A_e = C_t \times A_n = 0.90 \times 3.32 \text{ in.}^2 = 2.99 \text{ in.}^2 < 3.45 \text{ in.}^2 \quad \text{NG}$$

Try W 8 X 18

$$A = 5.26 \text{ in.}^2, \quad t_f = 0.33 \text{ in.}, \quad b_f = 5.25 \text{ in.}, \quad d = 8.14 \text{ in.}$$

$$A_n = 5.26 \text{ in.}^2 - 4 \times \left[\left(\frac{3}{4} \text{ in.} + \frac{1}{8} \text{ in.} \right) \times 0.33 \text{ in.} \right] = 4.11 \text{ in.}^2$$

$$C_t = 0.90 \left(b_f/d = \frac{2}{3} \right)$$

$$A_e = C_t \times A_n = 0.90 \times 4.11 \text{ in.}^2 = 3.69 \text{ in.}^2 > 3.45 \text{ in.}^2 \quad \text{ok}$$

$$l = 240 \text{ in.}; \quad r_{yy} = 1.23 \text{ in.}$$

$$\frac{l}{r} = \frac{240 \text{ in.}}{1.23 \text{ in.}} = 195 < 240 \quad \text{ok}$$

Use W 8 X 18.

Example 1.23. Design the most economical common single angle to carry a 50-kip tensile load. The angle is to be 15 ft long and is to be connected by four $\tfrac{3}{4}$-in. bolts in one line.

Solution.

$$\text{Gross area required} = \frac{50 \text{ k}}{22.0 \text{ ksi}} = 2.27 \text{ in.}^2$$

$$\text{Effective net area required} = \frac{50 \text{ k}}{29.0 \text{ ksi}} = 1.72 \text{ in.}^2 \quad (C_t = 0.85)$$

In determining the most economical angle, choose the angle with the least area that can carry the load. Therefore, the easiest method of investigating different angles is to make a chart.

Lightest Angles Available	Wt k/ft	Gross Area (in.^{2})	Area of One Hole (in.^{2})*	Effective Net Area (in.^{2})
$3\frac{1}{2} \times 3 \times \frac{3}{8}$	7.9	2.30	0.33	1.67
$3\frac{1}{2} \times 3\frac{1}{2} \times \frac{3}{8}$	8.5	2.48	0.33	1.83
$4 \times 4 \times \frac{5}{16}$	8.2	2.40	0.27	1.81
$5 \times 3 \times \frac{5}{16}$	8.2	2.40	0.27	1.81

*$(\frac{3}{4} + \frac{1}{8}$ in.) $\times$ thickness.

Try L $5 \times 3 \times \frac{5}{16}$

$$r_{zz} = 0.658 \text{ in.}$$

$$\frac{l}{r} = \frac{15 \text{ ft} \times 12 \text{ in./ft}}{0.658 \text{ in.}} = 273.6 \quad \text{NG}$$

Try L $4 \times 4 \times \frac{5}{16}$

$$r_{zz} = 0.791 \text{ in.}$$

$$\frac{l}{r} = \frac{15 \text{ ft} \times 12 \text{ in./ft}}{0.791 \text{ in.}} = 227.6 \quad \text{ok}$$

Use L $4 \times 4 \times \frac{5}{16}$ angle.

1.8 FLUCTUATIONS

Occasionally, it becomes necessary to design for stress fluctuations, if frequent variations or reversals in stress occur. An example of a system that encounters fluctuations is found in bridge structures. AISCS 1.7 covers fluctuations.

AISC Appendix B divides the loading into four categories, depending on the anticipated number of loading cycles. The appendix then lists the material subjected to stress fluctuations and provides a category classification for that member. The allowable range of stress is then determined by Table B3, based on the category classification and the loading category.

PROBLEMS TO BE SOLVED

1.1. Using AISCS, determine the tensile capacity for the following tension members connected by welds:

a) One angle $4 \times 6 \times \frac{3}{8}$, A36 steel, 6-in. outstanding leg
b) One angle $4 \times 6 \times \frac{3}{8}$, $F_y = 50$ ksi, 4-in. outstanding leg

c) One W 8 × 24, A36 steel

d) One C 8 × 11.5, 50 ksi steel

1.2. Determine the capacity of the members connected as shown. Use $\frac{7}{8}$-in. diameter bolts.

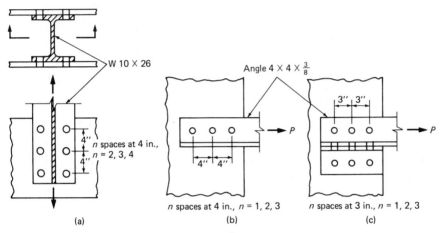

(a)

(b)

(c)

1.3. Determine the tensile capacity of $\frac{1}{2}$-in.-thick plates connected as shown. Use $\frac{7}{8}$-in. diameter bolts.

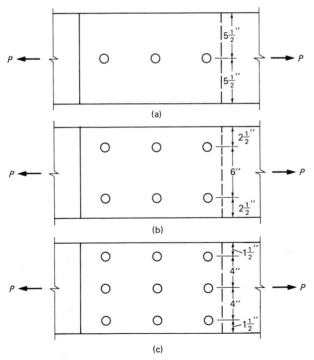

(a)

(b)

(c)

1.4. Calculate $P_{allowable}$ for the plates shown if F_y = 50 ksi, $t = \frac{1}{2}$ in., and bolt diamter = $\frac{3}{4}$ in.

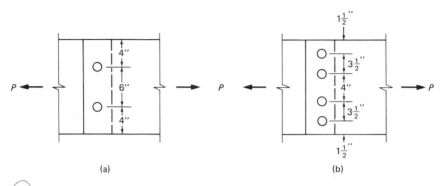

(a)   (b)

1.5. Determine the tensile capacity of the plate shown if F_y = 36 ksi, $t = \frac{1}{2}$ in., and bolt diameter is $\frac{3}{4}$ in.

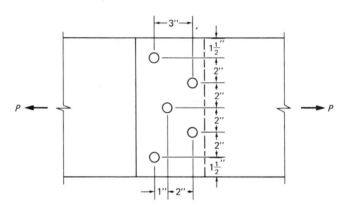

1.6. Determine the tensile capacity of a $7 \times 4 \times \frac{3}{4}$ angle as shown, if F_y = 50 ksi and bolt diameter = 1 in.

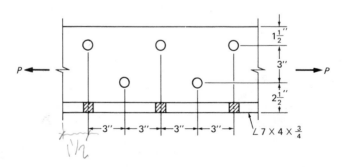

1.7. Calculate the pitch s, such that the net area is equal to the gross area less the area of two holes. What is the allowable tensile force if F_y = 50 ksi and bolt diameter is $\frac{7}{8}$ in.?

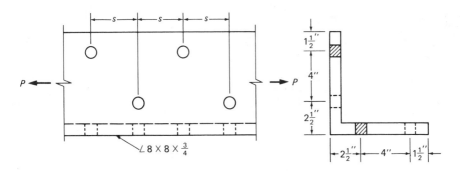

1.8. Determine the pitch s, such that the loss in area is equal to the loss of two holes. Use A36 steel and $\frac{3}{4}$-in. diameter bolts for the 7 × 4 × $\frac{3}{4}$ angle.

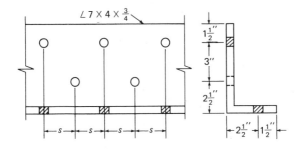

1.9. What is the maximum recommended length for a 4 × 4 × $\frac{1}{4}$ angle tension member?

1.10. A 30-ft-long angle of equal legs is to carry 150 kips. Design the tension member so that it is within the recommended slenderness ratio. Assume the angle will be welded at the ends.

1.11. Design a round tension bar to carry 100 kips if F_y = 50 ksi. Use threaded ends.

1.12. Design an eyebar to carry 200 kips. Use A36 steel.

1.13. Design a pin-connected plate to carry 200 kips. Assume 6-in. ϕ pin and A36 steel.

1.14. Design the 14 X 14-in. tension member shown to carry 400 kips if L = 50 ft, F_y = 36 ksi, and bolt diameter = $\frac{3}{4}$ in.

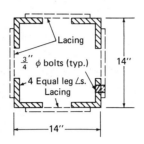

2
Members under Flexure: 1

2.1 MEMBERS UNDER FLEXURE

Flexural members are generally defined as structural members that support transverse loads. This chapter will cover the design of simple flexural members only. Flexural members, commonly called beams, can be categorized into the following:

joists, which are closely spaced beams supporting floors and roofs of buildings
lintels, which span openings in walls, such as doors and windows
spandrel beams, which support the exterior walls of buildings and, in some cases, part of the floor loads
girders, which are generally large beams carrying smaller ones
beams per se

Beams can be designed as simply supported, fixed-ended, partially fixed, or continuous. It is very important that the designer verify that the support conditions of the beam be detailed to satisfy design assumptions for the member to behave as predicted in its analysis.

2.2 DETERMINING THE ALLOWABLE BENDING STRESS

Bending stress in a beam is determined by the flexure formula

$$f_b = \frac{Mc}{I} \tag{2.1}$$

where M is the bending moment, c is the distance of the extreme fibers of the beam from the neutral axis, and I is the moment of inertia of the cross section. Because the section modulus of a beam is defined as the value I/c, the flexure

35

Fig. 2.1. A wide-flange beam and girder floor system.

formula becomes

$$f_b = \frac{M}{S} \qquad (2.2)$$

where S is the section modulus.

Example 2.1. Calculate the maximum bending stress f_b due to a 170-ft-k moment about the strong axis on a:

a) W 12 × 65 wide-flange beam
b) W 18 × 65 wide-flange beam

Solution.

a) For a W 12 × 65, d = 12.12 in.; I_x = 533 in.^{4}

Because the neutral axis is at the midpoint of the section,

$$c = \frac{12.12 \text{ in.}}{2} = 6.06 \text{ in.}$$

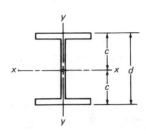

$$f_b = \frac{Mc}{I} = \frac{170 \text{ ft-k} \times 12 \text{ in./ft} \times 6.06 \text{ in.}}{533 \text{ in.}^4}$$

$$f_b = 23.19 \text{ ksi}$$

b) For a W 18 × 65, d = 18.35 in.; I_x = 1070 in.^{4}

$$c = \frac{d}{2} = \frac{18.35 \text{ in.}}{2} = 9.175 \text{ in.}$$

$$f_b = \frac{Mc}{I} = \frac{170 \text{ ft-k} \times 12 \text{ in./ft} \times 9.175 \text{ in.}}{1070 \text{ in.}^4}$$

$$f_b = 17.49 \text{ ksi}$$

Example 2.2. Determine the section modulus of a W 12 × 79 beam, and calculate the maximum bending stress according to

$$f_b = \frac{M}{S}, \quad M = 170 \text{ ft-k}$$

Solution.

$$S = \frac{I}{c}$$

For a W 12 × 79, d = 12.38 in.; I_x = 662 in.^{4}; $c = d/2$ = 6.19 in.

$$S = \frac{I}{c} = \frac{662 \text{ in.}^4}{6.19 \text{ in.}} = 106.9 \text{ in.}^3$$

From the AISCM, S = 107 in.^{3} for a W 12 × 79 (value checks)

$$f_b = \frac{M}{S} = \frac{170 \text{ ft-k} \times 12 \text{ in./ft}}{107 \text{ in.}^3} = 19.07 \text{ ksi}$$

The allowable bending stress, as determined by the AISC, can be found in AISCS 1.5.1.4. Under usual conditions, the allowable bending stress is F_b = 0.66 F_y (AISCS 1.5.1.4.1). To qualify, a member must meet the requirements of flange compactness and lateral support of the compression flange as specified. *Compact* flexural members are defined as those capable of developing their full plastic moment before localized buckling can occur (AISCS 1.5.1.4.1). The yield stresses beyond which a shape is not compact are labeled F_y' and F_y''' and are given in the properties tables. These limits are established from subparagraphs 2 and 4 of AISCS 1.5.1.4.1. Almost all of the W and S shapes are compact for A36 steel, and only some are not compact for 50-ksi steel.

Members bent about their major axis and having an axis of symmetry may fail by buckling of their compression flange and twisting about the longitudinal axis. To avoid this, the member must be "laterally braced" within certain intervals to resist such buckling.

In most situations, the compression flange is laterally supported, and therefore F_b = 0.66 F_y. At times, however, it is not possible to brace the compression flange. For such cases, F_b = 0.66 F_y can be used, provided the interval of lateral support is less than 76.0 b_f/F_y and 20,000/$(d/A_f)F_y$ (AISCS 1.5.1.4.1). If any of the above requirements cannot be satisfied, the allowable stress shall be reduced to 0.60 F_y or to a more restrictive value indicated below.

The AISC, in Section 1.5.1.4.5, specifies the stresses that can be used for certain unbraced lengths. Considering lateral instability, F_b = 0.66 F_y, provided the unsupported length of a beam is less than or equal to an established length denoted by L_c. The value F_b = 0.60 F_y may be used when the unsupported length falls between L_c and another established length L_u. For unsupported lengths greater than L_u, F_b must be the largest value computed from Eqs. 1.5-6a, 1.5-6b, or 1.5-7, but must not exceed 0.60 F_y (see Fig. 2.2).

In the case of plates and rolled shapes bent about their weak axis and square and round bars, the allowable stress is

$$F_b = 0.75 F_y \tag{2.3}$$

If a rolled shape is not compact, but the flange width-to-thickness ratio is

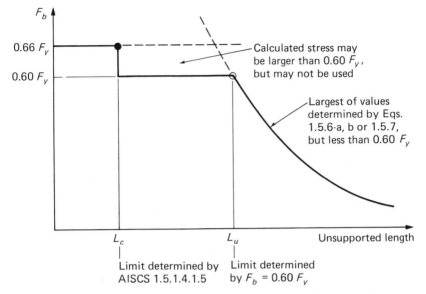

Fig. 2.2. Bending stress versus unsupported length. In the range of relatively short unsupported lengths, the allowable stress is given as a step function. For longer unsupported lengths, the stress is determined by a hyperbolic type function.

$65\sqrt{F_y} < b_f/2t_f < 95/\sqrt{F_y}$, then

$$F_b = F_y\left[1.075 - 0.005\left(\frac{b_f}{2t_f}\right)\sqrt{F_y}\right] \tag{2.4}$$

Otherwise, Appendix C must be consulted for the use of a reduction factor (AISCS 1.5.1.4.3).

2.3 LATERAL SUPPORT OF BEAMS

When a transverse load is applied to a beam, the compression flange behaves in the same manner as a column. As the length of the member increases, the flange tends to buckle. The resulting displacements in the weaker axis will induce torsion and may ultimately cause failure. Members bent solely about their minor axis need not be braced. However, they need to satisfy requirements of AISCS 1.9.2. Box members very seldom need to be braced, subject to requirements of AISCS 1.5.1.4.1 and 1.5.1.4.4. Members bent about their major axis may need their compression flanges braced to prevent lateral instability.

What constitutes lateral support is at times a matter of judgment. A beam encased in a concrete slab is fully laterally supported. Cross beams framing into

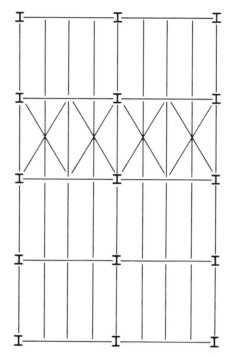

Fig. 2.3. Lateral bracing of a floor or roof framing system. Bracing in one bay can offer rigidity for several bays.

the sides of beams provide lateral support *if* an adequate connection is made to the compression flange. However, care must be taken to provide rigidity to the cross beams. It may be necessary to provide diagonal bracing in one section to resist movement in both directions. Bracing as shown in Fig. 2.3 will provide rigidity for several bays.

Metal decking, in some cases, does not constitute lateral bracing. With adequate connections, up to full lateral support may be assumed. Cases of partial support are usually transformed to full support by a multiple of the actual spacing. For instance, decking that is tack-welded every 4 ft may be considered to provide a third of the full lateral support, yielding an equivalent full lateral support every 12 ft.

2.4 DESIGN OF BEAMS FOR FLEXURE

The design of a beam for flexure is the selection of a bending member having a section modulus slightly larger than or equal to the one required. The section modulus is obtained by dividing the bending moment by the allowable bending

Fig. 2.4. Lateral bracing for a floor system made with channel sections.

stress, $S = M/F_b$, from Eq. 2.2. To select a member with adequate section modulus, use the properties tables or the allowable stress design selection tables. Because one of the objectives of good design is economy, selecting the lightest rolled section that has the required S is desirable. It is implicitly understood that the requirements of space limitations and deflections, generally governed by building codes, are satisfied. The S_x tables in Part 2 of AISCM are especially useful for selecting adequately braced rolled shapes of 36 and 50 ksi steels, for these tables are composed in groups where the top member in each group, printed in heavier print, has the largest S and the least weight in the group. These tables can be entered either by required section modulus or by resisting moment. Values of M_r are valid only for beams with unbraced lengths less than or equal to L_c, or L_u if no value of L_c is listed. The AISC has provided a design aid for unbraced lengths greater than allowable, which will be further discussed in Section 2.9.

Examples 2.3 through 2.7. Select the most economical W sections for the beams shown. Assume full lateral support and compactness. Neglect the weight of the beam.

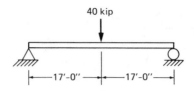

Solution 2.3.

$$M = \frac{PL}{4} = \frac{40 \text{ k} \times 34 \text{ ft}}{4} = 340 \text{ ft-k}$$

$F_b = 0.66 \, F_y$ for full lateral support   (AISCS 1.5.1.4.1)

$F_b = 0.66 \times 36 \text{ ksi} = 24.0 \text{ ksi}$ (for A36 steel)

$$S_{req} = \frac{M}{F_b} = \frac{340 \text{ ft-k} \times 12 \text{ in./ft}}{24 \text{ ksi}} = 170 \text{ in.}^3$$

W Shapes that are satisfactory (from properties tables)

W 12 × 136	$S = 186$ in.^{3}
W 14 × 109	$S = 173$
W 16 × 100	$S = 175$
W 18 × 97	$S = 188$
W 21 × 83	$S = 171$
W 24 × 76	$S = 176$
W 27 × 84	$S = 213$
W 30 × 99	$S = 269$
W 33 × 118	$S = 359$
W 36 × 135	$S = 439$

The last number represents the member weight per foot length. Therefore, W 24 × 76 is the lightest satisfactory section.

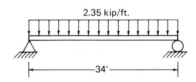

Solution 2.4.

$$M_{max} = \frac{wL^2}{8} = \frac{2.35 \text{ k} \times (34 \text{ ft})^2}{8} = 339.6 \text{ ft-k}$$

$$F_b = 0.66 \, F_y = 24.0 \text{ ksi}   \text{(AISCS 1.5.1.4.1)}$$

$$S_{req} = \frac{M}{F_b} = \frac{339.6 \text{ ft-k} \times 12 \text{ in./ft}}{24 \text{ ksi}} = 169.8 \text{ in.}^3$$

W shapes that are satisfactory are the same W shapes that are satisfactory for Example 2.3.

The lightest section is W 24 × 76, S_x = 176 in.³

Alternative approach

$$S_{req} = 169.8 \text{ in.}^3$$

From S_x tables (Part 2), W 24 × 76 (in bold print) has an S_x of 176 in.³ and for F_y = 36 ksi can carry a moment of 352 ft-k.
  Use W 24 × 76

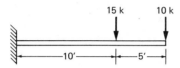

Solution 2.5.

$$M_{max} = (15 \text{ k} \times 10 \text{ ft}) + (10 \text{ k} \times 15 \text{ ft}) = 300 \text{ ft-k}$$

$$F_b = 24.0 \text{ ksi}$$

$$S_{req} = \frac{M}{F_b} = \frac{300 \text{ ft-k} \times 12 \text{ in./ft}}{24.0 \text{ ksi}} = 150 \text{ in.}^3$$

Through investigation, W 24 × 68 is the lightest section, S = 154 in.³

Alternative approach

$$M_{req} = 300 \text{ ft-k}$$

From the AISCM, M_R between 280 ft-k and 308 ft-k, W 24 × 68 (in bold print) has a moment capacity of 308 ft-k.

W 24 × 68 is satisfactory and is the lightest section.

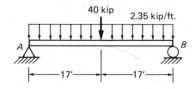

Solution 2.6. To determine the maximum moment, either (a) draw shear and moment diagrams, (b) use the method of sections, or (c) use superposition principles.

a) Total load = 40 k + (2.35 k/ft × 34 ft) = 119.9 k

$$R_A = R_B = \frac{119.9 \text{ k}}{2} = 59.95 \text{ k--say } 60 \text{ k}$$

Shear just to the left of the 40-k load

$$60 \text{ k} - (2.35 \text{ k/ft} \times 17 \text{ ft}) = 20 \text{ k}$$

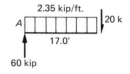

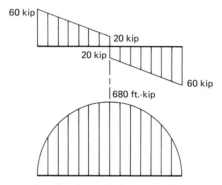

M_{max} at center = area under the shear diagram

$$M_{max} = \frac{60.0 \text{ k} + 20.0 \text{ k}}{2} \times 17 \text{ ft} = 680 \text{ ft-k}$$

b) $R_A = R_B = 60.0$ k

M_{max} at center = (60.0 k × 17 ft) - ((2.35 k/ft × 17 ft) × (17 ft/2)) = 680 ft-k

c) By inspection, the maximum moment is at the center.

Moment due to 40-kip load = $PL/4$

$$\frac{40 \text{ k} \times 34 \text{ ft}}{4} = 340 \text{ ft-k}$$

Moment due to distributed load = $wL^2/8$

$$\frac{2.35 \text{ k/ft} \times (34 \text{ ft})^2}{8} = 339.6 \text{ ft-k--say } 340 \text{ ft-k}$$

Total moment = 340 ft-k + 340 ft-k = 680 ft-k.

Using the AISCM beam chart, select a W 33 × 118 (in bold print) as the lightest satisfactory section with the moment capacity of 718 ft-k for A36 steel.

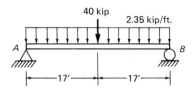

Solution 2.7. Redesign the beam in Example 2.6 for 50-ksi steel.

$$M_{max} = 680 \text{ ft-k} \quad \text{(from Example 2.6)}$$

$$F_b = 0.66 \, F_y = 0.66 \times 50 \text{ ksi} = 33.0 \text{ ksi}$$

$$S_{req} = \frac{M}{F_b} = \frac{680 \text{ ft-k} \times 12 \text{ in./ft}}{33.0 \text{ ksi}} = 247 \text{ in.}^3$$

Using the AISCM beam chart for F_y = 50 ksi, select a W 30 × 99 as the lightest satisfactory section, S = 269 in.^{3}

Example 2.8. Determine the most economical wide-flange section for the loading case shown. Assume full lateral support. Include the weight of the beam.

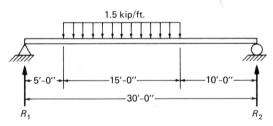

Solution. Reactions, shears, and moments can be determined from the beam diagrams and formulas (AISCM p. 2-114). Use diagram 1 for beam weight and diagram 4 for imposed loading.

Imposed Loading

$$M_{max} = R_1\left(a + \frac{R_1}{2w}\right) \text{ at } x = a + \frac{R_1}{w}$$

$$R_1 = \frac{wb}{2L}(2c + b)$$

$$a = 5 \text{ ft}, \quad b = 15 \text{ ft}, \quad c = 10 \text{ ft}, \quad L = 30 \text{ ft}$$

$$w = 1.5 \text{ k/ft}$$

$$R_1 = \frac{1.5 \text{ k/ft} \times 15 \text{ ft}}{2 \times 30 \text{ ft}} (2(10 \text{ ft}) + 15 \text{ ft}) = 13.13 \text{ k}$$

$$M_{max} = 13.13 \text{ k} \left(5 \text{ ft} + \frac{13.13 \text{ k}}{2(1.5 \text{ k/ft})}\right) = 123.12 \text{ ft-k}$$

$$x = 5 \text{ ft} + \frac{13.13 \text{ k}}{1.5 \text{ k/ft}} = 13.75 \text{ ft}$$

Beam Loading

Assume beam weight is 40 lb/ft

Determine moment at point of maximum applied moment.

$$x = 13.75 \text{ ft}$$

$$M_x = \frac{wx}{2}(L - x) = \frac{0.04 \text{ k/ft} \times 13.75 \text{ ft}}{2}(30.0 \text{ ft} - 13.75 \text{ ft}) = 4.47 \text{ ft-k}$$

Total Maximum Moment

$$123.12 \text{ ft-k} + 4.47 \text{ ft-k} = 127.6 \text{ ft-k}$$

Consulting the allowable stress design selection table (AISCM, Part 2), find the lightest beam with a resisting moment (M_R) greater than or equal to 127.6 ft-k $(F_y = 36 \text{ ksi})$.

Use W 16 $\times$ 40 $(M_R = 129 \text{ ft-k})$.

Example 2.9. Determine the distributed load that can be carried by a W 14 $\times$ 22 beam spanning 11 ft with an unsupported length of 5 ft, 6 in.

Solution. For a W 14 $\times$ 22,

$$L_c = 5.3 \text{ ft}; \quad L_u = 5.6 \text{ ft}; \quad S = 29.0 \text{ in.}^3 \quad \text{(from Part 2 of AISCM)}$$

Because the unsupported $L = 5.5$ ft is between L_c and L_u,

$$F_b = 0.6 \, F_y = 22.0 \text{ ksi} \quad \text{AISCS 1.5.1.4.5}$$

$$M = S \times F_b = 29.0 \text{ in.}^3 \times 22 \text{ ksi} \times \frac{1}{12 \text{ in./ft}} = 53 \text{ ft-k}$$

$$M = \frac{wL^2}{8}, \quad w = \frac{8M}{L^2} = \frac{8 \times 53 \text{ ft-k}}{(11.0)^2} = 3.5 \text{ k/ft}$$

Subtracting the weight of the beam,

$$w = 3.5 \text{ k/ft} - 0.022 \text{ k/ft} = 3.48 \text{ k/ft}$$

Example 2.10. Determine the distributed load that can be carried by a W 10 X 22 beam spanning 16 ft, 6 in. with an unsupported length of 5 ft, 6 in. $F_y = 50$ ksi.

Solution. For a W 10 X 22,

$$L_c = 5.2 \text{ ft}; \quad L_u = 6.8 \text{ ft}; \quad S = 23.2 \text{ in.}^3$$

Because the unsupported $L = 5.5$ ft is between L_c and L_u,

$$F_b = 0.6\,F_y = 0.6 \times 50 \text{ ksi} = 30 \text{ ksi} \quad \text{AISCS } 1.5.1.4.5$$

$$M = S \times F_b = 23.2 \text{ in.}^3 \times 30 \text{ ksi} \times \frac{1}{12 \text{ in./ft}} = 58.0 \text{ ft-k}$$

$$w = \frac{8\,M}{L^2} = \frac{8 \times 58.0 \text{ ft-k}}{(16.5)^2} = 1.70 \text{ k/ft}$$

Subtracting the weight of the beam,

$$w = 1.70 \text{ k/ft} - 0.02 \text{ k/ft} = 1.68 \text{ k/ft}$$

Example 2.11. Determine the distributed load that can be carried by a W 16 X 31 beam spanning 30 ft with an unsupported length of 15 ft, 0 in.

Solution. For a W 16 X 31,

$$L_c = 5.8 \text{ ft}; \quad L_u = 7.1 \text{ ft}; \quad S = 47.2 \text{ in.}^3$$

Because the unsupported $L = 15$ ft is greater than both L_c and L_u, F_b must be determined by AISCS 1.5.1.4.5

$$r_T = 1.39 \text{ in.}$$

$$\frac{l}{r_T} = \frac{15 \text{ ft} \times 12 \text{ in./ft}}{1.39 \text{ in.}} = 129.5$$

Referring to the inequalities of AISC Section 1.5.1.4.5.2a, we need to determine the relative magnitude of l/r_T. The value of C_b is

$$1.75 + 1.05\ (M_1/M_2) + 0.3\ (M_1/M_2)^2$$

which depends on the magnitude and sign of end moments and can be taken conservatively equal to 1 (AISCS 1.5.1.4.5.2.a footnote).

$$\sqrt{\frac{102 \times 10^3\ C_b}{F_y}} = \sqrt{\frac{102 \times 10^3}{36}} = 53.2$$

$$\sqrt{\frac{510 \times 10^3\ C_b}{F_y}} = \sqrt{\frac{510 \times 10^3}{36}} = 119.0$$

$$\frac{l}{r_T} > 119.0$$

$$F_b = \frac{170 \times 10^3\ C_b}{(l/r_T)^2} = 10.14\ \text{ksi} \qquad\qquad \text{AISCS (1.5-6b)}$$

or

$$F_b = \frac{12 \times 10^3\ C_b}{(l\ d/A_f)} = \frac{12 \times 10^3 \times 1}{180 \times (15.88/2.43)} = 10.21 \qquad \text{AISCS (1.5-7)}$$

The greater value is to be used.

$$F_b = 10.21\ \text{ksi}$$

$$M = S \times F_b = \frac{47.2\ \text{in.}^3 \times 10.21\ \text{ksi}}{12\ \text{in./ft}} = 40.2\ \text{ft-k}$$

$$w = \frac{8\,M}{l^2} = \frac{8 \times 40.2\ \text{ft-k}}{(30\ \text{ft})^2} = 0.36\ \text{k/ft}$$

Subtracting the beam weight to determine the beam capacity,

$$w = 0.36\ \text{k/ft} - 0.031\ \text{k/ft} = 0.33\ \text{k/ft}$$

2.5 SHEAR

For a beam subjected to a positive bending moment, the lower fibers of the member are elongated and the upper fibers are shortened, while, at the neutral axis, the length of the fibers remain unchanged (see Fig. 2.5). Due to these varying deformations, individual fibers have a tendency to slip on the adjacent

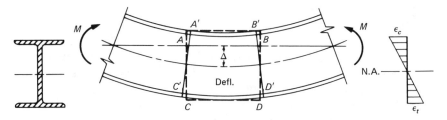

Fig. 2.5. Deformation of a beam in bending. Due to a moment, a section of beam in original position $ABCD$ will deflect a distance Δ. In doing so, compressive strains will develop on the loaded side of the neutral axis, and tensile strains on the opposite side. Under elastic deformation, the strains are linear, with no strain occurring at the neutral axis. The original section $ABCD$ assumes a new position $A'B'C'D'$.

ones. If a beam is built by merely stacking several boards on top of each other and then transversely loaded, it will take the configuration shown in Fig. 2.6(a). If the boards are connected, as in Fig. 2.6(b), the tendency to slip will be resisted by the shearing strength of the connectors. For single-element beams, the tendency to slip is resisted by the shearing strength of the material.

From mechanics of materials, the longitudinal shearing stress in a beam is given by the formula

$$f_v = \frac{VQ}{It} \qquad (2.5)$$

where V is the vertical shear force, Q is the moment of the areas on one side of the sliding interface taken about the centroid of the cross section, I is the moment of inertia of the section, and t is the width of the section where the shearing stress is investigated. For rectangular beams, the maximum shearing stress is given by $f_v = 3V/2A$, where A is the area of the cross section. For rolled and fabricated shapes, the AISC allows the longitudinal shearing stress to be determined by the formula

$$f_v = \frac{V}{A_{\text{web}}} \qquad (2.6)$$

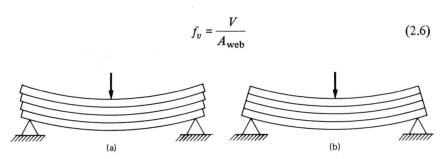

Fig. 2.6. Built-up beam in flexure: (a) transversely unconnected, (b) transversely connected.

where A_{web} is the product of the overall depth of the section d and the thickness of the web t_w (AISCS 1.5.1.2). However, this value is 10 to 15% less than the maximum shearing stress, as determined by the more accurate Equation 2.5. The allowable shearing stress, as determined by the AISC on the gross section of a member, is given by $F_v = 0.40\,F_y$ (AISCS 1.5.1.2). Except for short spans with heavy loading, or heavy concentrated loads close to supports, shear seldom governs in the design of beams.

Example 2.12. Determine the maximum shearing stress for the following sections when the external shear force $V = 75$ kips.

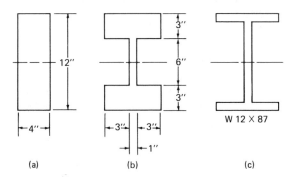

(a) (b) (c)

Solution.

a)
$$f_{v\,max} = \frac{3\,V}{2\,A}$$

$$A = 4 \text{ in.} \times 12 \text{ in.} = 48 \text{ in.}^2$$

$$f_{v\,max} = \frac{3\,V}{2\,A} = \frac{3 \times 75 \text{ k}}{2 \times 48 \text{ in.}^2} = 2.34 \text{ ksi}$$

b)
$$f_{v\,max} = \frac{VQ}{It}$$

By inspection, it is seen that the neutral axis is located at the center, 6 in. from the bottom.

$$I = \Sigma \frac{bh^3}{12} + (Ad^2)$$

$$= 2 \times \left[\frac{7 \times (3)^3}{12} + (3 \times 7)(6 - 1.5)^2 \right] + \frac{1 \times (6)^3}{12} = 900 \text{ in.}^4$$

Because $f_{v\,max}$ is at the neutral axis,

$$Q = (3 \times 7)(6 - 1.5) + (3 \times 1)(1.5) = 99.0 \text{ in.}^3$$

$$t = 1 \text{ in.}$$

$$f_{v\,max} = \frac{VQ}{It} = \frac{75 \text{ k} \times 99 \text{ in.}^3}{900 \text{ in.}^4 \times 1 \text{ in.}} = 8.25 \text{ ksi}$$

Verify with $f_v = \dfrac{V}{A_{web}}$ (AISCS 1.5.1.2.1)

$$f_v = \frac{75 \text{ k}}{12 \text{ in.} \times 1 \text{ in.}} = 6.25 \text{ ksi}$$

As the web is very thick, there is approximately one-third difference in shear values.

c)
$$f_v = \frac{V}{A_{web}}$$

$$A_{web} = d \times t = 12.53 \text{ in.} \times 0.515 \text{ in.} = 6.45 \text{ in.}^2$$

$$f_v = \frac{V}{A_{web}} = \frac{75 \text{ k}}{6.45 \text{ in.}^2} = 11.62 \text{ ksi}$$

Verify with $f_v = \dfrac{VQ}{It}$

$$Q = 0.81 \text{ in.} \times 12.125 \text{ in.} \times \left(\frac{12.53 \text{ in.}}{2} - \frac{0.810 \text{ in.}}{2} \right)$$

$$+ 0.515 \text{ in.} \times \left(\frac{10.91 \text{ in.}}{2} \right)^2 \times \frac{1}{2}$$

$$= 65.21 \text{ in.}^3$$

$$I = 740 \text{ in.}^4$$

$$f_v = \frac{75 \text{ k} \times 65.21 \text{ in.}^3}{740 \text{ in.}^4 \times 0.515 \text{ in.}} = 12.83 \text{ ksi}$$

A difference of 10% occurs between an exact approach and an estimated value. The code is aware of this error which seldom is greater than 15% and compensates in $F_v = 0.40 F_y$.

Example 2.13. Determine if the beam shown is satisfactory. Assume full lateral support. Check both flexural and shearing stresses.

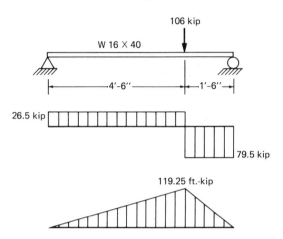

Solution.

$$d = 16.01 \text{ in.}; \quad t_w = 0.305 \text{ in.}; \quad S = 64.7 \text{ in.}^3$$

$$R_R = \frac{106 \text{ k} \times 4.5 \text{ ft}}{6.0 \text{ ft}} = 79.5 \text{ k}$$

$$R_L = 106 \text{ k} - 79.5 \text{ k} = 26.5 \text{ k}$$

M_{max} (from diagram) $= 119.25$ ft-k

The moment resisted by the section is

$$M_R = S \times F_b = 64.7 \text{ in.}^3 \times 24 \text{ ksi} \times \frac{1 \text{ ft}}{12 \text{ in.}} = 129.4 \text{ ft-k} > 119.25 \text{ ft-k} \quad \text{ok}$$

$$V_{max} = 79.5 \text{ k}$$

$$f_v = \frac{V}{A_{web}} = \frac{79.5 \text{ k}}{16.01 \text{ in.} \times 0.305 \text{ in.}} = 16.28 \text{ ksi}$$

$$F_v = 0.40 F_y = 0.4 \times 36 \text{ ksi} = 14.4 \text{ ksi} \quad \text{(AISCS 1.5.1.2)}$$

$$16.28 > 14.4 \text{ ksi} \quad\quad\quad\quad\quad\quad\quad\quad \text{N.G.}$$

W 16 × 40 is not satisfactory for resisting shear.

2.6 HOLES IN BEAMS

Whenever possible, holes in beams should be avoided. If drilling or cutting holes in a beam is absolutely necessary, it is good practice to avoid holes in the web at the locations of large shear and in the flange where the moment is large. Tests have shown that there is little difference in strengths between beams with no holes and beams with holes up to 15% of the gross area of either flange. For this reason, the AISC does not require the subtraction of holes from either flange when the area of the holes does not exceed 15% of the gross area of that flange, and then any deduction made is only for the hole area in excess of the 15% (AISCS 1.10.1). After the allowable net area of the flange is determined, the reduced net moment of inertia may be calculated by considering the net area of the flange.

Example 2.14. Calculate the design section modulus for a W 14 × 145:

a) for two 1-in. diameter holes in each flange
b) for two 2-in. diameter holes in each flange

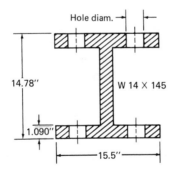

Solution.

$$S_x = 232 \text{ in.}^3; \quad b_f = 15.5 \text{ in.}; \quad t_f = 1.090 \text{ in.}; \quad I_x = 1710 \text{ in.}^4$$

$$A_{fl} = b_f \times t_f = 15.5 \text{ in.} \times 1.090 \text{ in.} = 16.90 \text{ in.}^2$$

a) Determine flange area loss due to two 1-in. diam holes.

$$(1.090 \text{ in.} \times 1 \text{ in.}) \times 2 = 2.18 \text{ in.}^2$$

$$15\% \text{ of flange area} = 0.15 \times 16.90 \text{ in.}^2 = 2.53 \text{ in.}^2$$

No deduction in flange area is necessary if the loss is less than 15% of the gross flange area (AISCS 1.10.1).

$$S = S_{\text{gross}} = 232 \text{ in.}^3$$

b) Determine flange area loss due to two 2-in. diam holes.

$$(1.090 \text{ in.} \times 2 \text{ in.}) \times 2 = 4.36 \text{ in.}^2$$

$$15\% \text{ of flange area} = 2.53 \text{ in.}^2$$

Only the excess of area loss need be deducted from the flange area (AISCS 1.10.1).

$$\text{Excess loss} = 4.36 \text{ in.}^2 - 2.53 \text{ in.}^2 = 1.83 \text{ in.}^2$$

$$\text{Allowable net area} = 16.90 \text{ in.}^2 - 1.83 \text{ in.}^2 = 15.07 \text{ in.}^2$$

$$I_{net} = 1710 \text{ in.}^4 - 1.83 \text{ in.}^2 \times \left(\frac{14.78 \text{ in.}}{2} - \frac{1.09 \text{ in.}}{2}\right)^2 = 1624 \text{ in.}^4$$

$$S_{all} = \frac{I_{net}}{d/2} = \frac{1624 \text{ in.}^4}{14.78 \text{ in.}/2} = 220 \text{ in.}^3$$

Beam end connections using high-strength bolts in relatively thin webs may create a condition of web tear-out. Failure can occur by the combination of shear through the line of bolts and tension across the bolt block. This condition should be considered when designing the bolt connection and should be investigated in situations when the flange is coped or when shear resistance is at a minimum. For further discussion see Chapter 5, Bolts and Rivets.

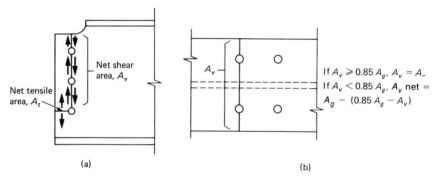

Fig. 2.7. Holes in beams. (a) Holes in beam webs should be avoided at locations of large shear. When holes are necessary, as in the end connection shown, the section must be checked for web tear-out. (b) Flange holes should be avoided at regions of large moments. If and when holes are provided in a beam flange, only the area of the holes in excess of 15% of the gross flange area must be deducted for computing net flange area.

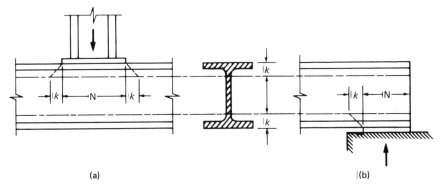

(a) |(b)

Fig. 2.8. Beam web crippling. The applied load or end reaction is transferred to the web from the flange at the web-flange intersection over a length extending a distance k on either side: (a) interior concentrated load, (b) end reaction.

2.7 WEB CRIPPLING

Beams may fail by web crippling at points of large stress concentrations due to concentrated loads or reactions. This condition can be obviated by the use of vertical web stiffeners. Failure occurs at the toe of the fillet where the beam transfers compression from a relatively wide flange to a narrow web. At this point, the AISC allows a maximum compressive stress of $0.75 F_y$ (AISCS 1.10.10).

Formulas have been developed by considering that the stress in the web spreads along the web with an angle of $45°$ and that the crippling stress is calculated at the end of the fillet, or the intersection of the web and flange. These formulas, as provided by AISC, are

$$\frac{R}{t(N + 2k)} \leqslant 0.75 F_y \qquad (2.7)$$

for interior loads (Fig. 2.8(a)) and

$$\frac{R}{t(N + k)} \leqslant 0.75 F_y \qquad (2.8)$$

for end reactions (Fig. 2.8(b)).

Example 2.15. Select the lightest section for the beam shown and determine the bearing length required for each support, in order not to use bearing stiffeners. Assume full lateral support and 50 ksi steel.

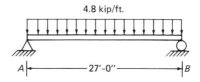

Solution.

$$W = 4.8 \text{ k/ft} \times 27 \text{ ft} = 130 \text{ k}$$

$$M = WL/8 = 437.4 \text{ ft-k}$$

Referring to AISC beam selection table,
Use W 24 × 76

$$R_A = \frac{W}{2} = \frac{130.0 \text{ k}}{2} = 65.0 \text{ k}$$

$$R_B = \frac{W}{2} = 65.0 \text{ k}$$

$$\frac{R}{t(N + k)} \leqslant 0.75 \, F_y \quad \text{(AISCS 1.10.10.1)}$$

Transposing, we obtain the minimum bearing length

$$N = \frac{R}{t_w \, (0.75 \, F_y)} - k$$

$$t_w = 0.440 \text{ in.}$$

$$k = 1\frac{7}{16} \text{ in.} = 1.44 \text{ in.} \quad (k \text{ can be found in W shapes dimensions})$$

Bearing length at A and B

$$N = \frac{65.0 \text{ k}}{0.44 \text{ in.} \, (0.75 \times 50.0 \text{ ksi})} - 1.44 \text{ in.} = 2.50 \text{ in.}$$

Provide a minimum of $2\frac{1}{2}$ in. end bearing.

Example 2.16. A column is supported by a transfer girder as shown.

a) Determine the lightest girder.
b) Determine safety against shear.
c) Determine lengths of bearing under the column and at the supports.

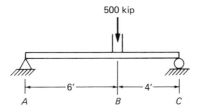

Solution.

a)
$$R_A = \frac{500 \text{ k} \times 4 \text{ ft}}{10 \text{ ft}} = 200.0 \text{ k}$$

$$R_B = 500 \text{ k} - 200 \text{ k} = 300 \text{ k}$$

$$M_{\max} = 200 \text{ k} \times 6 \text{ ft} = 1200 \text{ ft-k}$$

Use W 36 $\times$ 182 as the lightest available section.

b)
$$f_v = \frac{V}{A_{\text{web}}} = \frac{300 \text{ k}}{36.33 \text{ in.} \times 0.725 \text{ in.}} = 11.39 \text{ ksi}$$

$$F_v = 0.40 \, F_y = 14.4 \text{ ksi} > 11.39 \text{ ksi} \quad \text{ok}$$

W 36 $\times$ 182 is safe against shear.

c)
$$\frac{R}{t_w(N+2k)} \leqslant 0.75 \, F_y; \quad N \geqslant \frac{R}{t_w(0.75\,F_y)} - 2k \quad \text{for interior loads}$$

$$\frac{R}{t_w(N+k)} \leqslant 0.75 \, F_y; \quad N \geqslant \frac{R}{t_w(0.75\,F_y)} - k \quad \text{for end bearing}$$

At point A,

$$N = \frac{200.0 \text{ k}}{0.725 \text{ in.} \, (0.75 \times 36.0 \text{ ksi})} - 2.125 \text{ in.} = 8.1 \text{ in.} - \text{say } 8\frac{1}{4} \text{ in.}$$

At point B,

$$N = \frac{500.0 \text{ k}}{0.725 \text{ in.} \, (0.75 \times 36.0 \text{ ksi})} - (2 \times 2.125 \text{ in.}) = 21.3 \text{ in.} - \text{say } 21\frac{1}{2} \text{ in.}$$

At point C,

$$N = \frac{300.0 \text{ k}}{0.725 \text{ in.} \, (0.75 \times 36.0 \text{ ksi})} - 2.125 \text{ in.} = 13.2 \text{ in.} - \text{say } 13\frac{1}{4} \text{ in.}$$

2.8 DEFLECTIONS

Allowable deflections of beams are usually limited by codes and may need to be verified as part of the beam selection. The AISC sets the limit for live load deflection of beams supporting plastered ceilings at 1/360 of the span (AISCS 1.13.1). The deflection of a member is a function of its moment of inertia. To facilitate design, Part 2 of the AISCM has tabulated values of I_x and I_y for all W and M shapes, grouped in order of magnitude, where the shapes in bold type are the most economical in their respective groups.

Example 2.17. Show that beams of the same approximate depth have the same deflection when loaded to $f_b = F_b$ for the same span.

Solution.

$$M_1 = \frac{w_1 l^2}{8} = F_b S_1 = F_b \frac{I_1}{2d} \quad d = \text{depth}$$

$$M_2 = \frac{w_2 l^2}{8} = F_b S_2 = F_b \frac{I_2}{2d}$$

$$\Delta_1 = \frac{5 w_1 l^4}{384 EI_1} = \frac{5}{384} \frac{w_1 l^2}{8} \frac{8 l^2}{EI_1}$$

$$= CF_b \frac{I_1}{2d} \times l^2 \times \frac{1}{EI_1} = CF_b \frac{l^2}{2dE}, \quad C = \frac{40}{384}$$

Similarly, we find that Δ_2 is equal to the same quantity. Hence, we see the deflection is dependent on the depth and the span for the same F_b and E, and not on the load.

Example 2.18. Determine if the beam in Example 2.15 is adequate to limit deflection to $l/360$. Choose the next lightest satisfactory section if the maximum deflection is exceeded.

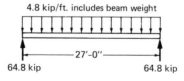

4.8 kip/ft. includes beam weight

27'-0"

64.8 kip   64.8 kip

Solution.

$$\Delta_{max} = \frac{5 wl^4}{384 EI}$$

$$E = 29{,}000 \text{ ksi}$$

For W 24 × 76, I_x = 2100 in.^{4}

$$\Delta_{max} = \frac{5(4.8/12)\text{k/in. } ((27 \times 12) \text{ in.})^4}{384 \,(29{,}000 \text{ ksi}) \, 2100 \text{ in.}^4} = 0.942 \text{ in.}$$

$$\frac{l}{360} = \frac{27 \times 12}{360} = 0.90 \text{ in.} < 0.942 \text{ in.} \qquad \text{N.G.}$$

Determine next lightest satisfactory section

$$\frac{5 \, wl^4}{384 \, EI} \leqslant \frac{l}{360}$$

$$I_x \geqslant \frac{5 \, wl^4}{384 \, E} \times \frac{1}{l/360}$$

$$I_x \geqslant 2199 \text{ in.}^4$$

Choosing from the moment of inertia selection table,

$$\text{W } 24 \times 84, \qquad I_x = 2370 \text{ in.}^4$$

$$\Delta_{max} = 0.835 \text{ in.} \qquad \text{ok}$$

Use **W** 24 × 84.

Example 2.19. Design the beam B and girder G for the framing plan shown. Assume full lateral support, A242Gr.50 steel, and the maximum depth of members to be 22 in. The floor is 5 in. reinforced concrete and carries a live load of 150 lb/ft^2. Assume there is no flooring, ceiling, or partition. Furthermore, assume that there is no restriction on deflection, but the value is requested.

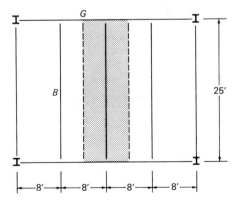

Solution. Because the weight of reinforced concrete is 150 lb/ft^3, the weight of

5 in. of concrete is

$$\frac{5 \text{ in.}}{12 \text{ in./ft}} \times 150 \text{ lb/ft}^3 = 62.5 \text{ lb/ft}^2$$

The tributary width of the beams is 8 ft.

$$DL_{conc} = 62.5 \text{ lb/ft}^2 \times 8 \text{ ft} = 500 \text{ lb/lin. ft of beam}$$

$$LL = 150 \text{ lb/ft}^2 \times 8 \text{ ft} = 1200 \text{ lb/lin. ft}$$

Assume weight of beam = 50 lb/lin. ft

Total distributed loading on the beams

$$w = (500 + 1200 + 50) \text{ lb/lin. ft} = 1.75 \text{ k/ft}$$

$$M_{max} = \frac{wL^2}{8} = \frac{1.75 \text{ k/ft} \times (25 \text{ ft})^2}{8} = 136.7 \text{ ft-k}$$

$$F_y = 50 \text{ ksi} \quad \text{(AISCS, Table 1)}$$

A W 18 × 35 may be selected as the lightest satisfactory section.

Because the beams frame into the girders, they are exerting their end reactions as concentrated loads on the girders. Each girder carries the end reactions of three beams on either side for a total of six concentrated loads.

The end reaction of each beam is the total load on the beam divided by 2.

$$R = \frac{1.75 \text{ lb/ft} \times 25 \text{ ft}}{2} = 21.875 \text{ k}$$

Assume weight of girder to be 150 lb/lin. ft.

End reactions of the girder

$$\frac{3 \times 2(21.875 \text{ k}) + (0.15 \text{ k/ft} \times 32 \text{ ft})}{2} = 68.0/k$$

Moment at midspan

$$(68 \text{ k} \times 16 \text{ ft}) - (43.75 \text{ k} \times 8 \text{ ft}) - (0.15 \text{ k/ft} \times 16 \text{ ft} \times 8 \text{ ft}) = 718.8 \text{ ft-k}$$

A W 30 X 99 may be selected as the lightest section.   However, the depth of W 30 X 99 = 29.64 in. > 22.00 in.   N.G.

$$S_{req} = \frac{M}{F_b} = \frac{718.8 \text{ ft-k} \times 12 \text{ in./ft}}{0.66 \times 50 \text{ ksi}} = 261.4 \text{ in.}^3$$

From the properties tables, select W 21 X 122 as the shallowest member with $S > 261.4 \text{ in.}^3$

$$S = 273 \text{ in.}^3, \quad d = 21.68 \text{ in.} \quad \text{ok}$$

Use W 21 X 122 girder.

Check deflection

Beam B

$$\Delta_{max} = \frac{5 wL^4}{384 EI}, \quad w = \frac{1.75 \text{ k/ft}}{12 \text{ in./ft}} = 0.15 \text{ k/in.}$$

$$I = 510 \text{ in.}^4$$

$$E = 29{,}000 \text{ ksi}$$

$$\Delta_{max} = \frac{5 \times 0.15 \text{ k/in.} \times (25 \text{ ft} \times 12 \text{ in./ft})^4}{384 \times 29{,}000 \text{ ksi} \times 510 \text{ in.}^4} = 1.04 \text{ in.}$$

Girder G

From tables 1, 7, and 9 of beam diagrams and formulas, AISCM p. 2-114

$$\Delta_{max} = \frac{5 wl^4}{384 EI} + \frac{Pl^3}{48 EI} + \frac{Pa}{24 EI} (3l^2 - 4a^2)$$

$$a = 8 \text{ ft} \times 12 \text{ in./ft} = 96 \text{ in.}$$

$$I = 2960 \text{ in.}^4$$

$$\Delta_{max} = \frac{5 \times 0.122 \times (32 \times 12)^4}{384 \times 29000 \times 2960} + \frac{43.75 \times (32 \times 12)^3}{48 \times 29000 \times 2960}$$

$$+ \frac{43.75 \times 96}{24 \times 29000 \times 2960} [3 \times (32 \times 12)^2 - 4 \times (96)^2]$$

$$= 0.402 + 0.601 + 0.827 = 1.83 \text{ in.}$$

2.9 USE OF AISCM DESIGN AIDS

In Part 2 of the eighth edition of the AISCM, beams tables are presented in a manner different from former editions. Total allowable uniformly distributed loads and deflections for laterally supported beams can be obtained by applying appropriate coefficients in equations dependent upon span length. The product of the total uniform design load and span length yields a constant W_c from which the beam size is selected. Tabulated values in the "Beams" tables include maximum web shear, maximum end reactions, and section modulus. Values of L_c and L_u are included to verify lateral support. The maximum span length for which beam web shear controls ($L_v = W_c/2\ V$) exemplifies the transition from the critical shear condition of heavily loaded short beams to the consideration of flexure in long spans. The constant D_c is given from which the maximum midspan deflection for a known span length can be obtained. The approximate deflection due to design loads can be calculated by multiplying the maximum deflection by the ratio of the total design load to the total allowable load. However, in cases where deflection is critical, it should be calculated manually.

Equally spaced, concentrated loads of equal magnitude can also be used with the "Beams" tables, provided appropriate coefficients are applied. The table of concentrated load equivalents on AISCM p. 2-113 provides the equivalent uniform load and deflection coefficients for various loading cases. If the span length is less than L_v, then web shear controls and the sum of the concentrated loads must be less than or equal to 2 V.

Example 2.20. Use the beams table to verify the sections chosen in Examples 2.4, 2.3, and 2.6. Beam weight is taken into account in the tables and should not be included in total weight.

Solution.

a) Verify the beam design of Example 2.4.

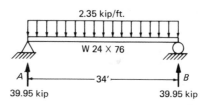

Total load = 79.9 k
Maximum allowable total load

$$W_{\text{all}} = \frac{W_c}{L}$$

$$W_c = 2820 \text{ ft-k} \quad \text{(from beams table)}$$

$$W_{\text{all}} = \frac{2820 \text{ ft-k}}{34.0 \text{ ft}} = 82.94 \text{ k} > 79.9 \text{ k} \quad \text{ok}$$

Maximum end reaction

$$R_{\text{all}} = 58.7 \text{ k for } 3\tfrac{1}{2} \text{ in. bearing} \quad \text{(from beams table)}$$

$$R_A = R_B = 39.95 \text{ k} < 58.7 \text{ k} \quad \text{ok}$$

Maximum deflection

$$\Delta_{\text{all}} = \frac{l}{360} = \frac{34.0 \text{ ft} \times 12 \text{ in./ft}}{360} = 1.13 \text{ in.}$$

$$\Delta_{\text{max}} = D_c \times L^2 / 1000$$

$$D_c = 1.0 \quad \text{(from beams table)}$$

$$\Delta_{\text{max}} = 1.0 \times (34.0)^2 / 1000 = 1.16 \text{ in. for } W_{\text{all}} = 82.94 \text{ k}$$

$$\Delta_{\text{max}} = \frac{79.9 \text{ k}}{82.94 \text{ k}} (1.16 \text{ in.}) = 1.12 \text{ in.}$$

The exact deflection is

$$\Delta_{\text{max}} = \frac{5 \, wl^4}{384 \, EI}$$

$$w = 2.35 \text{ k/ft} + 0.076 \text{ k/ft} = 2.43 \text{ k/ft}$$

$$I = 2100 \text{ in.}^4$$

$$\Delta_{\text{max}} = \frac{5\left(\dfrac{2.43 \text{ k/ft}}{12 \text{ in./ft}}\right)(34 \text{ in.} \times 12 \text{ in./ft})^4}{384 \, (29{,}000 \text{ ksi}) \, 2100 \text{ in.}^4} = 1.20 \text{ in.} > 1.13 \text{ in. allowable}$$

The approximate deflection from the beams table is 7% in error with exact deflection. Larger span deflections computed from the table may be underestimated. Deflection in critical cases should be checked by exact solution. In this case, the member chosen should not be used to support plastered ceilings.

b) Verify the beam design of Example 2.3.

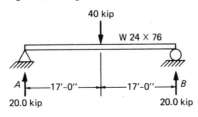

The equivalent uniform load for a concentrated load at the center is $(f \times P)$ (AISCM p. 2-113).

$$f = 2$$

$$W_{equiv} = (2 \times 40.0 \text{ k}) = 80.0 \text{ k}$$

Maximum allowable total load

$$W_{all} = \frac{W_c}{L}$$

$$W_{all} = \frac{2820 \text{ ft-k}}{34.0 \text{ ft}} = 82.94 \text{ k} > 80.0 \text{ k} \quad \text{ok}$$

Maximum end reaction

$$R_{all} = 58.7 \text{ k for } 3\tfrac{1}{2} \text{ in. bearing}$$

$$R_A = R_B = 20.0 \text{ k} < 58.7 \text{ k} \quad \text{ok}$$

Maximum deflection

The equivalent deflection coefficient for a concentrated load at the center is (g).

$$g = 0.80$$

$$\Delta_{all} = \frac{l}{360} = 1.13 \text{ in.}$$

$$\Delta_{max} = 0.8 \, (D_c \times L^2/1000) = 0.925 \text{ in. for } W_{all} = 82.94 \text{ k}$$

$$\Delta_{max} = \frac{80.0 \text{ k}}{82.94 \text{ k}} (0.925 \text{ in.}) = 0.892 \text{ in.}$$

c) Verify the beam design of Example 2.6.

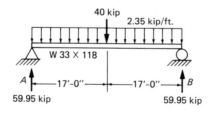

From part a)

$$W_t = 79.9 \text{ k}$$

From part b)

$$W_t = 80.0 \text{ k}$$

Total load on beam

$$W_{total} = 79.9 \text{ k} + 80.0 \text{ k} = 159.9 \text{ k (uniform load equivalent)}$$

Maximum allowable total load

$$W_{all} = \frac{W_c}{L} = \frac{5740 \text{ ft-k}}{34.0 \text{ ft}} = 168.82 \text{ k} > 159.9 \text{ k} \quad \text{ok}$$

Maximum end reaction

$$R_{all} = 75.2 \text{ k for } 3\frac{1}{2} \text{ in. bearing}$$

$$R_A = R_B = \left(\frac{79.9 \text{ k} + 40.0 \text{ k}}{2}\right) = 59.95 \text{ k} < 75.2 \text{ k} \quad \text{ok}$$

Maximum deflection

For deflection due to uniform load, $D_c = 0.76$

$$\Delta = \left(\frac{79.9 \text{ k}}{168.82 \text{ k}}\right) (0.76 \times (34.0)^2 / 1000) = 0.416 \text{ in.}$$

For deflection due to concentrated force, $D_c = 0.76, g = 0.8$

$$\Delta = \left(\frac{80.0 \text{ k}}{168.22 \text{ k}}\right) (0.8 \times 0.76 \times (34.0)^2 / 1000) = 0.333 \text{ in.}$$

$$\Delta_{total} = 0.416 \text{ in.} + 0.333 \text{ in.} = 0.749 \text{ in.}$$

Beams that are not fully laterally supported, cannot be designed using the above-mentioned tables, unless the interval of lateral support is less than the length L_c. For laterally unsupported beams, the allowable bending stress F_b must be first determined by AISCS 1.5.1.4.5. Because this procedure involves lengthy calculations and numerous checks, the AISC has established "Allowable Moments in Beams" charts to aid design. C_b has been taken as unity in the charts, which is an accepted conservative design assumption. To select a member for a given moment and unbraced length, enter an appropriate chart at the required resisting moment (ordinate) and proceed to the right to meet the vertical line corresponding to the unbraced length (abscissa). Any beam located above and to the right of the intersection is satisfactory, with the beam shown in solid line immediately to the right and above being the lightest satisfactory section. Beams indicated by broken lines satisfy the requirements of unbraced length and moment, but are not the lightest sections available for those conditions. The values corresponding to L_c are indicated in the charts by solid circles (dots), and values corresponding to L_u are indicated by open circles.

Example 2.21. Select the lightest wide-flange section for the beam shown if lateral support is provided at the ends and midpoint. Include the weight of the beam. Use the "Allowable Moments in Beams" graph.

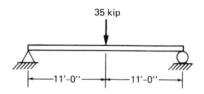

Solution.

Unsupported length $L = 11.0$ ft

Beam weight (assumed) = 60 lb/ft

$$M = \frac{PL}{4} + \frac{wL^2}{8} = \frac{35.0 \text{ k} \times 22.0 \text{ ft}}{4} + \frac{0.06 \text{ k/ft} \times (22.0 \text{ ft})^2}{8}$$

$$= 196.1 \text{ ft-k}$$

According to the graph "Allowable Moments in Beams," the solid line above the required moment and unsupported length intersection is the most economical section.

Use W 18 × 60, $S = 108$ in.^{3}

The bending stress is

$$f_b = \frac{196.1 \text{ ft-k} \times 12 \text{ in./ft}}{108 \text{ in.}^4} = 21.79 \text{ ksi}$$

Example 2.22. Select the lightest wide-flange section for the beam shown if lateral support is provided only at the point loads and beam ends. Use the "Allowable Moments in Beams" graph for F_y = 50 ksi.

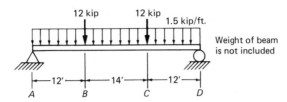

Solution.

$$\text{Maximum unsupported length } L = 14.0 \text{ ft}$$

$$\text{Beam weight (assumed)} = 90 \text{ lb/ft}$$

$$M_{max} = 1.59 \text{ k/ft} \times \frac{(38 \text{ ft})^2}{8} + (12 \text{ k} \times 12 \text{ ft}) = 431 \text{ ft-k}$$

Bending stress

$$f_b = \frac{431 \text{ ft-k} \times 12 \text{ in./ft}}{213 \text{ in.}^3} = 24.28 \text{ ksi}$$

From the "Allowable Moments in Beams" graph, W 27 X 84 is satisfactory.

Use W 27 X 84.

PROBLEMS TO BE SOLVED

2.1. Calculate the bending stress due to a bending moment of 105 ft-k (foot-kips) about the strong axis on the following wide-flange beams:

a) W 10 X 49
b) W 12 X 45
c) W 14 X 38

2.2. Calculate the section modulus of a W 10 X 68 wide-flange beam, and determine its allowable bending moment.

2.3. Select the most economical W section for a simply supported beam carrying a 25-k concentrated load at the center of a 30-ft span. Neglect the weight of the beam, and assume full lateral support.

2.4. Select the most economical W section for a simply supported beam loaded with a uniformly distributed load of 3.0 k/ft over a span of 35 ft. Neglect the weight of the beam, and assume full lateral support.

2.5 through 2.7. Select the most economical W section for the beams shown. Assume full lateral support, and neglect the weight of the beam.

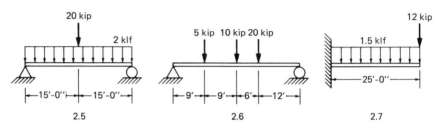

2.5   2.6   2.7

2.8. Redesign the beams in Problems 2.3 through 2.7 using the AISC beam tables. Determine the deflection in each case.

2.9. Assuming that a W 18 X 50 wide-flange beam was mistakenly installed in the situation described in Problem 2.3, determine the load that the new beam can carry.

2.10. Determine the distributed load that can be carried by a W 21 X 57 beam over a simply supported span of 25 ft with lateral supports every 5 ft.

2.11. Determine the concentrated load that a W 18 X 40 beam can carry if the span is 42 ft with the load and a lateral support located at midspan. Assume F_y = 50 ksi.

2.12. Compute the maximum shearing stress on the following sections due to an external shear force of 60 kips.

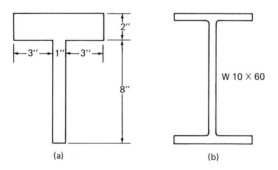

(a)   (b)

2.13. Select the most economical W 12 section that can carry a concentrated load of 85 kips at the third point of a 4-ft, 6-in. simple span. Check both flexural and shear stresses.

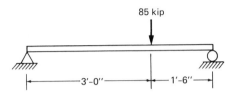

2.14. Calculate the design section modulus for a W 16 X 89 for:

a) two $\frac{1}{2}$-in. diameter holes in each flange
b) two $1\frac{1}{2}$-in. diameter holes in each flange

2.15. Assuming that the 85-k load on the W 12 beam in Problem 2.13 is due to a column bearing on the beam, calculate the bearing lengths required under the column and for each support, such that stiffeners are not necessary.

3
Members under Flexure: 2

3.1 COVER-PLATED BEAMS

The moment of inertia and section modulus of a structural steel member can be increased by the use of cover plates. In new steel work, height restrictions may limit the use of deep sections, making a cover-plated shallow beam necessary. Occasionally, the availability of rolled sections suggests the economical use of cover plates. Cover plates can also be added to increase the flexural capacity of existing beams of renovated or modified structures. Plates, in many cases, can be attached to beam flanges and cut off where they are no longer necessary.

The normal procedure for design of cover plates, where depth restrictions govern, is to select a wide-flange beam that has a depth less than the allowable depth, leaving room for the cover plates. The moments of inertia of symmetrical cover plates are added to that of the wide flange, making

$$I_{req} = I_{WF} + 2 A_{pl} \left(\frac{d}{2}\right)^2 \tag{3.1}$$

where d is the distance between the centers of gravity of the two cover plates. From Eq. 3.1, the plate area required is equal to

$$A_{pl} = 2\frac{I_{req} - I_{WF}}{d^2} \tag{3.2}$$

The use of an unsymmetrical flange addition requires the location of the section neutral axis and the calculation of the moment of inertia.

Example 3.1. Determine the maximum uniform load a cover-plated W 8 × 67 can support simply spanning 30 ft. The plates are $\frac{3}{4} \times 12$ in., and allowable stress $F_b = 24.0$ ksi.

70

Solution.

$$M = F_b \times S, \qquad M = \frac{wL^2}{8}, \qquad w = \frac{8\,F_b S}{L^2}$$

The section modulus is equal to the moment of inertia I of the section divided by the distance c from the neutral axis to the most extreme fiber of the section.

$$I_{\text{section}} = I_{\text{WF}} + 2\,I_{\text{pl}}$$

Neglecting I_0 for the cover plates,

$$I_{\text{section}} = I_{\text{WF}} + 2\,A_{\text{pl}}\left(\frac{d}{2}\right)^2 = 272.0 \text{ in.}^4 + 2 \times 9.0 \text{ in.}^2 \times (4.875 \text{ in.})^2$$

$$I_{\text{section}} = 699.8 \text{ in.}^4$$

$$c = \frac{9.0 \text{ in.}}{2} + 0.75 \text{ in.} = 5.25 \text{ in.}$$

$$S = \frac{699.8 \text{ in.}^4}{5.25 \text{ in.}} = 133.3 \text{ in.}^3$$

$$L = 30.0 \text{ ft}$$

$$w = \frac{8\,F_b S}{L^2} = \frac{8 \times 24.0 \text{ ksi} \times 133.3 \text{ in.}^3}{12 \text{ in./ft} \times (30.0 \text{ ft})^2} = 2.37 \text{ k/ft}$$

Subtracting the weight of the beam and cover plates,

$$w = 2.37 \text{ k/ft} - 0.067 \text{ k/ft} - 2(0.03 \text{ k/ft}) = 2.24 \text{ k/ft}$$

The maximum applied uniform load is 2.24 k/ft

Checking deflection,

$$\Delta = \frac{5\,Wl^3}{384\,EI} = \frac{5 \times (2.37 \times 30) \times (30 \times 12)^3}{384 \times 29{,}000 \times 699.8} = 2.13 \text{ in.}$$

The maximum deflection corresponds to a $l/169$ deflection. Note that in many cases this deflection is considered excessive and therefore unsatisfactory (AISCS 1.13.1). It is recommended that whenever practical, a minimum depth of $(F_y/800)l$ be provided (AISCS Commentary 1.13.1).

Example 3.2. An existing W 16 $\times$ 67, whose compression flange is laterally supported, needs to be reinforced to carry a maximum moment of 260 ft-k. Due to actual conditions, the top flange cannot be reinforced, and the bottom flange will be reinforced alone to carry the moment. What size 1-in. plate attached to the bottom flange with two lines of $\frac{3}{4}$-in. bolts will give the beam the required additional capacity?

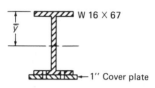

Solution.

$$S_{req} = \frac{260 \text{ ft-k} \times 12 \text{ in./ft}}{24.0 \text{ ksi}} = 130.0 \text{ in.}^3$$

The section modulus is determined from the new neutral axis.

Check for bolt hole reduction in flange of W 16 × 67

$$A_{hole} = 2(\tfrac{3}{4} + \tfrac{1}{8}) \times 0.665 = 1.16 \text{ in.}^2$$

$$A_{fl} = 10.24 \times 0.665 = 6.81 \text{ in.}^2$$

$$15\% \text{ of flange loss} = 0.15 \times 6.8 \text{ in.}^2 = 1.02 \text{ in.}^2$$

$$\text{excess loss} = 1.16 \text{ in.}^2 - 1.02 \text{ in.}^2 = 0.14 \text{ in.}^2 - \text{neglect loss}$$

Try plate width $w = 12$ in.

$$\bar{y} = \frac{\left(\dfrac{16.33}{2} \times 19.7\right) + ((16.33 + 0.50) \times 12.00)}{19.7 + 12.00} = 11.45 \text{ in.}$$

$$I = \Sigma I_0 + Ad^2 = 954 + (19.7 \times (11.45 - 8.165)^2) + (12.0 \times (16.83 - 11.45)^2)$$

$$= 1513.9 \text{ in.}^4$$

$$S = \frac{I}{c} = \frac{1513.9 \text{ in.}^4}{11.45 \text{ in.}} = 132.22 \text{ in.}^3$$

Try smaller plate width, $w = 9$ in.

$$\bar{y} = 10.88 \text{ in.}$$

$$I = 1417.0 \text{ in.}^4$$

$$S = 130.3 \text{ in.}^3 > 130.0 \text{ in.}^3$$

Bolt hole loss in plate $= 2 \times 0.875$ in. $\times 1.0$ in. $= 1.75$ in.^{2}

15% of plate loss $= 0.15 \times 9.0$ in.$^2 = 1.35$ in.^{2} but say ok

Use 1×9 in. bottom flange cover plate.

Example 3.3. A simply supported beam spanning 24.0 ft is to carry a uniform load of 4.2 k/ft. However, height restrictions limit the total beam depth to 12 in. Determine if a standard wide flange can be used, and design a section using cover plates if necessary. Assume full lateral support and $F_b = 24.0$ ksi.

Solution.

$$M = \frac{wL^2}{8}, \quad S = \frac{M}{F_b}$$

$$M = \frac{4.2 \text{ k/ft} \times (24.0 \text{ ft})^2}{8} = 302.4 \text{ ft-k}$$

$$S_{req} = \frac{302.4 \text{ ft-k} \times 12 \text{ in./ft}}{24.0 \text{ ksi}} = 151.2 \text{ in.}^3$$

Check wide flange sections that have a depth less than 12.0 in.

Largest section is W 10×112 $S_x = 126$ in.$^3 < 151.2$ in.^{3} N.G. Cover plated section is required.

Try W 10×68 $d = 10.4$ in., $S = 75.7$ in.^{3}, $I = 394.0$ in.^{4}

Assume cover plate thickness $= \frac{3}{4}$ in.

$$I_{req} - I_{WF} = 2 A_{pl} d^2$$

$$I_{req} = S \times c = 151.2 \text{ in.}^3 \times \left(\frac{11.9 \text{ in.}}{2}\right) = 899.6 \text{ in.}^4$$

$$I_{WF} = 394.0 \text{ in.}^4$$

$$I_{pl} = 899.6 \text{ in.}^4 - 394.0 \text{ in.}^4 = 505.6 \text{ in.}^4$$

$$d = \frac{10.4 \text{ in.}}{2} + \frac{0.75 \text{ in.}}{2} = 5.575 \text{ in.}$$

$$505.6 \text{ in.}^4 = 2 A_{pl} (5.575 \text{ in.})^2$$

$$A_{pl} = \frac{505.6 \text{ in.}^4}{62.2 \text{ in.}^2} = 8.13 \text{ in.}^2$$

$$A_{pl} = w \times t$$

$$w = \frac{8.13 \text{ in.}^2}{0.75 \text{ in.}} = 10.84 \text{ in.} \qquad \text{say } 11 \text{ in.}$$

Use W 10 × 68 beam with $\frac{3}{4}$ × 11-in. cover plates.

$$I = 906 \text{ in.}^4$$

$$d = 10.40 \text{ in.} + 2 \times 0.75 \text{ in.} = 11.90 \text{ in.} < 12 \text{ in.} \qquad \text{ok}$$

Checking deflection (for full-length cover plates),

$$\Delta = \frac{5 \, Wl^3}{384 \, EI} = \frac{5 \times (4.2 \times 24) \times (24 \times 12)^3}{384 \times 29000 \times 906} = 1.19 \text{ in.}$$

$$\frac{\Delta}{l} = \frac{1.19 \text{ in.}}{24 \text{ ft} \times 12 \text{ in./ft}} = \frac{1}{242}$$

Deflection-to-span ratio is within tolerable limits.

Cover plates to flanges and flanges to webs of built-up beams or girders are connected by bolts, rivets, or welds. The longitudinal spacing must be adequate to transfer the horizontal shear, calculated by dividing the strength of the fasteners acting across the section by the shear flow at the section. Maximum spacings for compression flange fasteners and tension flange fasteners are given in Figs. 3.2 and 3.3, as limited by AISCS 1.18.2.3 and 1.18.3.1. As with beam

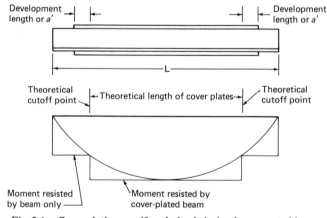

Fig. 3.1. Cover-plating a uniformly loaded, simply supported beam.

flanges, no plate area reduction is necessary for fastener holes if the reduction of flange gross area is less than 15% (AISCS 1.10.1).

Partial length cover plates are attached to rolled shapes to increase the section modulus where increased strength is required and are then discontinued where they are unnecessary (see Fig. 3.1). The plates are extended beyond the theoretical cutoff point and adequately fastened to develop the cover plate's portion of the flexural stresses in the beam or plate girder at the theoretical cutoff point (AISCS 1.10.4). Additionally, welded connections for the plate termination must be adequate to develop the cover plate's portion of the flexural stresses in the section at the distance a' from the actual end of the cover plate. In some cases (usually under fatigue loading) it may be required that the cover plate cutoff point be placed at a lower bending stress than the stress at the theoretical cutoff point. To assure adequate weld strength in all cases, it is required to extend the cover plate the minimum length a' past the theoretical cutoff point, even if the connection can be completed in a shorter development length. The

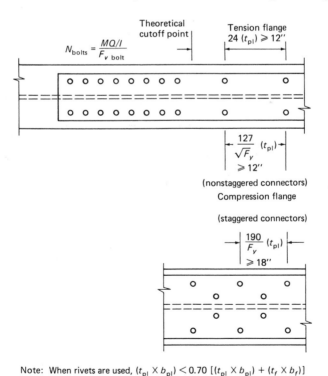

Note: When rivets are used, $(t_{pl} \times b_{pl}) < 0.70 \left[(t_{pl} \times b_{pl}) + (t_f \times b_f) \right]$

Fig. 3.2. Bolt and rivet requirements for cover plates. Fasteners in partial plate extensions must develop flexural stress at cutoff point. Maximum bolt or rivet spacing throughout plate must not exceed values shown.

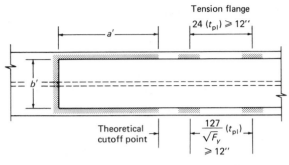

Development length a'

1) $a' = b_{pl}$ for weld leg $\geqslant \frac{3}{4} t_{pl}$ along b' side

2) $a' = 1\frac{1}{2} b_{pl}$ for weld leg $< \frac{3}{4} t_{pl}$ along b' side

3) $a' = 2 b_{pl}$ for no weld along end b'

Fig. 3.3. Fillet weld requirements for cover plates. The length a' is used as the minimum length to assure that all development requirements are met.

requirements for bolts or rivets are shown in Fig. 3.2, and weld requirements are shown in Fig. 3.3. For further discussion, see AISC Commentary Section 1.10.4.

The design of connectors for attaching cover plates to members is covered in Chapters 5 and 6.

Example 3.4. Design partial length cover plates for a W 27 × 94 loaded as shown. The plates will be connected to the flanges at the termination, with continuous fillet weld at the sides and across the end. The beam is fully laterally supported. Uniform load includes weight of beam.

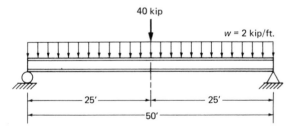

Solution.

$$V = \frac{40\,k}{2} + \left(2\,k/ft \times \frac{50\,ft}{2}\right) = 70\,k$$

$$M_{max} = \frac{40 \times 50}{4} + \frac{2 \times (50)^2}{8} = 1125\,ft\text{-}k$$

If b_c is the width and t_c the thickness of the cover plate,

$$S_{tot} = \frac{3270 + b_c t_c \times \left(\frac{26.92}{2} + \frac{t_c}{2}\right)^2 \times 2}{\left(\frac{26.92}{2} + t_c\right)}$$

$$S_{tot} \leqslant \frac{1125 \text{ ft-k} \times 12 \text{ in./ft}}{24.0 \text{ ksi}} = 562.5 \text{ in.}^3$$

$$2 \times 3270 + b_c t_c \times (26.92 + t_c)^2 \leqslant 562.5 \times (26.92 + 2 t_c)$$

Try

$$t_c = 1\tfrac{1}{2} \text{ in.} \quad b_c = 8.42 \text{ in.}$$
$$t_c = 1\tfrac{3}{8} \text{ in.} \quad b_c = 9.22 \text{ in.}$$
$$t_c = 1\tfrac{5}{8} \text{ in.} \quad b_c = 7.88 \text{ in.}$$

Select cover plates to be $1\tfrac{5}{8} \times 8$ in.

$$I = 3270 + 8 \times 1.625 \times \left(\frac{26.92}{2} + \frac{1.625}{2}\right)^2 \times 2 = 8566 \text{ in.}^4$$

$$S = \frac{8566}{\frac{26.92}{2} + 1.625} = 567.9 \text{ in.}^3 > 562.5 \text{ in.}^3 \quad \text{ok}$$

Moment at cutoff point is determined from wide flange capacity.

$$M = \frac{F_b \times S}{12 \text{ in./ft}} = \frac{24.0 \text{ ksi} \times 243 \text{ in.}^3}{12 \text{ in./ft}} = 486 \text{ ft-k}$$

Determine the theoretical cutoff points.

2 kip/ft.   486 ft.-kip   x   70 kip

$$70 X - 2 \times \frac{X^2}{2} - 486 = 0$$

Solving for X in

$$X^2 - 70X + 486 = 0$$

$$X = 7.82 \text{ ft}$$

Determine connection of cover plates to flanges.

$$V_{cutoff} = 70 \text{ k} - (7.82 \text{ ft} \times 2 \text{ k/ft}) = 54.36 \text{ k}$$

$$Q = 1.625 \times 8 \times \left(\frac{26.92}{2} + \frac{1.625}{2}\right) = 185.5 \text{ in.}^3$$

$$q = \frac{VQ}{I} = \frac{54.36 \text{ k} \times 185.5 \text{ in.}^3}{8566 \text{ in.}^4} = 1.18 \text{ k/in.}$$

Minimum size weld is $\frac{5}{16}$ in. (AISCS 1.17.2).

Use E 60 welds. Capacity of weld is

$$0.30 \times \frac{\sqrt{2}}{2} \times 60 \text{ ksi} \times \frac{5}{16} \text{ in.} = 3.98 \text{ k/in.}$$

Capacity of base metal is

$$0.40 \times 36 \text{ ksi} \times \tfrac{5}{16} \text{ in.} = 4.5 \text{ k/in.} \quad \text{(AISCS 1.5.3)}$$

Using intermittent fillet welds at 12 in. on center (AISCS 1.18.2.3, 1.18.3.1)

$$l_w = \frac{1.18 \text{ k/in.} \times 12 \text{ in.}}{2 \times 3.98 \text{ k/in.}} = 1.78 \text{ in.} \quad \text{Use } 1\frac{3}{4} \text{ in. weld length}$$

Termination welds

$$M_{cutoff} = 486 \text{ ft-k} \quad \text{(AISCS 1.10.4, AISCC 1.10.4)}$$

Force to be carried in termination length

$$H = \frac{MQ}{I} = \frac{486 \text{ ft-k} \times 12 \text{ in./ft} \times 185.5 \text{ in.}^3}{8566 \text{ in.}^4} = 126.3 \text{ k}$$

$$a' = 1.5 \times 8 \text{ in.} = 12 \text{ in.}$$

Total length of termination weld = 12 in. $\times$ 2 + 8 in. = 32 in.

H_w = 32 in. $\times$ 3.98 k/in. = 127.4 k > 126.3 k   ok

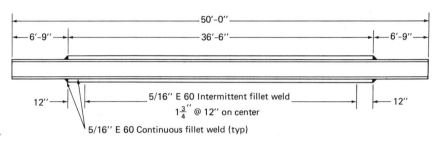

Example 3.5. A W 16 $\times$ 100 with $\frac{7}{8} \times$ 10 in. cover plates spans a distance of 32 ft with a uniform load of 3.6 k/ft (including weight of beam) and point loads of 15 kips located 10 ft from each end. Determine the length of the partial length cover plates and the connections to the flanges. Use $\frac{3}{4}$-in. ϕ A 325-N bolts throughout and type F bolts for the plate termination. Assume full lateral support.

Solution.

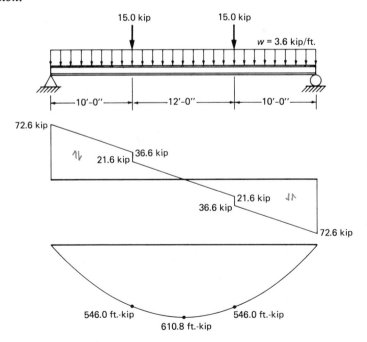

$$I = 1490 \text{ in.}^4 + 2 \times \frac{7}{8} \text{ in.} \times 10 \text{ in.} \times \left(\frac{16.97 \text{ in.}}{2} + \frac{0.875 \text{ in.}}{2}\right)^2 = 2883 \text{ in.}^4$$

$$S = \frac{I}{c} = \frac{2883 \text{ in.}^4}{\dfrac{16.97 \text{ in.}}{2} + 0.875 \text{ in.}} = 308.0 \text{ in.}^3$$

$$M_{all} = F_b \times S = 7392 \text{ in.-k} = 616 \text{ ft-k} > 610.8 \text{ ft-k} \quad \text{ok}$$

Moment at termination of cover plates

$$M = 175 \text{ in.}^3 \times \frac{24 \text{ ksi}}{12 \text{ in./ft}} = 350 \text{ ft-k}$$

Distance to termination point from end of beam

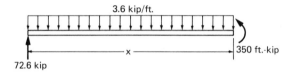

From free body diagram,

$$72.6 \, X - 3.6 \times \frac{X^2}{2} - 350 = 0$$

Solving for X,

$$X = 5.60 \text{ ft}$$

Shear at cutoff point of cover plates

$$V = 72.6 \text{ k} - 5.6 \text{ ft} \times 3.6 \text{ k/ft} = 52.44 \text{ k}$$

Shear flow at cutoff point

$$Q = \frac{7}{8} \text{ in.} \times 10 \text{ in.} \times \left(\frac{16.97 \text{ in.}}{2} + \frac{0.875 \text{ in.}}{2}\right) = 78.07 \text{ in.}^3$$

$$q = \frac{52.44 \text{ k} \times 78.07 \text{ in.}^3}{2883 \text{ in.}^4} = 1.42 \text{ k/in.}$$

Shear capacity of $\frac{3}{4}$-in. A 325-N bolts

$$r_v = 9.3 \text{ k} \qquad \text{(AISCS Table I-D)}$$

Spacing of bolts throughout plated region

Minimum spacing

$$s \leqslant 0.875 \text{ in.} \times \frac{127}{\sqrt{F_y}} = 18.5 \text{ in.}$$

$$s \leqslant 12 \text{ in.}$$

Spacing of bolts for top plate must be minimum 12 in. (AISCS 1.18.2.3).

$$s \leqslant 24 \times 0.875 \text{ in.} = 21 \text{ in.}$$

$$s \leqslant 12 \text{ in.}$$

Spacing of bolts for bottom plate must be minimum 12 in. (AISCS 1.18.3.1).

Spacing of bolts in termination region

$$H = \frac{MQ}{I} = \frac{350 \text{ ft-k} \times 12 \text{ in./ft} \times 78.07 \text{ in.}^3}{2883 \text{ in.}^3} = 113.7 \text{ k}$$

Capacity of termination bolts ($\frac{3}{4}$-in. A 325-F)

$$r_v = 7.7 \text{ k} \qquad \text{(AISCS Table I-D)}$$

Number of bolts needed

$$N = \frac{H}{r_v} = \frac{113.7 \text{ k}}{7.7 \text{ k}} = 14.77 \qquad \text{use 16 bolts}$$

To reduce length of plate termination, stagger bolts.

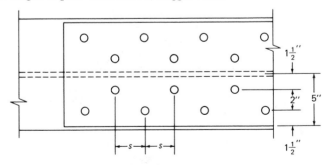

Determine s such that loss of area is less than or equal to two bolt holes.

$$10 \text{ in.} - 4 \times \left(\frac{3}{4} \text{ in.} + \frac{1}{8} \text{ in.}\right) + 2 \times \frac{s^2}{4 \times 2 \text{ in.}} = 10 \text{ in.} - 2 \times \left(\frac{3}{4} \text{ in.} + \frac{1}{8} \text{ in.}\right)$$

$s^2 = 7.0$ in. $s = 2.65$ in.—say $2\frac{3}{4}$ in.

3.2 PLATE GIRDERS

The use of plate girders is necessary when the flexural capacity of rolled sections becomes inadequate. Welded plate girders usually consist of a web with flange plates connected to the two edges of the web. Cover plates are sometimes used on the flanges and are cut off where unnecessary. Web stiffeners are attached to one or both sides of the web when high shear stresses occur in the web. Today, almost all plate girders have their flanges welded to the webs.

Generally the depth of plate girders varies from $\frac{1}{6}$ to $\frac{1}{15}$ of the span, with $\frac{1}{10}$ to $\frac{1}{12}$ being the most common. The AISC Commentary in Section 1.13 suggests that the depth not exceed $F_y/800$ times the span for floors and $F_y/1000$ times the span for roof purlins. These are offered only as guides, and the requirements for shear, flexure, and deflection are of primary concern.

There is no unique solution for the design of a plate girder. Although minimum cost is the most desired result, a girder can be designed for any number of

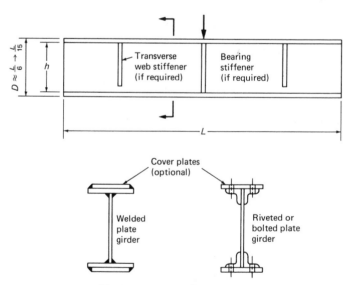

Fig. 3.4. A typical plate girder.

combinations of web and flange dimensions and stiffener arrangements (when necessary). A number of trials may be necessary for cost optimization. The basic approach, however, is to make assumptions on the size of web and flanges and to modify the section as deemed necessary.

The usual design procedure for plate girders is to select a trial cross section by the flange-area method and then to check by the moment of inertia method. The flange-area method assumes that the flanges will carry most of the bending moment and the web will carry all the shear forces. Required web area is then

$$A_w = \frac{V}{F_v} \tag{3.3}$$

where V is the vertical shear force at the section, and F_v is the allowable shear stress on the web. The required flange area is then determined from

$$A_f = \frac{M}{F_b h} - \frac{A_w}{6} \tag{3.4}$$

where M is the bending moment at the section, F_b is the allowable bending stress, and h is the clear distance between flanges.

Verification by the moment of inertia method requires calculating the moment of inertia of the girder cross section. The allowable bending stress in the compression flange is determined from AISCS 1.5.1.4. Stress reduction is required if $(h/t_w) > (760/\sqrt{F_b})$ (AISCS 1.10.6).

When transverse stiffeners of adequate strength are provided at required spacings, they act as compression members. The web then behaves as a membrane that builds up diagonal tension fields consisting of shear forces greater than those associated with the theoretical buckling load. The result, commonly referred to as *tension field action*, is similar to a Pratt truss. From this additional stiffness, shear capacity is gained in the web, unaccountable to linear buckling theory. Applying flat plate theory, it is understandable that the additional shear strength is dependent on the web plate section and stiffener spacing. Tension field action is not considered when the panel ratio a/h exceeds 3.0 (AISCS Commentary 1.10.5).

Hybrid girders are sections in which the web plate is of a different grade of steel than that of the flange plates. Hybrid plate girders may be used, provided they are not required to resist any axial force greater than 15% of the flange yield stress times the area of the gross section. Hybrids cannot be designed for tension field action, and the bending stress is limited to $0.60 F_y$ (AISCS 1.10.1).

When choosing the web plate dimensions, one must consider the ratio of the web height to thickness (h/t_w). Common ratios and their significance are shown in Table 3.1. If a plate girder with no stiffeners is desired, the ratio must not

Table 3.1. Web Plate Ratios.

		Values of Ratio (F)	
If $h/t_w \leqslant$	Then	F_y = 36 ksi	F_y = 50 ksi
$\dfrac{640}{\sqrt{F_y}}$	Member is compact $(F_b = 0.66\,F_y)$	107	91
$\dfrac{760}{\sqrt{F_b}}$	No flange stress reduction $F_b = 0.60\,F_{y\,\text{flange}}$	162	139
$\dfrac{14{,}000}{\sqrt{F_y(F_y + 16.5)}}$	No web stiffeners	322	243
$\dfrac{2000}{\sqrt{F_y}}$	Web stiffeners spaced no more than $1\frac{1}{2}\,d$ apart	333	283
260	No intermediate stiffeners with $f_v \leqslant F_v$	260	260

Yield stress value of web plate unless noted otherwise.

exceed

$$\frac{h}{t_w} \leqslant \frac{14{,}000}{\sqrt{F_y\,(F_y + 16.5)}} \tag{3.5}$$

(AISCS 1.10.2). The limit of the ratio can be increased to

$$\frac{h}{t_w} \leqslant \frac{2000}{\sqrt{F_y}} \tag{3.6}$$

when stiffeners are provided at a spacing not exceeding $1\frac{1}{2}$ times the girder height.

Flange proportions are limited in AISCS 1.9.1.2 by the following flange width-to-thickness ratio

$$\frac{b_f}{2\,t_f} \leqslant \frac{95}{\sqrt{F_y}} \tag{3.7}$$

If it is found necessary to increase the section's moment of inertia, the flange area should be increased. By increasing slightly the flange thickness, the ratio will remain below the limiting, value, while increasing the I value.

Bearing stiffeners must be placed in pairs at unframed girder ends and at points of concentrated loads when required. Load stiffeners are required to prevent web crippling when

$$\frac{R}{t_w(N+2k)} > 0.75\, F_y \tag{3.8}$$

where k is the distance from the outer face of the flange to the web toe of the fillet. The stiffener must be designed as columns (AISCS 1.10.5.1) and satisfy requirements for unstiffened elements under compression (AISCS 1.9.1). Figure 3.5 demonstrates the design criteria for bearing stiffeners (AISCS 1.10.5).

Intermediate stiffeners may be required in otherwise unstiffened panels to increase the web plate's shear capacity. When stiffeners are required, the panel ratio a/h shall not exceed $260/(h/t)^2$ nor 3.0.

If web shear stress is less than

$$F_v = \frac{F_y}{2.89}\,(C_v) \leqslant 0.40\, F_y \tag{3.9}$$

and the web height-to-thickness ratio is less than 260, no intermediate stiffeners are required in the web panel. C_v is a function of the web dimensions and transverse stiffeners as shown in Fig. 3.6.

Girders other than hybrid may be designed for tension field action, for which the allowable shear stress is

$$F_v = \frac{F_y}{2.89}\left[C_v - \frac{1-C_v}{1.15\sqrt{1+(a/h)^2}}\right] \leqslant 0.40\, F_y \tag{3.10}$$

In addition, combined shear and tension stress must be considered such that the flexural tensile stress is limited to

$$F_b \leqslant \left(0.85 - 0.375\,\frac{f_v}{F_y}\right)F_y \tag{3.11}$$

but not greater than $0.60\, F_y$.

Allowable shear stress values are tabulated in the AISC Appendix A as a function of the ratios a/h and h/t. Tables 10-36 and 10-50 may be used when tension field action is not considered, and Tables 11-36 and 11-50 include the effect of the tension field. The values "36" and "50" in the table designation refer to yield stress.

When the web depth-to-thickness ratio exceeds $760\sqrt{F_b}$, the maximum bending

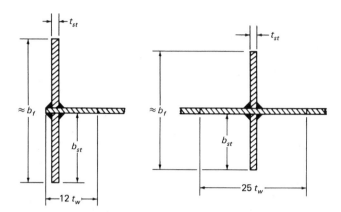

Effective area of end-bearing stiffeners. Provide at unframed ends

Effective area of load-bearing stiffeners. Provide under concentrated loads when

$$\frac{R}{t_w(N + 2 k)} > 0.75 F_y$$

Design requirements

1. $\dfrac{b_{st}}{t_{st}} < \dfrac{95}{\sqrt{F_y}}$

2. Struts designed as columns. Determine allowable axial stress

End bearing

$$A_{\text{eff}} = 2 (b_{st} \times t_{st}) + 12 t_w^2$$

Interior bearing

$$A_{\text{eff}} = 2 (b_{st} \times t_{st}) + 25 t_w^2$$

$$I_{st} = \frac{t_{st} (2b_{st} + t_w)^3}{12}$$

$$r = \sqrt{\frac{I}{A}}$$

$Kh, K = 1.0$, assuming joint translation possible at top and bottom (conservative).

Allowable axial stress is found in AISC Tables 3-36 and 3-50 for value of (Kh/r). If stiffener yield stress is different from web yield stress, use lesser grade of the two.

3. Actual bearing stress

$$f_a = \frac{V}{A_{\text{eff}}}$$

Fig. 3.5. Requirements for girder bearing stiffeners.

Known: $a/h, h/t, F_y$ (web steel)
Is $a/h < 1.0$

$$\text{Yes} \rightarrow k = 4.00 + \frac{5.34}{(a/h)^2}$$

$$\text{No} \rightarrow k = 5.34 + \frac{4.00}{(a/h)^2}$$

$$C_v = \frac{45,000 \, k}{F_y \, (h/t)^2}$$

Is $C_v > 0.8$

$$\text{Yes} \rightarrow C_v = \frac{190}{h/t} \sqrt{\frac{k}{F_y}}$$

$$\text{No} \rightarrow C_v = C_v$$

Fig. 3.6. Computation of critical web stress ratio C_v.

stress in the compression flange shall not exceed

$$F_b' \leqslant F_b \left[1.0 - 0.0005 \frac{A_w}{A_f} \left(\frac{h}{t} - \frac{760}{\sqrt{F_b}}\right)\right] \tag{3.12}$$

where F_b is the allowable bending stress in accordance with AISCS 1.5.1.4. Further, the maximum stress in either flange of a hybrid girder shall not exceed the value determined from Eq. 3.12 nor

$$F_b' \leqslant F_b \left[\frac{12 + (A_w/A_f)(3\alpha - \alpha^3)}{12 + 2(A_w/A_f)}\right] \tag{3.13}$$

where $\alpha = F_{y \, \text{web}}/F_{y \, \text{flange}}$.

Intermediate stiffeners, when provided, should be proportioned and detailed as shown in Fig. 3.8. In addition, stiffeners provided for tension field action must be connected for total shear transfer for a minimum of

$$f_{vs} \geqslant h \sqrt{\left(\frac{F_y}{340}\right)^3} \tag{3.14}$$

Shear transfer stress may be reduced in the same proportion as the largest computed shear stress to allowable shear stress in adjacent panels.

Fig. 3.7. Fascia, or edge girder of a steel frame building with vertical web stiffeners.

Due to complexity of design and the numerous options available to the designer, it is advantageous to organize the design process into a flow chart, making decisions when called for. It must be remembered though that the result leads to one possible solution. To optimize the design, it would be necessary to view the solution critically and make revisions where they seem advantageous. In some cases, the flow chart may be reentered with a different decision and continued to determine another solution.

Example 3.6. A plate girder framed between two columns 65 ft apart supports a uniform load of 4.8 kips per foot. Design trial web and flange plates that will not reduce allowable flange stress and will not require transverse stiffeners. Check by the moment of inertia method. Beam is fully laterally supported.

1. $A_{st\,min} = \dfrac{1-C_v}{2}\left[\dfrac{a}{h} - \dfrac{(a/h)^2}{1+9(a/h)^2}\right] YDht\dfrac{f_v}{F_v}$

(total area for stiffener pairs)
C_v, as defined in Fig. 3.6

$Y = \dfrac{F_{y\,web}}{F_{y\,st}}$

$D = 1.0$ stiffeners in pairs
  $= 1.8$ single-angle stiffeners
  $= 2.4$ single-plate stiffeners

f_v = maximum shear stress in panel under investigation
F_v = maximum allowable shear stress in panel under investigation

$\dfrac{b_{st}}{t_{st}} < \dfrac{95}{\sqrt{F_y}}$

2. $I_{st} \geqslant \left(\dfrac{h}{50}\right)^4$

3. Stiffener length

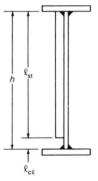

$\dfrac{D}{16}$ weld $+ 6\,t_w \geqslant l_{cl} \geqslant \dfrac{D}{16}$ weld $+ 4\,t_w$

$l_{st} = h - l_{cl}$

Fig. 3.8. Requirements for intermediate girder stiffeners.

Solution.

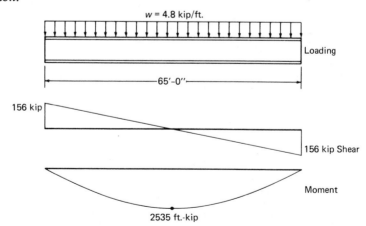

Assume $h = l/10 = 78$ in.

Web trial

$$t_w \geqslant \frac{h}{F}$$

For no flange stress reduction with $F_y = 36.0$ ksi, $F = 162$ (see Table 3.1)

$$t_w \geqslant \frac{78}{162} = 0.48$$

Try $\frac{1}{2}$-in. web

$$\frac{h}{t_w} = \frac{78}{0.5} = 156$$

From AISC Table 10-36 (Appendix A), $F_v = 3.7$ ksi

$$f_v = \frac{V}{A_w} = \frac{156.0\text{ k}}{0.5\text{ in.} \times 78\text{ in.}} = 4.0\text{ ksi} > 3.7\text{ ksi} \qquad \text{N.G.}$$

Try $\frac{9}{16}$ in. web

$$\frac{h}{t_w} = 139; \qquad F_v = 4.2\text{ ksi}$$

$$f_v = \frac{156.0\text{ k}}{0.56\text{ in.} \times 78\text{ in.}} = 3.57\text{ ksi} < 4.2\text{ ksi} \qquad \text{ok}$$

Because $162 > h/t_w = 139 > 107$

$$F_b = 0.60\,F_y = 22.0\text{ ksi} \qquad \text{(AISCS 1.5.1.4.5)}$$

$$S_{\text{req}} = \frac{M}{F_b} = \frac{2535\text{ ft-k} \times 12\text{ in./ft}}{22.0\text{ ksi}} = 1383\text{ in.}^3$$

Preliminary flange size

$$A_f = \frac{M}{F_b h} - \frac{A_w}{6} = \frac{2535 \text{ ft-k} \times 12 \text{ in./ft}}{22.0 \text{ ksi} \times 78 \text{ in.}} - \frac{0.56 \text{ in.} \times 78 \text{ in.}}{6} = 10.45 \text{ in.}^2$$

$$b_f \times t_f = 10.45$$

Try

$$t_f = \frac{3}{4} \text{ in.}, \quad b_f = 14.0 \text{ in.}$$

$$\frac{14 \text{ in.}}{2 \times 0.75 \text{ in.}} = 9.33 < \frac{95}{\sqrt{F_y}} = 15.83 \quad (\text{AISCS } 1.9.1.2)$$

Check by moment of inertia

$$I = I_{0 \text{ web}} + 2(Ad^2)_{fl} = \frac{9}{16} \text{ in.} \times \frac{(78 \text{ in.})^3}{12} + 2\left[10.5 \text{ in.}^2 \times \left(\frac{78.75 \text{ in.}}{2}\right)^2\right]$$

$$= 54,803 \text{ in.}^4$$

$$S = \frac{I}{c} = \frac{54,803 \text{ in.}^4}{39.75 \text{ in.}} = 1379 \text{ in.}^3 < 1383 \text{ in.}^3 \quad \text{say ok}$$

Use $\frac{9}{16} \times 78$ in. web plate with $\frac{3}{4} \times 14$ in. flange plates.

NOTE: This may not be the most economic solution, as a solution with web stiffeners, much thinner web, and heavier flanges could give much less poundage, even including the extra labor cost, as its equivalent in steel weight.

Weight of web + flanges

$$
\begin{array}{llll}
\text{web} & 23 \text{ psf} \times 6.5 \text{ ft} & = 149.5 \text{ lb/ft} \\
\text{flg} & 35.6 \text{ plf} \times 2 & = \underline{71.2 \text{ lb/ft}} \\
& & 220.7 \text{ lb/ft} \times 65 \text{ ft} = 14,346 \text{ lb}
\end{array}
$$

Example 3.7. Design bearing stiffeners and intermediate stiffeners in pairs for a plate girder with loading as shown, $56 \times \frac{5}{16}$-in. web plate, and flange of $1\frac{1}{4} \times$

20-in. plate. The member is on the building exterior and is unframed at girder ends.

Solution.

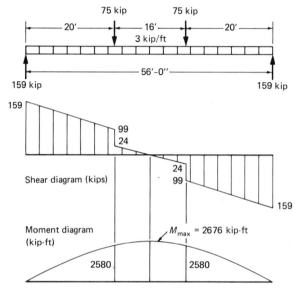

Check for bearing stiffeners under concentrated loads (AISC 1.10.5.1).

To remain very conservative, assume $\frac{1}{4}$-in. flange to web fillet welds and point loading ($N = 0$).

$$\frac{R}{t_w(N + 2k)} = \frac{75.0\,k}{\dfrac{5}{16}(2 \times (1.25 + 0.25))} = 80\,ksi > 0.75 \times 36\,ksi$$

Provide bearing stiffeners.

Design stiffeners to act as compression members (see Fig. 3.5)

$$A_{\text{eff}} = 2(b_{st} \times t_{st}) + 25\,t_w^2$$

$$\frac{b_{st}}{t_{st}} \leqslant \frac{95}{\sqrt{F_y}} = 15.83 \quad \text{(AISCS 1.9.1.2)}$$

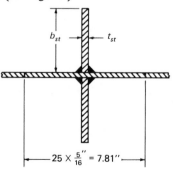

Try $\frac{5}{8} \times 9\frac{3}{4}$-in. bearing stiffeners

$$\frac{b_{st}}{t_{st}} = \frac{9.75\,\text{in.}}{.625\,\text{in.}} = 15.60 < 15.83 \quad \text{ok}$$

$$A_{eff} = 2(9.75 \times .625) + 25(0.31)^2 = 14.59 \text{ in.}^2$$

$$I_{eff} = \frac{t_{st}(2\,b_{st} + t_w)^3}{12} = \frac{.625\,(2 \times 9.75 + .31)^3}{12} = 404.9 \text{ in.}^4$$

$$r = \sqrt{\frac{I}{A}} = \sqrt{\frac{404.9}{14.59}} = 5.27 \text{ in.}$$

$$\frac{Kh}{r} = \frac{0.75 \times 56}{5.27} = 8.0 \quad F_a = 21.25 \text{ ksi} \quad K = 0.75 \quad (\text{AISCS } 1.10.5.1)$$

$$f_a = \frac{R}{A_{eff}} = \frac{75.0 \text{ k}}{14.59 \text{ in.}^2} = 5.14 \text{ ksi} < 21.25 \text{ ksi} \quad \text{extremely overdesigned}$$

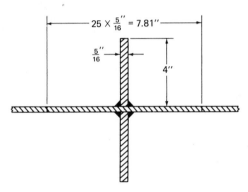

Try $\frac{5}{16} \times$ 4-in. stiffeners

$$A_{eff} = 2 \times \frac{5}{16} \text{ in.} \times 4 \text{ in.} + 7.81 \text{ in.} \times 0.31 \text{ in.} = 4.92 \text{ in.}^2$$

$$I_{eff} = \frac{\dfrac{5}{16} \text{ in.} \times (2 \times 4.0 \text{ in.} + 0.31 \text{ in.})^3}{12} = 14.95 \text{ in.}^4$$

$$r = \sqrt{\frac{14.95 \text{ in.}^4}{4.92 \text{ in.}^2}} = 1.74 \text{ in.}$$

$$\frac{Kh}{r} = \frac{0.75 \times 56}{1.74} = 24.1 \quad F_a = 20.34 \text{ ksi}$$

$$f_a = \frac{R}{A_{eff}} = \frac{75.0 \text{ k}}{4.92 \text{ in.}^2} = 15.24 \text{ ksi} < 20.34 \text{ ksi} \quad \text{ok}$$

$$\frac{b_{st}}{t_{st}} = \frac{4 \text{ in.}}{\dfrac{5}{16} \text{ in.}} = 12.8 < 15.83 \quad \text{ok} \quad (\text{AISCS } 1.9.1.2)$$

Use two 4 in. $\times \frac{5}{16}$ in. $\times$ 4 ft, 8 in. interior bearing plates.

End bearing plates (for unframed ends)

$$A_{eff} = 2(b_{st} \times t_{st}) + 12\, t_w^2$$

Try $\frac{1}{2} \times$ 7-in. end bearing stiffeners

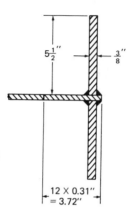

$$\frac{b_{st}}{t_{st}} = \frac{7 \text{ in.}}{\frac{1}{2} \text{ in.}} = 14.0 < 15.83 \qquad \text{ok}$$

$$A_{eff} = 2 \times \frac{1}{2} \text{ in.} \times 7 \text{ in.} + 12 \times (0.31 \text{ in.})^2 = 8.15 \text{ in.}^2$$

$$I_{eff} = \frac{\frac{1}{2} \text{ in.} \times (2 \times 7.0 \text{ in.} + 0.31 \text{ in.})^3}{12} = 122.1 \text{ in.}^4$$

$$r = \sqrt{\frac{122.1 \text{ in.}^4}{8.15 \text{ in.}^2}} = 3.87 \text{ in.}$$

$$\frac{Kh}{r} = \frac{0.75 \times 56}{3.87} = 10.85 \qquad F_a = 21.11 \text{ ksi}$$

$$f_a = \frac{159.0 \text{ k}}{8.15 \text{ in.}^2} = 19.51 \text{ ksi} < 21.11 \text{ ksi} \qquad \text{ok}$$

Use two $7 \times \frac{1}{2}$ in. $\times$ 4 ft, 8 in. bearing plates.

Intermediate stiffener spacing

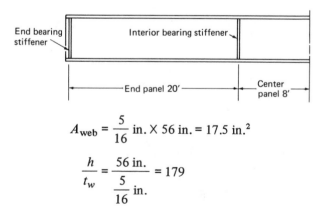

$$A_{web} = \frac{5}{16} \text{ in.} \times 56 \text{ in.} = 17.5 \text{ in.}^2$$

$$\frac{h}{t_w} = \frac{56 \text{ in.}}{\frac{5}{16} \text{ in.}} = 179$$

Because $h/t < 260$, stiffeners are not required when $f_v \leqslant 2.6$ ksi (AISCS 1.10.5.3).

Stiffeners are not required when shear is less than $A_w \times f_v$.

17.5 in.$^2 \times 2.6$ ksi $= 45.5$ k—Refer to shear diagram for $V < 45.5$ k

No intermediate stiffeners required in center panel of girder.

End panel intermediate stiffener spacing—no tension field permitted.

$$V = 159.0 \text{ k}$$

$$f_v = \frac{159.0 \text{ k}}{17.5 \text{ in.}^2} = 9.1 \text{ ksi}$$

For $h/t = 180$ and $F_v = 9.1$ ksi,

$$a/h = 0.6 \quad \text{(AISC Table 10-36)}$$

$$h = 56 \text{ in.}, \quad a = 0.6 \times 56 \text{ in.} = 33.6 \text{ in.}$$

Provide first intermediate stiffener 2 ft, 9 in. from end.

$$\text{At } a = 2.75 \text{ ft}, \quad V = 159.0 - (2.75 \times 3.0) = 150.75 \text{ k}$$

$$f_v = \frac{150.75 \text{ k}}{17.5 \text{ in.}^2} = 8.61 \text{ ksi}$$

Tension field is allowed beyond the first intermediate stiffener.

For $h/t = 180$ and $F_v = 8.7$ ksi,

$$a/h = 1.2 \quad \text{(AISC Table 11-36)}$$

$$a = 1.2 \times 56 \text{ in.} = 67 \text{ in.}\text{—say 5 ft, 9 in. for even spacing}$$

Provide other intermediate stiffeners at same spacing.

Check combined shear and tension in tension field panels.

$$I = \frac{0.31 \, (56)^3}{12} + 2 \times (1.25 \times 20)(28.625)^2 = 45,543 \text{ in.}^4$$

$$c = 29.25 \text{ in.}$$

$$S = \frac{I}{c} = \frac{45{,}543 \text{ in.}^4}{29.25 \text{ in.}} = 1557 \text{ in.}^3$$

$$f_b < \left(0.825 - 0.375\,\frac{f_v}{F_v}\right) F_y \quad \text{(AISCS 1.10.7)}$$

Shear

Moment

Pt A

$$f_v = \frac{V}{A_w} = \frac{150.8 \text{ k}}{17.5 \text{ in.}^2} = 8.62 \text{ ksi}$$

$$f_b < \left(0.825 - 0.375\,\frac{8.62}{8.7}\right) 36 = 16.32 \text{ ksi}$$

$$f_b = \frac{M}{S} = \frac{425.9 \text{ ft-k} \times 12 \text{ in./ft}}{1557 \text{ in.}^3} = 3.28 \text{ ksi} < 16.32 \text{ ksi} \quad \text{ok}$$

Pt D

$$f_v = \frac{V}{A_w} = \frac{99.0 \text{ k}}{17.5 \text{ in.}^2} = 5.66 \text{ ksi}$$

$$f_b < \left(0.825 - 0.375\,\frac{5.66}{8.7}\right) \times 36 \text{ ksi} = 20.92 \text{ ksi}$$

$$f_b = \frac{M}{S} = \frac{2580.0 \text{ ft-k} \times 12 \text{ in./ft}}{1557 \text{ in.}^3} = 19.88 \text{ ksi} < 20.92 \text{ ksi} \quad \text{ok}$$

Check web stability under uniform compressive loads. Because member is spandrel girder (exterior exposure), assume flange is not restrained against rotation in the middle panel (AISCS 1.10.10.2).

$$F_p = \left[2 + \frac{4}{(a/h)^2}\right] \frac{10,000}{(h/t)^2}$$

$$(a/h) = \frac{16 \text{ ft} \times 12 \text{ in./ft}}{56 \text{ in.}} = 3.43$$

$$(h/t) = \frac{56 \text{ in.}}{\frac{5}{16} \text{ in.}} = 179.2$$

$$F_p = \left[2 + \frac{4}{(3.43)^2}\right] \frac{10,000}{(179.2)^2} = 0.73 \text{ ksi}$$

$$f_p = \frac{3 \text{ k/ft}}{12 \text{ in./ft} \times \frac{5}{16} \text{ in.}} = 0.80 \text{ ksi} > 0.73 \text{ ksi} \qquad \text{N.G.}$$

Provide one additional stiffener at center of girder.

$$(a/h) = \frac{8 \text{ ft} \times 12 \text{ in./ft}}{56 \text{ in.}} = 1.71$$

$$F_p = \left[2 + \frac{4}{(1.71)^2}\right] \frac{10,000}{(179.2)^2} = 1.05 \text{ ksi} > 0.80 \text{ ksi} \qquad \text{ok}$$

Intermediate stiffener sizes (in pairs) (AISCS 1.10.5.4)

$$A_{st\,min} = \frac{1 - C_v}{2} \left[\frac{a}{h} - \frac{(a/h)^2}{1 + (a/h)^2}\right] YDht \frac{f_v}{F_v}$$

$$\frac{b_{st}}{t_{st}} < \frac{95}{\sqrt{F_y}}$$

$$I_{st\,min} = \left(\frac{h}{50}\right)^4 = \left(\frac{56 \text{ in.}}{50}\right)^4 = 1.57 \text{ in.}^4$$

Design stiffeners for end panels, and provide identical pair at midspan.

$$a/h = \frac{5.75 \text{ ft} \times 12 \text{ in./ft}}{56 \text{ in.}} = 1.23$$

$$k = 5.34 + \frac{4.00}{(1.23)^2} = 7.98$$

$$C_v = 0.311 \text{ (see Fig. 3.6)}$$

$$Y = \frac{F_{y \text{ web}}}{F_{y \text{ st}}} = 1.0$$

$$D = 1.0 \text{ for stiffeners in pairs}$$

$$f_{v \text{ max}} = 8.62 \text{ ksi–say } \frac{f_v}{F_v} = 1.0$$

$$A_{st \text{ min}} = \frac{1 - 0.311}{2} \left[1.23 - \frac{(1.23)^2}{1 + (1.23)^2} \right] 56 \text{ in.} \times \frac{5}{16} \text{ in.} = 3.78 \text{ in.}^2$$

Try $\frac{3}{8} \times 5\frac{1}{2}$-in. intermediate stiffeners

$$A = 2(0.375 \times 5.50) = 4.13 \text{ in.}^2 > 3.78 \text{ in.}^2 \quad \text{ok}$$

$$\frac{b_{st}}{t_{st}} = \frac{5.50}{0.375} = 14.67 < 15.83 \quad \text{ok}$$

$$I_{st} = \frac{0.375 \times (2 \times 5.5 + 0.31)^3}{12} = 45.2 \text{ in.}^4 > 1.57 \text{ in.}^4 \quad \text{ok}$$

Stiffener may be stopped between 4 to 6 times web thickness from web toe of fillet weld ($\frac{1}{4}$ in.).

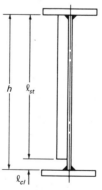

$$\tfrac{1}{4} \text{ in.} + 6(\tfrac{5}{16} \text{ in.}) \geqslant l_{cl} \geqslant \tfrac{1}{4} \text{ in.} + 4(\tfrac{5}{16} \text{ in.})$$

$$2.125 \text{ in.} \geqslant l_{cl} \geqslant 1.50 \text{ in.–say } l_{cl} = 2 \text{ in.}$$

$$l_{st} = h - l_{cl} = 56 \text{ in.} - 2 \text{ in.} = 54 \text{ in.}$$

Use two $5\frac{1}{2} \times \frac{3}{8}$ in. $\times$ 4 ft, 6 in. intermediate stiffeners.

Use plates as shown.

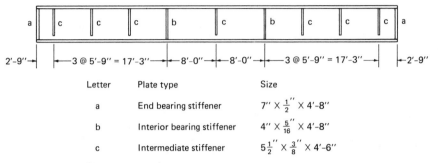

Letter	Plate type	Size
a	End bearing stiffener	$7'' \times \frac{1}{2}'' \times 4'-8''$
b	Interior bearing stiffener	$4'' \times \frac{5}{16}'' \times 4'-8''$
c	Intermediate stiffener	$5\frac{1}{2}'' \times \frac{3}{8}'' \times 4'-6''$

Note: All stiffeners to be supplied in pairs.

Example 3.8. A plate girder loaded as shown is not framed at the supports. It is laterally supported at the points of applied concentrated load. Design this girder and compare results

a) without any intermediate stiffeners
b) with intermediate stiffeners.

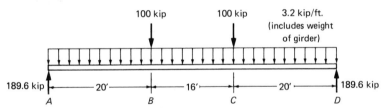

Solution.

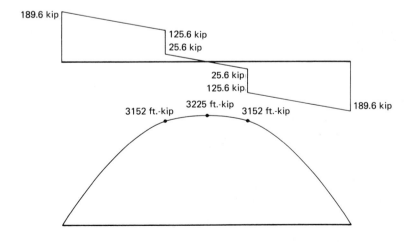

Depth of girder

The usual depth of plate girders varies between $\frac{1}{6}$ to $\frac{1}{15}$ of the span, with a general average depth of $\frac{1}{10}$.

Select for depth $l/10$

$$\frac{l}{10} = \frac{56 \text{ ft} \times 12 \text{ in./ft}}{10} = 67.2 \text{ in.–say } 69 \text{ in.}$$

a) Design without intermediate stiffeners.

Choose web thickness t_w for no flange reduction

$$\frac{h}{t_w} \leqslant \frac{640}{\sqrt{F_y}} \text{ for } f_a = 0 \qquad \text{(AISCS 1.5.1.4.1.4)}$$

$$\frac{t_f}{2\,b_f} \leqslant \frac{65}{\sqrt{F_y}} \qquad \text{(AISCS 1.5.1.4.1.2)}$$

For $F_b = 0.6\,F_y = 22$ ksi

$$\frac{h}{t_w} \leqslant \frac{760}{\sqrt{F_y}} \qquad \text{(AISCS 1.10.6)}$$

The ratio

$$\frac{t_f}{2\,b_f} < \frac{95}{\sqrt{F_y}}$$

must always be observed in order to have nonslender compression members (AISCS 1.9.1.2).

$$t_w = \frac{h \times \sqrt{F_y}}{640} = \frac{69 \text{ in.} \times \sqrt{36}}{640} = 0.647 \text{ in.} \qquad \text{for } F_b = 24 \text{ ksi}$$

$$t_w = \frac{h \times \sqrt{F_y}}{760} = \frac{69 \text{ in.} \times \sqrt{36}}{760} = 0.545 \text{ in.} \qquad \text{for } F_b = 22 \text{ ksi}$$

Allowable shear stress in web

The first panel must not have tension field action used.

After some trials, we obtain

$$f_v = \frac{189.6 \text{ k}}{(69 \text{ in.} + 2 \text{ in.}) \times \frac{9}{16} \text{ in.}} = 4.74 \text{ ksi} < F_b = 5.5 \text{ ksi}$$

ok (Table 10-36, AISC Appendix A)

AISCS 1.5.1.2.1 is used for the calculated shear stress of the section.

$$\text{Overall depth} = h + t_f, \qquad h = \text{web height}$$

The overall depth is used in AISCS 1.5.1.2.1, and h is used in AISCS 1.10.5.2.

Assume web thickness of $\frac{5}{8}$ in. and 1-in. flange thickness.

Based on this web thickness (for $F_b = 22$ ksi), we calculate the area of the flanges

$$\left(\frac{A_w d}{6} + A_f d \right) F_b = M$$

$$A_f = \frac{M}{F_b h} - \frac{A_w}{6}$$

$$A_w = \frac{5}{8} \text{ in.} \times 69 \text{ in.} = 43.1 \text{ in.}^2$$

$$A_f = \frac{3225 \text{ ft-k} \times 12 \text{ in./ft}}{22 \text{ ksi} \times 69 \text{ in.}} - \frac{43.1 \text{ in.}^2}{6} = 18.31 \text{ in.}^2 - 1 \text{ in.} \times 19 \text{ in.}$$

$$I = \frac{\frac{5}{8} \times (69)^3}{12} + 2 \times (1 \times 19) \times \left(\frac{69.0}{2} + 0.5 \right)^2 = 63660 \text{ in.}^4$$

$$S = \frac{63660 \text{ in.}^4}{\frac{69}{2} \text{ in.} + 1 \text{ in.}} = 1793 \text{ in.}^3$$

$$f_b = \frac{3225 \text{ ft-k} \times 12 \text{ in./ft}}{1793 \text{ in.}^3} = 21.58 \text{ ksi} < 22 \text{ ksi} \qquad \text{ok}$$

Plate girder section

$$\text{web} = \tfrac{5}{8} \text{ in.} \times 69 \text{ in.}$$

flanges = 1 in. $\times$ 19 in.

$$I = 63660 \text{ in.}^4, \quad S = 1793 \text{ in.}^3, \quad f_b = 21.58 \text{ ksi}$$

Investigate lateral stability for allowable flange stresses (AISCS 1.5.1.4.5)

$$I_T = \frac{1.0 \text{ in.} \times (19 \text{ in.})^3}{12} = 571.6 \text{ in.}^4$$

$$A_T = 1.0 \text{ in.} \times 19 \text{ in.} + \frac{43.1 \text{ in.}^2}{6} = 26.18 \text{ in.}^2$$

$$r_T = \sqrt{\frac{571.6 \text{ in.}^4}{26.18 \text{ in.}^2}} = 4.67 \text{ in.}$$

Calculate F_b in region BC and AB.

F_b in BC

$$\frac{l}{r_T} = \frac{16 \text{ ft} \times 12 \text{ in./ft}}{4.67 \text{ in.}} = 41.1$$

$$C_b = 1$$

$$\sqrt{\frac{102 \times 10^3}{F_y}} \, C_b = 53 \quad \text{(AISCS 1.5.1.4.5.2, or Table 6, AISC Appendix A)}$$

Because $41.1 < 53$,

$$F_b = 22 \text{ ksi}$$

F_b in AB

$$\frac{l}{r_T} = \frac{20 \text{ ft} \times 12 \text{ in./ft}}{4.67 \text{ in.}} = 51.4$$

$$C_b = 1.75 + 1.05 \times \frac{M_1}{M_2} + 0.30 \times \left(\frac{M_1}{M_2}\right)^2$$

For $M_1 = 0$,

$$C_b = 1.75$$

$$53\sqrt{C_b} = 70.1$$

Because $51.4 < 70.1$,

$$F_b = 22 \text{ ksi}$$

Tentative selection of web and flanges is satisfactory

End-bearing stiffeners (AISCS 1.10.5.1)

Try two $\frac{5}{8}$-in. $\times$ 9-in. bars

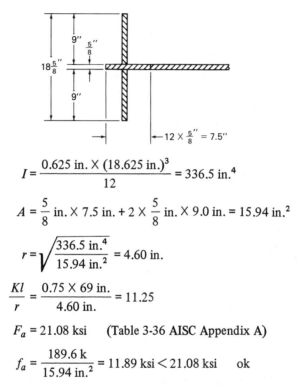

$$I = \frac{0.625 \text{ in.} \times (18.625 \text{ in.})^3}{12} = 336.5 \text{ in.}^4$$

$$A = \frac{5}{8} \text{ in.} \times 7.5 \text{ in.} + 2 \times \frac{5}{8} \text{ in.} \times 9.0 \text{ in.} = 15.94 \text{ in.}^2$$

$$r = \sqrt{\frac{336.5 \text{ in.}^4}{15.94 \text{ in.}^2}} = 4.60 \text{ in.}$$

$$\frac{Kl}{r} = \frac{0.75 \times 69 \text{ in.}}{4.60 \text{ in.}} = 11.25$$

$$F_a = 21.08 \text{ ksi} \quad \text{(Table 3-36 AISC Appendix A)}$$

$$f_a = \frac{189.6 \text{ k}}{15.94 \text{ in.}^2} = 11.89 \text{ ksi} < 21.08 \text{ ksi} \quad \text{ok}$$

The large difference in stresses indicates that the stiffeners may be reduced considerably.

Try $\frac{1}{2}$-in. $\times$ 7-in. plates

$$I = 130.3 \text{ in.}^4, \quad A = 11.69 \text{ in.}^2, \quad r = 3.34 \text{ in.}$$

$$\frac{Kl}{r} = \frac{0.75 \times 69 \text{ in.}}{3.34 \text{ in.}} = 15.5 \quad F_a = 20.86 \text{ ksi}$$

$$f_a = \frac{P}{A} = \frac{189.6 \text{ k}}{11.69 \text{ in.}^2} = 16.22 \text{ ksi} < 20.86 \text{ ksi}$$

Verify also AISCS 1.9.1.2

$$\frac{b}{t} = \frac{7}{.5} = 14 < 15.8 \quad ok$$

Investigating the need for stiffeners under columns (AISCS 1.10.10.2),

$$f_v = \left(\frac{100\,k}{69\,in.} + \frac{3.2\,k/ft}{12\,in./ft}\right) \times \frac{1}{\frac{5}{8}\,in.} = 2.75\,ksi$$

$$\frac{a}{h} = \frac{20\,ft \times 12\,in./ft}{69\,in.} = 3.48$$

$$\frac{h}{t} = \frac{69\,in.}{\frac{5}{8}\,in.} = 110.4$$

$$\left(2.0 + \frac{4}{(a/h)^2}\right) \times \frac{10000}{(h/t)^2} = 1.91\,ksi \quad (AISCS\ 1.10.10.2)$$

$$f_v = 2.75\,ksi > F_{all} = 1.91\,ksi \quad N.G.$$

Use load stiffeners

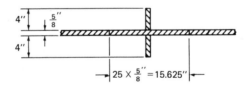

$$I = \frac{0.25 \times (8.625)^3}{12} = 13.37\,in.^4$$

$$A = 0.25\,in. \times 4\,in. \times 2 + 15.625\,in. \times \frac{5}{8}\,in. = 11.77\,in.^2$$

$$r = \sqrt{\frac{13.37\,in.^4}{11.77\,in.^2}} = 1.07\,in.$$

$$\frac{Kl}{r} = \frac{0.75 \times 69\,in.}{1.07\,in.} = 48.36 \quad F_a = 18.48\,ksi \quad (AISCS\ Table\ 3\text{-}36\ Appendix\ A)$$

Bearing stiffeners and stiffeners under concentrated loads will be connected to web by $\frac{1}{4}$-in. continuous fillet weld

$$\frac{h}{t_w} = \frac{4}{.25} = 16 > \frac{95}{\sqrt{F_y}} = 15.8 \quad \text{(AISCS 1.9.1.2)}$$

Checking the h/t_w ratio of stiffeners, we see that thickness must be increased to $\frac{5}{16}$ in.

No need to verify web crippling, as stiffeners are already provided.

Connection to flanges to web

$$Q = (1 \text{ in.} \times 19 \text{ in.}) \times \left(\frac{69 \text{ in.}}{2} + \frac{1.0 \text{ in.}}{2}\right) = 665 \text{ in.}^3$$

Connection in panel AB

$$q = \frac{VQ}{I} = \frac{189.6 \text{ k} \times 665 \text{ in.}^3}{63660 \text{ in.}^4} = 1.98 \text{ k/in.} = 23.77 \text{ k/ft}$$

Use $\frac{5}{16}$-in. E 70 welds (AISCS 1.17.2).

Considering AISCS 1.5.1.4.1.1, intermittent fillet welds may be used with $F_b = 22$ ksi.

Capacity of weld (AISCS 1.5.3)

$$W_c = 4.64 \text{ k/in.}$$

$$l_w = \frac{23.77 \text{ k/ft}}{2 \times 4.64 \text{ k/in.}} = 2.56 \text{ in. per ft} \quad \text{say } 2\frac{3}{4} \text{ in.}$$

Connection in panel BC

$$V = 25.6 \text{ k}$$

$$l_w = 2.75 \text{ in.} \times \frac{25.6}{189.6} = 0.37 \text{ in.}$$

Use $l_w = 1\frac{1}{2}$ in. (minimum) (AISCS 1.17.5).

For connection of bearing stiffeners to web, use $\frac{1}{4}$-in. continuous fillet weld with close bearing against bottom flange (AISCS 1.17.2).

Similar for stiffeners under concentrated loads, but close bearing against top flange.

Assume that full capacity of load is taken by the stiffener and carried to web by welds (very conservative assumption).

Use $\frac{1}{4}$-in. welds for end stiffeners.

$$W_c = 4 \times 0.928 \text{ k/in.} = 3.71 \text{ k/in.}$$

$$l_w = \frac{189.6 \text{ k}}{3.7 \text{ k/in.}} = 51.1 \text{ in.}$$

$$\frac{69 \text{ in.}}{51.1 \text{ in./2}} \times 1.5 \text{ in.} = 4.05 \text{ in.}$$

Use $1\frac{1}{2}$-in. intermittent fillet weld at 4 in. on center.

For bearing stiffeners

$$l_w = \frac{100 \text{ k}}{2.78 \text{ k/in.}} = 35.97 \text{ in.} \left(\frac{3}{16} \text{ in. fillet weld}\right)$$

$$\frac{69 \text{ in.}}{35.97 \text{ in./2}} \times 1.5 \text{ in.} = 5.75 \text{ in.}$$

Use $1\frac{1}{2}$-in. intermittent fillet weld at $5\frac{1}{2}$ in. on center.

Maximum distance between two intermittent welds

$$\frac{5}{16} \text{ in.} \times \frac{127}{\sqrt{36}} = 6.61 \text{ in.} > 5\frac{1}{2} \text{ in.} \quad \text{ok} \quad (\text{AISCS } 1.18.2.3)$$

Total weight

web	$1 - \dfrac{5}{8}$ in. $\times$ 69 in. $\times$ 56 ft-0 in. $\times \dfrac{490 \text{ pcf}}{144 \text{ in.}^2/\text{ft}^2} =$	8218 lb
flanges	$2 - 1$ in. $\times$ 19 in. $\times$ 56 ft-0 in. $\times (490/144) \quad =$	7241 lb
stiffeners	$4 - \dfrac{1}{2}$ in. $\times$ 7 in. $\times$ 69 in. $\times (490/1728) \quad =$	274 lb
	$4 - \dfrac{5}{16}$ in. $\times$ 4 in. $\times$ 69 in. $\times (490/1728) \quad =$	98 lb
		15831 lb

b) Design with stiffeners.

Minimum thickness

$$\frac{h}{t_w} < 333$$

$$t_w > \frac{69 \text{ in.}}{333} = 0.207 \text{ in.} \quad \text{use } \frac{5}{16} \text{ in.}$$

$$\frac{h}{t_w} = \frac{69 \text{ in.}}{0.3125 \text{ in.}} = 221$$

$$A_w = 0.3125 \text{ in.} \times 69 \text{ in.} = 21.56 \text{ in.}^2$$

Shear stress

Assume $1\frac{1}{4}$-in. thick flanges

$$f_v = \frac{189.6 \text{ k}}{(69 \text{ in.} + 2.5 \text{ in.}) \times 0.3125 \text{ in.}} = 8.49 \text{ ksi}$$

Area of flange needed

Try $F_b = 21$ ksi, considering flange stress reduction

$$A_f = \frac{3225 \text{ ft-k} \times 12 \text{ in./ft}}{21 \text{ ksi} \times 69 \text{ in.}} - \frac{21.56 \text{ in.}^2}{6} = 23.11 \text{ in.}^2$$

Try $1\frac{3}{8}$ in. $\times$ 17 in.

$$A_f = 23.375 \text{ in.}^2$$

$$\frac{b_f}{2 \, t_f} = \frac{17 \text{ in.}}{2 \times 1.375 \text{ in.}} = 6.18 < 15.8 \quad \text{ok} \quad \text{(AISCS 1.9.1.2)}$$

$$I = \frac{\frac{5}{16} \text{ in.} \times (69 \text{ in.})^3}{12} + 2 \times (1.375 \text{ in.} \times 17 \text{ in.}) \left(\frac{69 \text{ in.}}{2} + \frac{1.375 \text{ in.}}{2} \right)^2$$

$$= 66440 \text{ in.}^4$$

$$S = \frac{66440 \text{ in.}^4}{\dfrac{69 \text{ in.}}{2} + 1.375 \text{ in.}} = 1852 \text{ in.}^3$$

$$f_b = \frac{3225 \text{ ft-k} \times 12 \text{ in./ft}}{1852 \text{ in.}^3} = 20.90 \text{ ksi}$$

To obtain the reduction in flange stress, use AISCS 1.10.6.

Calculate first F_b by AISCS 1.5.1.4.5.2

$$I_T = \frac{1.375 \text{ in.} \times (17 \text{ in.})^3}{12} = 562.9 \text{ in.}^4$$

$$A_T = 1.375 \text{ in.} \times 17 \text{ in.} + \frac{21.56 \text{ in.}^2}{6} = 26.968 \text{ in.}^2$$

$$r_T = \sqrt{\frac{562.9 \text{ in.}^4}{26.968 \text{ in.}^2}} = 4.57 \text{ in.}$$

Stress in panel BC

$$C_b = 1, \quad \frac{l}{r_T} = \frac{16 \text{ ft} \times 12 \text{ in./ft}}{4.57 \text{ in.}} = 42.0 < 53$$

$$F_b = 22 \text{ ksi}$$

Stress in panel AB

$$C_b = 1.75, \quad \frac{l}{r_T} = \frac{20 \text{ ft} \times 12 \text{ in./ft}}{4.57 \text{ in.}} = 52.52 < 53\sqrt{1.75} \quad \text{ok}$$

$$F_b = 22 \text{ ksi}$$

$$F_b' = F_b\left[1.0 - 0.0005\frac{A_w}{A_f}\left(\frac{h}{t_w} - \frac{760}{\sqrt{F_b}}\right)\right] \quad \text{(AISCS 1.10.6)}$$

$$A_w = 21.56 \text{ in.}^2, \quad A_f = 23.375 \text{ in.}^2, \quad h/t_w = 221, \quad 760/\sqrt{F_b} = 162.0$$

$$F_b' = 22 \times \left[1.0 - 0.0005 \times \frac{21.56}{23.375} \times (221 - 162.0)\right] = 21.40 \text{ ksi}$$

$$f_b = 20.90 \text{ ksi} < F_b = 21.40 \text{ ksi} \quad \text{ok}$$

Location of bearing stiffeners

Bearing stiffeners are required at end of unframed girders (AISCS 1.10.5.1).

Under concentrated loads, compressive stress is

$$\frac{100 \text{ k}}{0.3125 \text{ in.} \times 69 \text{ in.}} + \frac{3.20 \text{ k/ft}}{12 \text{ in./ft} \times 0.3125 \text{ in.}} = 5.49 \text{ ksi}$$

Allowable compression stress

$$a = 16 \text{ ft} \times 12 \text{ in./ft} = 192 \text{ in.}$$

$$\left(2 + \frac{4}{\left(\frac{a}{h}\right)^2}\right)\left(\frac{10000}{\left(\frac{h}{t_w}\right)^2}\right) = 0.52 \text{ ksi} < 5.49 \text{ ksi}$$

Check web crippling

$$\frac{R}{t(N+2\text{k})} = \frac{100 \text{ k}}{\frac{5}{16} \text{ in. } (8 \text{ in.} + 2 \times 1.375 \text{ in.})} < 0.75 \, F_y = 27.0 \text{ ksi}$$

(AISCS 1.10.10.1)

$$29.76 \text{ ksi} > 27.0 \text{ ksi} \quad \text{N.G.}$$

Must use stiffeners under concentrated loads.

Location of intermediate stiffeners

$$\frac{h}{t} = \frac{69 \text{ in.}}{\frac{5}{16} \text{ in.}} = 221$$

$$f_v = \frac{189.6}{(69 + 2.5) \times 0.3125} = 8.49 \text{ ksi}$$

Panel AB

$$\frac{a}{h} = \frac{20 \text{ ft} \times 12 \text{ in./ft}}{69 \text{ in.}} = 3.5$$

No tension field action for end panel (AISCS 1.10.5.2)

$$F_v = 1.7 \text{ ksi} \quad \text{(Table 10-36)}$$

Must use stiffeners.

Intermediate stiffeners in end panels

Calculate by hand, because f_v is outside the range of Table 10-36.

Try for $C_v < 0.8$

$$F_v = 8.49 \text{ ksi} = \frac{F_y}{2.89} \times C_v$$

$$C_v = 0.68 = \frac{45000 \text{ k}}{F_y \left(\frac{h}{t}\right)^2} = \frac{45000 \text{ k}}{36 \times (221)^2}$$

$$k = 26.57$$

$$k = 26.57 = 4.00 + \frac{5.34}{(a/h)^2} \text{ for } a/h \text{ less than } 1.0$$

$$a = 69 \text{ in.} \times \sqrt{\frac{5.34}{26.57 - 4.00}} = 33.56 \text{ in.} \quad \text{say } 33 \text{ in. (2 ft-9 in.)}$$

Spacing to next stiffener (with tension field action)

$$V = 189.6 \text{ k} - 2.75 \text{ ft} \times 3.2 \text{ k/ft} = 180.8 \text{ k}$$

$$f_v = \frac{180.8 \text{ k}}{71.5 \text{ in.} \times \frac{5}{16} \text{ in.}} = 8.09 \text{ ksi}$$

Entering Table 11-36 for $h/t = 221$ and $f_v = 8.09$ ksi,

$$\frac{a}{h} = 1.2, \text{ with } 11\% \text{ of web area required for gross area of stiffener pair}$$

$$a = 1.2 \, h = 1.2 \times 69 \text{ in.} = 828 \text{ in. (6 ft-10 in.)} \quad \text{say 6 ft-9 in.}$$

Spacing to next stiffener (with tension field action)

$$V = 108.8 \text{ k} - 6.75 \text{ ft} \times 3.2 \text{ k/ft} = 87.2 \text{ k}$$

$$f_v = \frac{87.2 \text{ k}}{71.5 \text{ in.} \times \frac{5}{16} \text{ in.}} = 3.90 \text{ ksi}$$

Entering Table 11-36 for $h/t = 221$ and $f_v = 3.90$ ksi

$$\frac{a}{h} = 1.4, \text{ with } 10.6\% \text{ of web area required for gross area of stiffener pair}$$

$$a = 1.4 \, h = 1.4 \times 69 \text{ in.} = 96.6 \text{ in. (8 ft-0 in.)}$$

No intermediate stiffeners are required when $a/h < [260/(h/t)]^2$ (AISCS 1.10.5.3)

$$a < h \times \left(\frac{260}{h/t}\right)^2 = 69 \text{ in.} \times \left(\frac{260}{221}\right)^2 = 96 \text{ in. (8 ft-0 in.)}$$

Stiffeners must be provided at a maximum spacing of 8 ft-0 in.

Provide stiffeners at points shown.

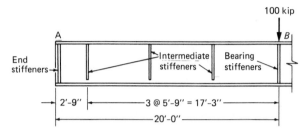

Stiffeners in panel BC (AISCS Appendix A, Table 11-36)

$$f_v = \frac{25.6 \text{ k}}{71.5 \text{ in.} \times \frac{5}{16} \text{ in.}} = 1.14 \text{ ksi} < F_v = 1.7 \text{ ksi}$$

No stiffeners are needed between points B and C.

Design of stiffeners

End stiffener

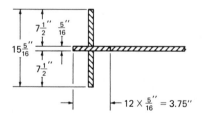 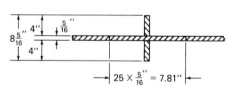

Try $\frac{9}{16}$ in. $\times$ $7\frac{1}{2}$ in. end stiffeners

$$I = \frac{0.5625 \text{ in.} \times (15.3125)^3}{12} = 168.30 \text{ in.}^4$$

$$A = 3.75 \text{ in.} \times \frac{5}{16} \text{ in.} + 2 \times 7.5 \text{ in.} \times \frac{9}{16} \text{ in.} = 9.61 \text{ in.}^2$$

$$r = \sqrt{\frac{168.30 \text{ in.}^4}{9.61 \text{ in.}^2}} = 4.18 \text{ in.}$$

$$\frac{Kl}{r} = \frac{0.75 \times 69 \text{ in.}}{4.18 \text{ in.}} = 12.4 \quad F_a = 21.03 \text{ ksi} \quad \text{(AISCS Appendix A Table 3-36)}$$

$$f_a = \frac{189.6 \text{ k}}{9.61 \text{ ksi}} = 19.72 \text{ ksi} < 21.03 \text{ ksi} \quad \text{ok}$$

Stiffeners under concentrated load

Try $\frac{3}{8}$ in. $\times$ 4 in. stiffeners

$$I = \frac{0.375 \text{ in.} \times (8.3125 \text{ in.})^3}{12} = 17.95 \text{ in.}^4$$

$$A = 7.81 \text{ in.} \times \frac{5}{16} \text{ in.} + 2 \times 4.0 \text{ in.} \times \frac{3}{8} \text{ in.} = 5.44 \text{ in.}^2$$

$$r = \sqrt{\frac{17.95 \text{ in.}^4}{5.44 \text{ in.}^2}} = 1.82 \text{ in.}$$

$$\frac{Kl}{r} = \frac{0.75 \times 69 \text{ in.}}{1.82 \text{ in.}} = 28.4 \quad F_a = 20.05 \text{ ksi}$$

$$f_a = \frac{100 \text{ k}}{5.44 \text{ in.}^2} = 18.38 \text{ ksi} < 20.05 \text{ ksi} \quad \text{ok}$$

$$\frac{l}{t} = \frac{4 \text{ in.}}{0.375 \text{ in.}} = 10.67 < 15.8 \quad \text{ok}$$

Intermediate stiffeners

$$A_{st} \geq 11\% \, A_w = 0.11 \times \tfrac{5}{16} \text{ in.} \times 69 \text{ in.} = 2.37 \text{ in.}^2$$

Try $\frac{1}{4}$ in. $\times$ 4 in. stiffeners

$$A_{st} = 2 \times 0.25 \text{ in.} \times 4 \text{ in.} = 2.00 \text{ in.}^2$$

Check width/thickness ratio (AISCS 1.9.1.2)

$$\frac{4}{\frac{1}{4}} = 16 > 15.8 \quad \text{N.G.}$$

Use $\frac{3}{8}$ in. $\times$ 4 in. stiffeners

$I = 17.95$ in.^{4}

$\left(\dfrac{h}{50}\right)^4 = 3.63$ in.^{4} (AISCS 1.10.5.4)

17.95 in.$^4 > 3.63$ in.^{4} ok

Connection of flanges to web

Use E 70 electrodes, $\frac{5}{16}$-in. welds (AISCS 1.17.2)

$$W_c = 5 \times 0.928 \text{ k/in.} = 4.64 \text{ k/in.}$$

Shear flow

$$Q = (1.375 \text{ in.} \times 17 \text{ in.}) \times \left(\frac{69 \text{ in.}}{2} + \frac{1.375 \text{ in.}}{2}\right) = 823 \text{ in.}^3$$

$$q_{max} = \frac{VQ}{I} = \frac{189.6 \text{ k} \times 823 \text{ in.}^3}{66440 \text{ in.}^4} = 2.35 \text{ k/in.} = 28.17 \text{ k/ft}$$

Using $\frac{5}{16}$-in. intermittent fillet weld at 12-in. centers,

$$l_w = \frac{28.17 \text{ k/ft}}{2 \times 4.64 \text{ k/in.}} = 3.04 \text{ k/in.} \quad \text{say } 3\frac{1}{4} \text{ in.}$$

Determine location at which length of weld may be $1\frac{1}{2}$ in.

$$\frac{V \times 823 \text{ in.}^3 \times 12 \text{ in./ft}}{66440 \text{ in.}^4 \times 4.64 \text{ k/in.} \times 2} = 1.5 \text{ in./ft}$$

$$V = 93.6 \text{ k} \quad \text{say 90 k}$$

Use intermittent fillet welds of length $1\frac{1}{2}$ in on 12-in. centers between points B and C.

Connection of stiffeners

Bearing stiffeners

Consider stiffener loaded with full load and the load transferred to web (conservative approach).

End stiffener

$$A = 7.5 \text{ in.} \times \tfrac{9}{16} \text{ in.} = 4.22 \text{ in.}^2$$

$$P = 19.72 \text{ ksi} \times 4.22 \text{ in.}^2 = 83.2 \text{ k}$$

Use $\frac{1}{4}$-in. intermittent fillet weld

$$W_c = 4 \times 0.928 \text{ k/in.} = 3.71 \text{ k/in.}$$

$$l_w = \frac{83.2 \text{ k}}{2 \times 3.71 \text{ k/in.}} = 11.21 \text{ in.}$$

Using $1\frac{1}{2}$ in. weld length, spacing is

$$\frac{69 \text{ in.}}{11.21 \text{ in.}} \times 1.5 \text{ in.} = 9.23 \text{ in.} \qquad \text{say 9 in. on center}$$

Use $\frac{1}{4}$-in. intermittent fillet weld of $1\frac{1}{2}$-in. length at 9-in. centers.

Stiffener under load

$$A = 4 \text{ in.} \times \tfrac{3}{8} \text{ in.} = 1.5 \text{ in.}^2$$

$$P = 18.38 \text{ ksi} \times 1.5 \text{ in.}^2 = 27.6 \text{ k}$$

Use $\frac{3}{16}$-in. intermittent fillet welds

$$W_c = 3 \times 0.928 \text{ k/in.} = 2.78 \text{ k/in.}$$

Total length needed on one side is

$$l_w = \frac{27.6 \text{ k}}{2 \times 2.78 \text{ k/in.}} = 4.96 \text{ in.}$$

Using $1\frac{1}{2}$ in. weld length

$$\frac{69 \text{ in.}}{4.96 \text{ in.}} \times 1.5 \text{ in.} = 20.87 \text{ in.}$$

Use $\frac{3}{16}$-in. intermittent fillet weld of $1\frac{1}{2}$-in. length at 12-in. centers.

Intermediate stiffeners (AISCS 1.10.5.4)

$$f_{vs} = h\sqrt{\left(\frac{F_y}{340}\right)^3} = 2.38 \text{ k/in.}$$

Total force for two stiffeners

$$V = \frac{69 \text{ in.} \times 2.38 \text{ k/in.}}{2} = 82.11 \text{ k}$$

Using $\frac{3}{16}$-in. weld

$$l_w = \frac{82.11 \text{ k}}{2 \times 2.78 \text{ k/in.}} = 14.77 \text{ in. (on one side)}$$

Using $1\frac{1}{2}$ in. intermittent fillet weld, spacing is

$$\frac{69 \text{ k}}{14.77 \text{ ksi}} = 7.0 \text{ in.}$$

Use $\frac{3}{16}$-in. intermittent fillet weld of $1\frac{1}{2}$-in. length at 7-in. centers.

Total weight

web	$1 - \frac{5}{16}$ in. $\times$ 69 in. $\times$ 56 ft-0 in. $\times \dfrac{490 \text{ pcf}}{144 \text{ in.}^2/\text{ft}^2}$ =	4109 lb
flanges	$2 - 1\frac{3}{8}$ in. $\times$ 17 in. $\times$ 56 ft-0 in. $\times$ (490/144) =	8908 lb
stiffeners	$4 - \frac{9}{16}$ in. $\times$ $7\frac{1}{2}$ in. $\times$ 69 in. $\times$ (490/1728) =	330 lb
	$4 \times 4 - \frac{3}{8}$ in. $\times$ 4 in. $\times$ 69 in. $\times$ (490/1728) =	470 lb
		13817 lb

Comparing weight of 15,831 lb without stiffeners and 13,817 lb with stiffeners, and considering 1 ft of weld is approximately equal to the cost of 5-$\frac{1}{2}$ lb of steel, we must add to the weight with stiffeners the equivalent weight of the length of intermediate stiffener weld required

$$l_w = 2 \times 14.77 \text{ in. (say 15 in.)} = 30 \text{ in.} = 2.5 \text{ ft}$$

equivalent weight = 2.5 ft $\times$ 5.5 lb/ft $\times$ 16 plates = 220 lb

The total equivalent weight is

$$Wt = 13,817 \text{ lb} + 220 \text{ lb} = 14,037 \text{ lb}$$

$$\frac{14,037 \text{ lb}}{15,831 \text{ lb}} = .89$$

We see that the design with stiffeners is approximately 10% less expensive than the design without stiffeners. Other advantages are realized in the weight reduction, such as for shipping and building dead load.

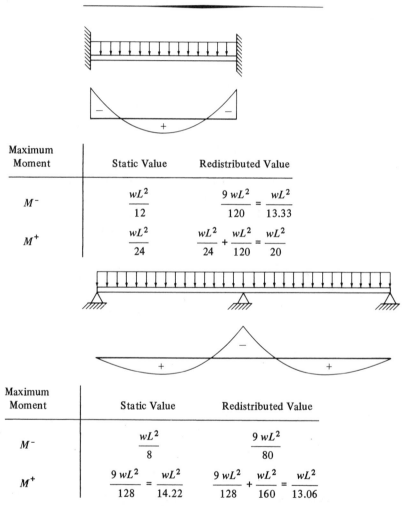

Maximum Moment	Static Value	Redistributed Value
M^-	$\dfrac{wL^2}{12}$	$\dfrac{9\,wL^2}{120} = \dfrac{wL^2}{13.33}$
M^+	$\dfrac{wL^2}{24}$	$\dfrac{wL^2}{24} + \dfrac{wL^2}{120} = \dfrac{wL^2}{20}$

Maximum Moment	Static Value	Redistributed Value
M^-	$\dfrac{wL^2}{8}$	$\dfrac{9\,wL^2}{80}$
M^+	$\dfrac{9\,wL^2}{128} = \dfrac{wL^2}{14.22}$	$\dfrac{9\,wL^2}{128} + \dfrac{wL^2}{160} = \dfrac{wL^2}{13.06}$

Fig. 3.9. Moment redistribution of rigid and continuous beams.

3.3 CONTINUOUS BEAMS

The AISC allows beams that span continuously over an interior support or are rigidly framed to columns to assume a slight moment redistribution. AISCS 1.5.1.4.1 states that, except for hybrid girders and members of A 514 steel, continuous or rigidity framed compact beams and girders may be designed for $\frac{9}{10}$ of the negative moments due to gravity loading if the positive moments are increased by $\frac{1}{10}$ of the average negative moments (see Fig. 3.9). These provisions do not apply for cantilevers. If negative moment is resisted by a rigid column-beam connection, the $\frac{1}{10}$ reduction may be used in proportioning the column for the combined axial and bending loads, provided that the stress f_a does not exceed $0.15 F_a$.

Example 3.9. A beam spanning two 30-ft bays supports a live load of 1.2 k/ft and a dead load of 0.5 k/ft. Assuming full lateral support, design the beam

a) for calculated positive and negative moments
b) using the moment redistribution allowed by AISCS.

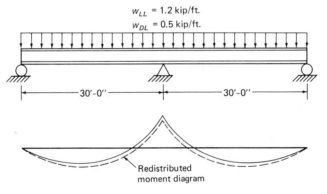

Solution.

$$M^+ = \tfrac{9}{128} wL^2 \text{ at 11 ft-3 in. from outside supports}$$

$$M^- = \tfrac{1}{8} wL^2 \text{ at interior support}$$

$$M^+ = \tfrac{9}{128} (1.2 \text{ k/ft} + 0.5 \text{ k/ft}) \times (30.0 \text{ ft})^2 = 107.6 \text{ ft-k}$$

$$M^- = \tfrac{1}{8} (1.2 \text{ k/ft} + 0.5 \text{ k/ft}) \times (30.0 \text{ ft})^2 = 191.25 \text{ ft-k}$$

a)
$$M_{\max} = 191.25 \text{ ft-k}$$

$$S = M_{\max}/F_b = \frac{191.25 \text{ ft-k} \times 12 \text{ in./ft}}{24.0 \text{ ksi}} = 95.6 \text{ in.}^3$$

From the allowable stress design selection table

$$\text{Use W 18} \times 55 \qquad S_x = 98.3 \text{ in.}^3$$

b) For AISC redistribution, the negative moment becomes $\frac{9}{10}$ that produced by gravity, and the positive moment increases $\frac{1}{10}$ of the average negative moments

$$M^- = \frac{9}{10} \times 196.9 \text{ ft-k} = 177.2 \text{ ft-k}$$

$$M^+ = 107.6 \text{ ft-k} + \left(\frac{1}{10} \times \frac{196.9 \text{ ft-k}}{2}\right) = 117.4 \text{ ft-k}$$

$$S = M_{max}/F_b = \frac{177.2 \text{ ft-k} \times 12 \text{ in./ft}}{24.0 \text{ ksi}} = 88.6 \text{ in.}^3$$

From the allowable stress design selection table

$$\text{Use W 18} \times 50 \qquad S_x = 88.9 \text{ in.}^3$$

Example 3.10. Design a continuous beam for the condition shown.

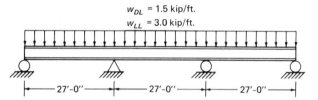

$w_{DL} = 1.5 \text{ kip/ft.}$
$w_{LL} = 3.0 \text{ kip/ft.}$

27'-0'' 27'-0'' 27'-0''

Solution. The analysis can be done easily with the aid of AISC beam diagrams and formulas. Design the beam to carry the dead load uniformly plus the maximum moment that can occur according to variable live load conditions (AISCM Part 2).

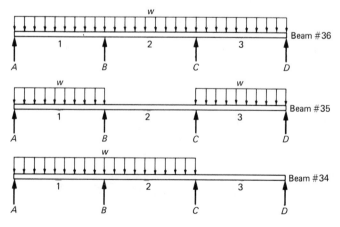

For dead load, use beam No. 36 (beam diagrams and formulas, AISCM Part 2).

$$M_1 = M_3 = 0.08 \ (1.5 \text{ k/ft} \times (27.0 \text{ ft})^2) = 87.5 \text{ ft-k}$$
$$M_B = M_C = -0.1 \ (1.5 \text{ k/ft} \times (27.0 \text{ ft})^2) = -109.4 \text{ ft-k}$$

It can be seen that the maximum midspan moment occurs in the outer bays, and that the maximum interior support moment occurs with one bay unloaded.

For midspan live load moment, use beam No. 35 (greater value)

$$M_1 = 0.1013 \ (3.0 \text{ k/ft} \times (27.0 \text{ ft})^2) = 221.5 \text{ ft-k}$$

For support moment, use beam No. 34 (greater value)

$$M_B = -0.1167 \ (3.0 \text{ k/ft} \times (27.0 \text{ ft})^2) = -255.2 \text{ ft-k}$$

The beam is symmetrical, and moments would be the same for the opposite side.

Adding dead load moment and live load moment

$$M_1 = 87.5 \text{ ft-k} + 221.5 \text{ ft-k} = 309.0 \text{ ft-k}$$
$$M_B = -109.4 \text{ ft-k} + -255.2 \text{ ft-k} = -364.6 \text{ ft-k}$$

Using moment redistribution

$$M_B = 0.9 \times -364.6 \text{ ft-k} = -328.1 \text{ ft-k}$$

$$M_1 = 309.0 \text{ ft-k} + \frac{36.46 \text{ ft-k}}{2} = 327.2 \text{ ft-k} \qquad \text{nearly equal moments}$$

$$S = M/F_b = \frac{328.1 \text{ ft-k} \times 12 \text{ in./ft}}{24.0 \text{ ksi}} = 164.05 \text{ in.}^3$$

Use W 24 × 76 $S_x = 176.0 \text{ in.}^3$

3.4 BIAXIAL BENDING

All cross sections have two principal axes passing through the centroid for which, about one axis, the moment of inertia is maximum and, about the other axis, the moment of inertia is minimum. If loads passing through the shear center are perpendicular to one axis, simple bending occurs about that axis. Members loaded such that bending occurs simultaneously about both principal axes are subject to biaxial bending. The total bending stress at any point in the cross section of such a member is

$$f_b = \frac{M_1}{I_1} m + \frac{M_2}{I_2} n \qquad (3.15)$$

where the subscripts refer to the principal axes, m is the distance to the point measured perpendicular to axis 1, and n is the distance to the point measured perpendicular to axis 2. The moment of inertia I_1 and I_2 about the principal axes can be determined by

$$I_1 = \frac{I_x + I_y}{2} + \sqrt{\left(\frac{I_x - I_y}{2}\right)^2 + I_{xy}^2}$$

$$I_2 = \frac{I_x + I_y}{2} - \sqrt{\left(\frac{I_x - I_y}{2}\right)^2 + I_{xy}^2}$$

where I_x and I_y are the moments of inertia through the centroid about the x and y axes, respectively, and I_{xy} is the product of inertia of the cross section. The product of inertia of a section is the geometrical characteristic of the section defined by the integral $I_{xy} = \int_A xy\, dA$. If orthogonal axes x and y, or one of them, are axes of symmetry, the product of inertia with respect to such axes is equal to zero. In such cases, the principal axes coincide with the x and y centroidal axes of the member.

In the case of loads applied perpendicular to symmetrical sections, the stresses and deflections may be calculated separately for bending about each axis and superimposed. Loads applied that are not perpendicular to either principal axis can be broken down into components that are perpendicular to the principal axes. The extreme bending stress due to biaxial bending is then

$$f_b = \frac{M_x y}{I_x} + \frac{M_y x}{I_y} \qquad (3.16)$$

and the resultant deflection is

$$\Delta_T = (\Delta_1)^2 + (\Delta_2)^2 \qquad (3.17)$$

Shapes that do not have an axis of symmetry have the principal axes inclined to the x and y centroidal axes. For these nonsymmetrical shapes, the total stress at any point in the cross section can be determined by

$$f_b = \frac{M_x - M_y (I_{xy}/I_y)}{I_x - (I_{xy}^2/I_y)} y + \frac{M_y - M_x (I_{xy}/I_y)}{I_y - (I_{xy}^2/I_x)} x$$

where M_x and M_y are the bending moments caused by loads perpendicular to the x and y axes, and I_{xy} is the product of inertia of the cross section, referred to the x and y axes.

In most cases, steel flexural members have different allowable bending stresses with respect to their major and minor axes. The use of the interaction equation below is then required to limit the stress levels in each plane

$$\frac{f_{bx}}{F_{bx}} + \frac{f_{by}}{F_{by}} \leqslant 1.0 \tag{3.18}$$

The allowable stresses, as discussed in Chapter 2, are applicable (AISCS 1.6.1).

Example 3.11. The wide-flange beam shown is loaded through the center subjecting the beam to a 150-ft-k moment. Design the lightest W 14 section by breaking down the moment to components of the principal axes and solving for biaxial bending. Assume full lateral support.

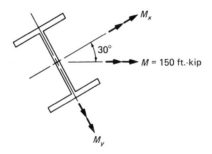

Solution. The moment is broken down by the components as shown

$$M_x = 150 \text{ ft-k } (\cos 30°) = 130.0 \text{ ft-k}$$
$$M_y = 150 \text{ ft-k } (\sin 30°) = 75.0 \text{ ft-k}$$

For member under combined stress

$$\frac{f_{bx}}{F_{bx}} + \frac{f_{by}}{F_{by}} \leqslant 1.0 \quad \text{(AISCS 1.6.1)}$$

Try W 14 × 109 $S_x = 173 \text{ in.}^3$ $S_y = 61.2 \text{ in.}^3$

$$F_{bx} = 24.0 \text{ ksi}$$
$$F_{by} = 27.0 \text{ ksi}$$

$$f_{bx} = M_x/S_x = \frac{130.0 \text{ ft-k} \times 12 \text{ in./ft}}{173 \text{ in.}^3} = 9.02 \text{ ksi}$$

$$f_{by} = M_y/S_y = \frac{75.0 \text{ ft-k} \times 12 \text{ in./ft}}{61.2 \text{ in.}^3} = 14.71 \text{ ksi}$$

$$\frac{9.02 \text{ ksi}}{24.0 \text{ ksi}} + \frac{14.71 \text{ ksi}}{27.0 \text{ ksi}} = 0.376 + 0.545 = 0.921 \leqslant 1.0 \quad \text{ok}$$

If the next lightest section, W 14 $\times$ 99, were checked, the interaction equation would yield $1.018 > 1.0$.

Therefore, W 14 $\times$ 109 is the lightest W 14 section.

Example 3.12. Design a C 12 channel spanning 20 ft with full lateral support as purlins subject to biaxial bending. Assume the live load of 100 lb/ft, dead load of 40 lb/ft, and wind load of 80 lb/ft to act through the shear center, thus eliminating torsion in the beam.

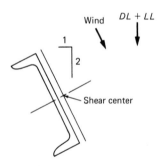

Solution. The total wind load is carried in the x axis of the channel, because the wind acts perpendicular to the roof surface. The live load and dead load will be broken down into x and y components.

$$\text{Gravity loads w} = 40 \text{ lb/ft (DL)} + 100 \text{ lb/ft (LL)}$$
$$= 140 \text{ lb/ft}$$

$$w_x = 80 \text{ lb/ft (wind)} + \frac{2}{\sqrt{5}} \, 140 \text{ lb/ft} = 205 \text{ lb/ft}$$

$$w_y = \frac{1}{\sqrt{5}} \, 140 \text{ lb/ft} = 63 \text{ lb/ft}$$

$$M_x = w_x L^2/8 = \frac{0.205 \text{ k/ft} \times (20.0 \text{ ft})^2}{8} = 10.25 \text{ ft-k}$$

$$M_y = w_y L^2/8 = \frac{0.063 \text{ k/ft} \times (20.0 \text{ ft})^2}{8} = 3.15 \text{ ft-k}$$

From members subject to combined stresses, the bending stresses must be proportioned such that

$$\frac{f_{bx}}{F_{bx}} + \frac{f_{by}}{F_{by}} \leqslant 1.0 \quad \text{(AISCS 1.6.1)}$$

$$F_{bx} = 24.0 \text{ ksi}$$

$$F_{by} = 27.0 \text{ ksi}$$

Try C 12 × 25 $\quad S_x = 24.1 \text{ in.}^3 \quad S_y = 1.88 \text{ in.}^3$

$$f_{bx} = M_x/S_x = \frac{10.25 \text{ ft-k} \times 12 \text{ in./ft}}{24.1 \text{ in.}^3} = 5.1 \text{ ksi}$$

$$f_{by} = M_y/S_y = \frac{3.15 \text{ ft-k} \times 12 \text{ in./ft}}{1.88 \text{ in.}^3} = 20.1 \text{ ksi}$$

$$\frac{5.1 \text{ ksi}}{24.0 \text{ ksi}} + \frac{20.1 \text{ ksi}}{27.0 \text{ ksi}} = 0.95 < 1.0 \quad \text{ok}$$

Use C 12 × 25 channel.

Fig. 3.10. Long-span joist roof system of the American Royal Arena in Kansas City, Missouri. (Courtesy of U.S. Steel Corp.)

3.5 OPEN-WEB STEEL JOISTS

In common usage are shop-fabricated, lightweight truss members referred to as *open-web steel joists* (see Fig. 3.11). Charts are readily available that provide

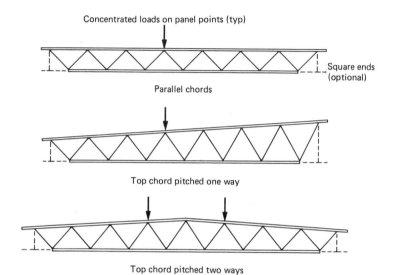

Fig. 3.11. Types of open-web steel joists. Generally, chords are double angles, and the web members are round bars. Parallel chord joists are most commonly used.

Fig. 3.12. View of a steel truss made from wide flange sections during fabrication.

standard fabricated sections for desired span and loading conditions. When the use of open-web joists is desired, the designer should refer to any one of the numerous design tables available from joist suppliers and the Steel Joist Institute.

PROBLEMS TO BE SOLVED

3.1. A W 16 × 100 beam has a simple span of 45 ft. If the member is fully laterally supported, determine

a) the maximum distributed load the beam can carry and
b) the maximum distributed load that can be carried by the beam if both flanges are cover-plated with $1\frac{1}{2}$ × 12 in. plates.

3.2. Determine the width of $1\frac{1}{2}$-in. cover plates necessary for a W 12 × 65 beam to resist a moment of 320 ft-k. Initially, design plates for top and bottom flanges, and rework for only bottom flange. Use F_b = 22.0 ksi.

3.3. The loading of an existing W 12 × 65 beam spanning 25 ft is to be increased to 2.75 k/ft. Because of existing conditions only the bottom flange can be reinforced. Determine the thickness of a 12-in. wide cover plate, for the bottom flange, necessary to carry the increased load.

3.4.[1] A simple beam spanning of 30 ft is to carry a uniform distributed load of 1.8 k/ft. However, height restrictions limit the total depth to $10\frac{1}{2}$ in. Determine the width of $\frac{3}{4}$-in. partial length cover plates necessary to increase a W 8 × 67 section to carry the load. Assuming bolted plates with two rows of $\frac{3}{4}$-in. A 325-N or F bolts as applicable, calculate the required spacing of bolts. CuT OFF PoiNTS

3.5.[1] Design partial length cover plates for Problem 3.4 assuming $\frac{3}{4}$ × 11-in. plates welded continuously along the sides and across the ends with $\frac{5}{16}$-in. welds.

3.6.[1] A W 10 × 112 beam with $\frac{3}{4}$ × 12-in. cover plates spans a distance of 35 ft with loads as shown. Assuming F_b = 22.0 ksi, design partial length cover plates bolted to the flanges with two rows of $\frac{3}{4}$-in. A 325-N bolts, using an allowable load of 9 kips per bolt, and determine how many bolts are required in the development length.

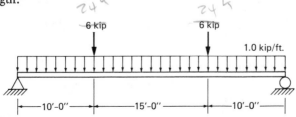

---

[1] Knowledge of connection design (Chapters 5, 6, and 7) is required.

3.7. Show that the required plate girder flange area equation $A_f = M/(F_b h) - A_w/6$ is derived from the gross moment of inertia equation. Hint: Assume flange thickness to be small compared with the web plate height.

3.8.[1] A plate girder framed between two columns 70 ft apart supports a uniform load 4.5 k/ft. Design trial web and flange plates that will not require transverse stiffeners. Assume the girder depth to be $L/12$. Check by moment of inertia method. Calculate welds.

3.9. Design bearing and intermediate stiffeners, in pairs, and their connections for a plate girder with loading as shown. The girder section consists of a 56 $\times$ $\frac{5}{16}$-in. web plate and $1\frac{1}{4} \times$ 20-in. flange plates and is unframed at girder ends. Assume the girder is fully laterally supported.

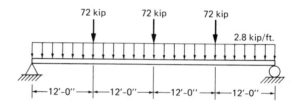

3.10. Completely design the plate girder shown. The girder is laterally supported at points A, B, C, and D, the compressive flange is not prevented from rotating, and the ends are not framed. Design such that intermediate stiffeners are not required.

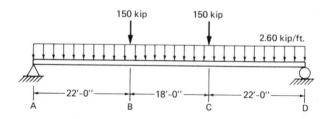

3.11. Redesign the plate girder shown in Problem 3.10 using minimum thickness web plate and intermediate web stiffeners.

3.12. A beam spanning continuously, as shown, supports a live load of 1.50 k/ft and a dead load of 0.50 k/ft. Assuming full lateral support, design the beam as a rolled section without cover plates for

---

[1] Knowledge of connection design (Chapters 5, 6, and 7) is required.

a) calculated positive and negative moments
b) positive and negative moments using redistribution allowed by AISCS

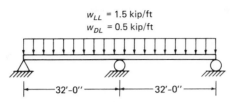

3.13. Design a continuous beam for the condition shown. Assume full lateral support.

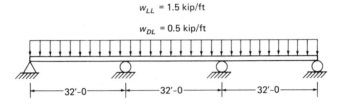

3.14. Assuming full lateral support for the loading shown, design a beam first for calculated moments with simple spans, then for continuous beam moments, and finally for redistributed moments.

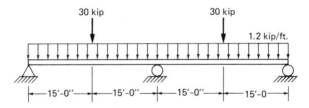

3.15. The wide-flange beam shown is loaded through the shear center, subjecting the member to a moment of 90 ft-k. Design the lightest W 12 section by breaking down the moment into components along the principal axes and considering biaxial bending. Assume full lateral support.

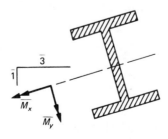

3.16. Design a C 10 channel purlin with full lateral support spanning 16 ft and subject to biaxial bending. Assume the live load of 90 lb/ft, dead load of 40 lb/ft, and wind load of 60 lb/ft to act through the shear center.

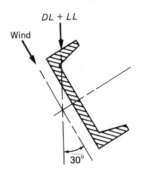

4
Columns

4.1 COLUMNS

Straight members that are subject to compression by axial forces are known as *columns*. The strength of a column is governed by the yielding of the material for short ones, by elastic buckling for long ones, and by inelastic (plastic) buckling for ones of intermediate lengths. A "perfect column," that is, one made of isotropic material, free of residual stresses, loaded at its centroid, and perfectly straight, will shorten uniformly due to uniform compressive strain on its cross section. If the load on the column is gradually increased, it will eventually cause the column to deflect laterally and fail in a bending mode. This load, called the *critical load*, is considered the maximum load that can be safely carried by the column.

Example 4.1. For a column with pin-connected ends and subjected to its critical load P_{cr}, show that

a) $y'' + k^2 y = 0$ for $k^2 = \dfrac{P}{EI}$

b) $kl = \pi$ and that $P_{cr} = \dfrac{\pi^2 EI}{l^2}$.

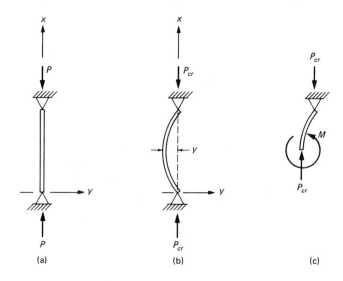

(a)   (b)   (c)

Solution.

a) From mechanics of materials, the moment in a beam is given as

$$M = EIy''$$

for the bending and axes indicated in figure *b*. From figure *c*

$$M = -Py$$

Equating the two

$$EIy'' = -Py$$
$$EIy'' + Py = 0$$
$$y'' + (P/EI)y = 0$$

Using $P/EI = k^2$

$$y'' + k^2 y = 0$$

b) For the differential equation $y'' + k^2 y = 0$, the end conditions are

at $x = 0, y = 0$
at $x = (\frac{1}{2}), y' = 0$ (slope of beam)

The solution of the differential equation $y'' + k^2 y = 0$ is

$$y = A \cos kx + B \sin kx$$

at $x = 0, y = A \cos 0 + B \sin 0 = 0$
$\cos 0 = 1, \sin 0 = 0$. Hence, $A = 0$

$$y = B \sin kx$$
$$y' = kB \cos kx$$

At $x = \frac{1}{2}, y' = 0$. Hence the second condition can be satisfied only if $\cos kl/2 = 0$ or $kl/2 = \pi/2$, which gives

$$k^2 l^2 = \pi^2$$

or

$$P = P_{\mathrm{cr}} = \frac{\pi^2 EI}{l^2}$$

As developed above for a long column with pinned ends

$$P_{\mathrm{cr}} = \frac{\pi^2 EI}{l^2} \qquad (4.1)$$

where P_{cr} is the critical load, called the *Euler load*. Because axial stress $f_a = P/A$

$$F_{a\ \mathrm{cr}} = \frac{\pi^2 EI}{l^2 A} = F_e \qquad (4.2)$$

and using $\sqrt{I/A} = r$ (radius of gyration), we find

$$F_e = \frac{\pi^2 E}{(l/r)^2} \qquad (4.3)$$

where F_e is Euler's stress.

When we graphically represent the ideal failure stresses (F_F) of a column versus the ratio l/r (called slenderness) we obtain a curve made of the branches

Fig. 4.1. Detail of a wide-flange beam to pipe column connection.

AB and BCD (see Fig. 4.2). The branch AB is where a column can be expected to fail by yielding $(F_F = F_y)$ and $l/r \leqslant (l/r)^* = \pi\sqrt{E/F_y}$ where $(l/r)^*$ is the slenderness for which the Euler stress is equal to the yielding stress. On branch BCD, the column can be expected to fail by elastic buckling with a failure stress $F_F = F_e = \pi^2 E/(l/r)^2 < F_y$ where $l/r > (l/r)^* = \pi\sqrt{E/F_y}$.

However tests have shown that columns fail in the zone shaded in Fig. 4.2. This variation in failure stresses of test samples was originally attributed to imperfections in the columns; but it has been proven that residual stresses created this condition. The residual stresses are created by uneven cooling of rolled sections such as in wide flange sections where the tips of the flanges and the middle portion of the web cools much faster than the juncture of the web and flanges. As a result of these observations, the Column Research Council (CRC) has adopted the slenderness C_c to separate the elastic from the non-elastic buckling. C_c is the slenderness corresponding to $F_F = F_y/2$ and is equal to $\pi\sqrt{2E/F_y}$. The branch AGC is a quadratic curve which fits the test results, has a value of $F_F = F_y$ and a horizontal tangent at $l/r = 0$, and matches Euler's curve at point C by having the same ordinate and the same tangent for $l/r = C_c$.

The branch CD is Euler's curve. Hence the failure stresses can be given by

$$F_F = F_y \left[1 - \frac{1}{2}\left(\frac{l/r}{C_c}\right)^2 \right] \tag{4.4}$$

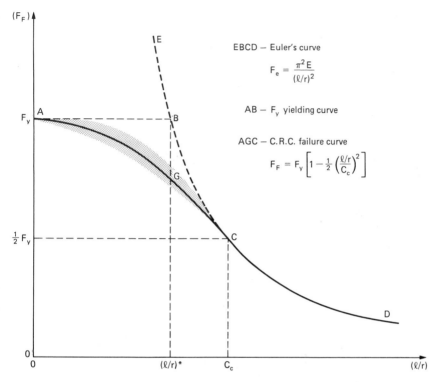

EBCD — Euler's curve

$$F_e = \frac{\pi^2 E}{(\ell/r)^2}$$

AB — F_y yielding curve

AGC — C.R.C. failure curve

$$F_F = F_y\left[1 - \tfrac{1}{2}\left(\frac{\ell/r}{C_c}\right)^2\right]$$

Fig. 4.2. Slenderness versus Failure Stress Curve.

for $l/r \leqslant C_c$ where $C_c = \sqrt{2\pi^2 E/F_y}$ and by

$$F_F = \frac{\pi^2 E}{(l/r)^2} \tag{4.5}$$

for $l/r \geqslant C_c$.

Example 4.2. Determine the Euler stress and critical load for a pin-connected W 8 $\times$ 31 column with a length of 16 ft. $E = 29,000$ ksi.

Solution. For a W 8 $\times$ 31

$A = 9.13$ in.^{2}, $I_x = 110$ in.^{4}, $I_y = 37.1$ in.^{4}, $r_x = 3.47$ in., $r_y = 2.02$ in.

I_y and r_y govern.

$$F_e = \frac{\pi^2 E}{(l/r)^2} = \frac{\pi^2 \times 29,000}{\left(\frac{16 \times 12}{2.02}\right)^2} = 31.68 \text{ ksi}$$

$$P_{cr} = A \times F_e = 9.13 \times 31.68 = 289.25 \text{ k}$$

Example 4.3. Determine the Euler stress and critical load for the pin-connected column as shown, with a length of 30 ft. $E = 29,000$ ksi.

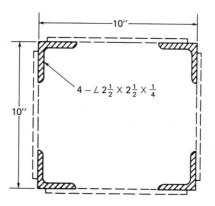

Solution. For a $2\frac{1}{2} \times 2\frac{1}{2} \times \frac{1}{4}$ in. angle

$$A = 1.19 \text{ in.}^2, \quad I = 0.703 \text{ in.}^4, \quad x = y = 0.717 \text{ in.}$$

Due to symmetry, $\bar{y}$ is 5 in. from the top or bottom.

$$I = 4 \times (I_0 + Ad^2) = 4(0.703 + 1.19(5 - .717)^2)$$

$$I = 90.1 \text{ in.}^4$$

$$r = \sqrt{\frac{I}{A}} = \sqrt{\frac{90.1}{4 \times 1.19}} = 4.35 \text{ in.}$$

$$F_e = \frac{\pi^2 E}{(l/r)^2} = \frac{\pi^2 \times 29,000}{\left(\frac{30 \times 12}{4.35}\right)^2} = 41.80 \text{ ksi}$$

$$P_{cr} = A \times F_e = (4 \times 1.19) \times 41.80 = 199.0 \text{ k}$$

The critical load equation can also be obtained by transforming the Euler stress equation

$$P_{cr} = A \times F_e = A \times \frac{\pi^2 E}{(l/\sqrt{I/A})^2} = \frac{\pi^2 EI}{l^2}$$

$$P_{cr} = \frac{\pi^2 \times 29,000 \times 90.1}{(30 \times 12)^2} = 199.0 \text{ k}$$

NOTE: Euler stress and critical load are theoretical column limits. For actual allowable stresses of columns, see examples beginning with 4.6.

Example 4.4. Determine the Euler stress and critical load for the pin-connected column as shown with a length of 30 ft. $E = 29,000$ ksi.

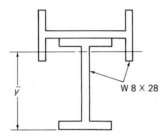

W 8 X 28

Solution. For a W 8 X 28

$A = 8.25 \text{ in.}^2, \quad I_x = 98.0 \text{ in.}^4, \quad I_y = 21.7 \text{ in.}^4, \quad d = 8.06 \text{ in.}, \quad t_w = 0.285 \text{ in.}$

$$\bar{y} = \frac{\left(8.25 \times \frac{8.06}{2}\right) + 8.25 \times \left(8.06 + \frac{0.285}{2}\right)}{2 \times 8.25} = 6.12 \text{ in.}$$

$I_x = \Sigma I_0 + Ad^2 = 98.0 + 8.25(4.03 - 6.12)^2$
$\qquad + 21.7 + 8.25(8.20 - 6.12)^2 = 191.4 \text{ in.}^4$

$I_y = \Sigma I_0 = 98.0 + 21.7 = 119.7 \text{ in.}^4 \text{ governs}$

$$\text{least } r = \sqrt{\frac{I}{A}} = \sqrt{\frac{119.7}{2(8.25)}} = 2.69 \text{ in.}$$

$$F_e = \frac{\pi^2 E}{(l/r)^2} = \frac{\pi^2 \times 29{,}000}{\left(\dfrac{30 \times 12}{2.69}\right)^2} = 15.98 \text{ ksi}$$

$$P_{cr} = A \times F_e = (2 \times 8.25) \times 15.98 \text{ ksi} = 263 \text{ k}$$

or

$$P_{cr} = \frac{\pi^2 EI}{l^2} = \frac{\pi^2 \times 29{,}000 \times 119.7}{(30 \times 12)^2} = 263 \text{ k}$$

4.2 EFFECTIVE LENGTHS

Columns with supporting conditions other than pinned at both ends have critical loads different from Euler columns.

Example 4.5. For the column conditions shown, show that

a) P_{cr} is four times the P_{cr} for the same column pin supported.
b) P_{cr} is one-fourth that for the same column pin supported.
c) P_{cr} is two times the P_{cr} of the same pin-supported column.

Solution.

a)

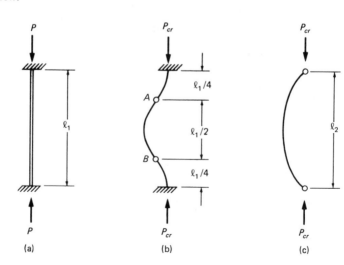

Considering the deflected shape of a column fixed at both ends, we see that contraflexure points occur at A and B. That portion of the column between A and B is the same as for a pin-ended column. Hence from figure b

$$P_{cr} = \frac{\pi^2 EI}{(l_1/2)^2} = \frac{\pi^2 EI}{(Kl_1)^2} = \frac{\pi^2 EI}{\frac{1}{4}(l_1)^2}$$

which gives 0.5 as the effective length factor. Due to the term in the denominator, direct load factor is 4 times the critical load of the pin-supported column.

b)

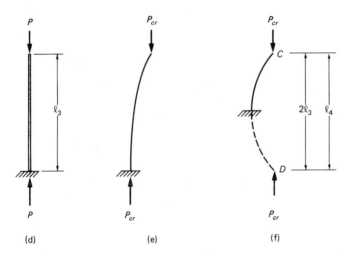

(d)   (e)   (f)

By taking the symmetry of the deflected column in figure f, the column behaves as an Euler column of length $2\,l_3$. The column of length C to D is the same as for a pin-ended column. Hence

$$P_{cr} = \frac{\pi^2 EI}{(2\,l_3)^2} = \frac{\pi^2 EI}{(Kl_3)^2} = \frac{\pi^2 EI}{4\,(l_3)^2}$$

with 2 for the effective length factor. Due to the term in the denominator, the direct load factor is one-fourth the critical load of the pin-supported column.

c)

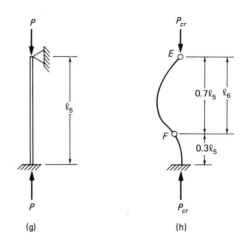

(g)   (h)

The distance EF is approximately $0.7\,l_5$, where F is the point of contraflexure (zero moment location, behaving like a hinge). We then have

$$P_{cr} = \frac{\pi^2 EI}{(0.7\,l_5)^2} = \frac{2\,\pi^2 EI}{(l_5)^2}$$

with the effective length factor $K = 0.7$. The direct load factor is approximately 2 times the critical load of the pin-supported column.

The values of K (effective length factor) above agree with the values found in Table 4.1.

As seen in the above example, for a column with both ends fixed the critical load is

$$F_{cr} = \frac{\pi^2 E}{(0.5\,l/r)^2} \tag{4.6}$$

and with one end fixed and the other pinned

$$F_{cr} = \frac{\pi^2 E}{(2\,l/r)^2} \tag{4.7}$$

This leads to the development of the expression

$$F_{cr} = \frac{\pi^2 E}{(Kl/r)^2} \tag{4.8}$$

Fig. 4.3. Erection of prefabricated, two-story column-and-beam system for the Sears Tower, Chicago. Note reinforced openings in the beam webs for heating and cooling ducts. (Courtesy of U.S. Steel Corp.)

where K, called the *effective length factor*, is based on the end connections of the column.

Because the allowable axial stress for a column depends upon the slenderness ratio (l/r), the AISC allows the slenderness ratio to be modified by a factor K, so that the allowable axial stress also depends upon the end conditions of the column. The effective length factor K can be obtained from Table C.1.8.1 of the AISC *Commentary* (reproduced as Table 4.1 here), and the value r for rolled sections can be found in the properties for designing tables.

The AISC code in Specifications 1.8.2 and 1.8.3 allows for laterally braced frames the use of $K = 1$ as a conservative simplification, while for unbraced frames the use of analysis is suggested. This can be done by interpolation between appropriate cases of Table 4.1 or by a more precise approach, (see AISCS 1.8, AISCM pages 5-123 through 5-127 and AISCS Fig. C.1.8.2). Determination of K for laterally unbraced frames is beyond the scope of this book and will not be discussed here.

Table 4.1. Effective Length Factor K Based on AISCS 1.8. (AISCS Table C1.8.1, reprinted with permission).

	(a)	(b)	(c)	(d)	(e)	(f)
Buckled shape of column is shown by dashed line						
Theoretical K value	0.5	0.7	1.0	1.0	2.0	2.0
Recommended design value when ideal conditions are approximated	0.65	0.80	1.2	1.0	2.10	2.0
End condition code		Rotation fixed and translation fixed				
		Rotation free and translation fixed				
		Rotation fixed and translation free				
		Rotation free and translation free				

Whether a column connection is pinned or fully fixed can not always be readily determined. Quite often, connections meant to be fully fixed do not have full fixity due to improper detailing. In general, pinned column base connections have the base plate anchored to the foundation with bolts located at or near the centerline of the base plate, and pinned column-beam connections have the column connected to the beam web only (see Fig. 4.4, left). Column base plates for fixed columns usually have the anchor bolts near the column flanges, as far away from the centerline as practical and in fixed column-beam connections the column flange is rigidly connected to the beam flanges (see Fig. 4.4, right).

4.3 ALLOWABLE AXIAL STRESS

AISCS 1.5.1.3.1 specifies the allowable axial compressive stress for members whose cross sections meet AISC Section 1.9 (width-thickness ratios). When Kl/r, the largest effective slenderness ratio of any unbraced segment is less than C_c

$$F_a = \frac{\left[1 - \dfrac{(Kl/r)^2}{2\,C_c^2}\right] F_y}{\dfrac{5}{3} + \dfrac{3\,(Kl/r)}{8\,C_c} - \dfrac{(Kl/r)^3}{8\,C_c^3}} \tag{4.9}$$

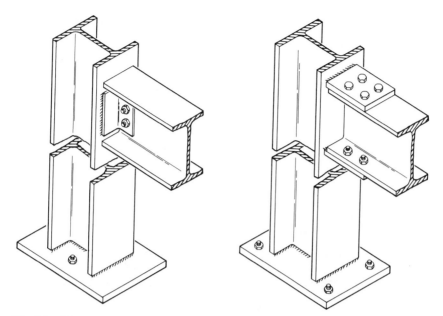

Fig. 4.4. Column connections. Left, pinned column base and pinned column-to-beam connection. Right, fixed column base and fixed column-to-beam connection.

where $C_c = \sqrt{2\pi^2 E/F_y}$. Referring to Section 4.1, the value C_c is slenderness corresponding in Euler's curve to the stress $F_y/2$, as at a stress close to this one Euler's curve and the CRC curve match. When Kl/r exceeds C_c

$$F_a = \frac{12\,\pi^2 E}{23\,(Kl/r)^2} \qquad (4.10)$$

It should be noted that the denominator in Eq. 4.9, and the factor $\frac{23}{12}$ in 4.10, are the safety factors for the allowable compressive stresses, and the numerators are the failure stresses for the particular slendernesses. For very short columns ($Kl/r \cong 0$), the safety factor is $\frac{5}{3}$ (same as for tension members) and is increased by 15% to $\frac{23}{12}$, as slenderness increases to when Kl/r equals C_c and remains at $\frac{23}{12}$ for Kl/r greater than C_c. The safety factor is increased as slenderness increases because slender columns are more sensitive to eccentricities in loading and flaws in the steel itself than short columns.

In the case of bracing and secondary members of slenderness larger than 120, the values of F_a obtained from Eqs. 4.9 and 4.10 are divided by $(1.6 - l/200\,r)$ to obtain more realistic F_a values. These values take into account for fixity developed at the members' ends by their connections and that secondary members can use a smaller safety factor. In view of the above, K should not be taken less than 1, as the advantage of the end restraints has already been taken into consideration.

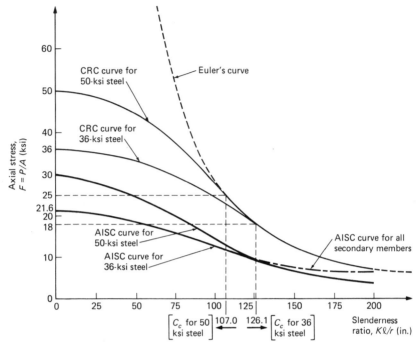

Fig. 4.5. Slenderness curves for the most commonly used steels. Euler's curve for theoretical stress and CRC curve for failure stress are also shown for comparison with the AISC specified curves for allowable stress.

To determine the allowable axial compressive stress F_a without lengthy calculations, the AISCS have established tables that give the allowable stress F_a as functions of Kl/r values and yield stress of the steel. AISCM Tables 3-36 and 3-50 are provided for the most commonly used steels and can be found in Appendix A. For other grades of steels, Tables 4 and 5 in Appendix A and AISCS 1.5.1.3 are needed. Figure 4.5 shows a graphical representation of the failure stresses and Fig. 4.5 the allowable axial stress for $F_y = 36$ ksi and $F_y = 50$ ksi as a function of slenderness of the member.

Example 4.6. Calculate $F_{\text{allowable}}$ and $P_{\text{allowable}}$ for a W 8 × 48 column with fixed ends and a length of 16 ft, 6 in.

Solution. The effective length concept is used to equate framed compression member length to that of an equivalent pin-ended member. For this purpose theoretical K values and suggested design values are tabulated in the AISCC.

$$K = 0.65 \qquad \text{(AISCS Table C1.8.1)}$$

$$r_y = 2.08 \text{ in.}$$

$$\frac{Kl}{r} = \frac{0.65 \times (16.5 \times 12)}{2.08} = 61.88$$

$$C_c{}^1 = \sqrt{\frac{2\pi^2 E}{F_y}} = \sqrt{\frac{2\pi^2\, 29{,}000}{36}} = 126.1 \qquad \text{(AISCS 1.5.1.3.1)}$$

$$61.88 < 126.1$$

Because the slenderness ratio is less than C_c, AISC equation 1.5-1 is to be used

$$F_a = \frac{\left[1 - \dfrac{(Kl/r)^2}{2C_c^2}\right] F_y}{\dfrac{5}{3} + \dfrac{3(Kl/r)}{8C_c} - \dfrac{(Kl/r)^3}{8C_c^3}}$$

$$F_a = \frac{\left[1 - \dfrac{(61.88)^2}{2(126.1)^2}\right] \times 36.0}{\dfrac{5}{3} + \dfrac{3(61.88)}{8(126.1)} - \dfrac{(61.88)^3}{8(126.1)^3}} = 17.25 \text{ ksi}$$

$$A = 14.1 \text{ in.}^2$$

$$P_{\text{all}} = F_a \times A = 17.25 \text{ ksi} \times 14.1 \text{ in.}^2 = 243.2 \text{ k}$$

Example 4.7. Calculate the allowable stress F_a and allowable load for a W 8 × 48 column with one end fixed and one end pinned with a length of 35 ft.

Solution.

$$K = 0.80$$

$$r_y = 2.08 \text{ in.}$$

$$C_c = 126.1 \text{ (from Example 4.6)}$$

$$\frac{Kl}{r} = \frac{0.80 \times (35 \times 12)}{2.08} = 161.54$$

$$161.54 > 126.1$$

$^1 C_c$ can also be obtained from Table 5 in Appendix A of the AISCM.

For slenderness ratios greater than C_c, AISC equation 1.5-2 is to be used (AISCS 1.5.1.3.2).

$$F_a = \frac{12\pi^2 E}{23\left(\frac{Kl}{r}\right)^2}$$

$$F_a = \frac{12 \times \pi^2 \times 29{,}000}{23 \times (161.54)^2} = 5.72 \text{ ksi}$$

$$A = 14.1 \text{ in.}^2$$

$$P = F_a \times A = 5.72 \text{ ksi} \times 14.1 \text{ in.}^2 = 80.7 \text{ k}$$

Example 4.8. What load can a bracing made up of two angles $3 \times 3 \times \frac{1}{4}$ in. separated by a $\frac{1}{4}$ in. plate carry, when

a) member length $L = 12$ ft, 6 in.
b) member length $L = 7$ ft, 6 in.

Solution.

a) For two angles $3 \times 3 \times \frac{1}{4}$ in. separated by $\frac{1}{4}$ in.

$$r_{min} = r_x = 0.930 \text{ in.} \qquad \text{AISCM pp. 1-73}$$

$$A = 2.88 \text{ in.}^2$$

$$\frac{Kl}{r_{min}} = \frac{12.5 \text{ ft} \times 12 \text{ in./ft}}{0.930 \text{ in.}} = 161.3 \ (K = 1 \text{ for bracing members})$$

$$161.3 > C_c = 126.1$$

$$F_a = \frac{12\pi^2 E}{23(Kl/r)^2} = \frac{12 \times \pi^2 \times 29{,}000}{23 \ (161.3)^2} = 5.74 \text{ ksi} \qquad \text{(AISCS 1.5.1.3.2)}$$

$$F_{as} = \frac{F_a}{1.6 - \dfrac{l}{200 \, r_{min}}} = \frac{5.74 \text{ ksi}}{1.6 - \dfrac{12.5 \times 12}{200 \times 0.930}} = 7.23 \text{ ksi} \qquad \text{(AISCS 1.5.1.3.3)}$$

$$P_{all} = F_{as} \times A = 7.23 \text{ ksi} \times 2.88 \text{ in.}^2 = 20.82$$

b) For member length $L = 7$ ft, 7 in.

$$\frac{Kl}{r_{min}} = \frac{7.5 \text{ ft} \times 12 \text{ in./ft}}{0.930 \text{ in.}} = 96.77$$

$$96.77 < C_c = 126.1$$

$$F_a = \frac{\left[1 - \dfrac{(Kl/r)^2}{2C_c^2}\right] F_y}{\dfrac{5}{3} + \dfrac{3(Kl/r)}{8C_c} - \dfrac{(Kl/r)^3}{8C_c^3}} = \frac{\left[1 - \dfrac{(96.77)^2}{2(126.1)^2}\right] 36.0}{\dfrac{5}{3} + \dfrac{3(96.77)}{8(126.1)} - \dfrac{(96.77)^3}{8(126.1)^3}} = 13.38 \text{ ksi}$$

$$\text{(AISCS 1.5.1.3.1)}$$

As $l/r < 120$, no factor is applied.

$$P_{all} = F_a \times A = 13.38 \text{ ksi} \times 2.88 \text{ in.}^2 = 38.53 \text{ k}$$

Example 4.9. Using the AISC table for allowable stress for compression members of 36-ksi yield stress, determine the allowable load a W 10 × 77 can support with a length of 16 ft and pinned ends.

Solution.

$$K = 1.0$$

$$r_y = 2.60 \text{ in.}$$

$$\frac{Kl}{r} = \frac{1.0 \times 16.0 \times 12}{2.60} = 73.85$$

Table 3-36 is in Appendix A of the AISCS.

$$F_a = 16.01 \text{ ksi} \quad \left(\frac{Kl}{r} = 74\right) \quad \text{(AISC Table 3-36)}$$

$$P_{all} = F_a \times A = 16.01 \times 22.6 = 361.8 \text{ k}$$

Example 4.10. Determine the allowable load a W 12 × 65 can support with a length of 20 ft and pinned ends.

Solution.

$$K = 1.0$$

$$r = 3.02 \text{ in.}$$

$$\frac{Kl}{r} = \frac{1.0 \times 20 \text{ ft} \times 12 \text{ in./ft}}{3.02 \text{ in.}} = 79.47$$

$$F_a = 15.42 \text{ ksi} \quad \text{(AISC Table 3-36)}$$

$$P_{all} = F_a \times A = 15.42 \text{ ksi} \times 19.1 \text{ in.}^2 = 294.5 \text{ k}$$

Example 4.11. Determine the allowable load a W 10 × 49 can support with a length of 24 ft, 6 in. One end is fixed and the other pinned. Steel yield stress F_y = 50 ksi.

Solution.

$$K = 0.80$$

$$r_y = 2.54 \text{ in.}$$

$$\frac{Kl}{r} = \frac{0.80 \times 24.5 \times 12}{2.54} = 92.6$$

$$F_a = 16.37 \text{ ksi} \quad \text{(AISC Table 3-50)}$$

$$P_{all} = F_a \times A = 16.37 \text{ ksi} \times 14.4 \text{ in.}^2 = 235.73 \text{ k}$$

Example 4.12. Determine the allowable load a WT 3 × 7.5 can carry if it is a bracing element and is 10 ft, 6 in.

Solution. For a WT 3 × 7.5, A = 2.21 in.^{2}

$$r_x = 0.797 \text{ in.}$$

$$r_y = 1.45 \text{ in.}$$

Use $r_{min} = r_x$ = 0.797 in.

$$K = 1.0$$

$$\frac{Kl}{r_{min}} = \frac{10.5 \text{ ft} \times 12 \text{ in./ft}}{0.797 \text{ in.}} = 158.1$$

$$F_{as} = 7.38 \text{ ksi} \quad \text{(AISC Table 3-36)}$$

$$P_{all} = F_{as} \times A = 7.38 \text{ ksi} \times 2.21 \text{ in.}^2 = 16.3 \text{ k}$$

Example 4.13. Find the allowable load a 10-in. standard steel pipe can support with a length of 20 ft and pinned ends.

Solution.

$$K = 1.0$$

$$r = 3.67 \text{ in.}$$

$$A = 11.9 \text{ in.}^2$$

$$\frac{Kl}{r} = \frac{1.0 \times 20 \times 12}{3.67} = 65.4$$

$$F_a = 16.9 \text{ ksi}$$

$$P_{all} = F_a \times A = 16.9 \times 11.9 = 201 \text{ k}$$

Example 4.14. A truss member consists of two $8 \times 8 \times \frac{3}{4}$-in. angles separated by $\frac{3}{4}$ in. back-to-back, a length of 24 ft, 6 in., pinned at both ends. Determine the allowable axial load.

Solution. Use r_x as least value of r.

$$K = 1.0$$

$$r_x = 2.47 \text{ in.} \qquad r_y = 3.62 \text{ in.}$$

$$A = 22.9 \text{ in.}^2$$

$$\frac{Kl}{r} = \frac{1.0 \times 24.5 \times 12}{2.47 \text{ in.}} = 119.0$$

$$F_a = 10.43 \text{ ksi}$$

$$P_{all} = F_a \times A = 10.43 \text{ ksi} \times 22.9 \text{ in.}^2 = 238.8 \text{ k}$$

4.4 DESIGN OF COLUMNS

The following procedure can be used to design an axially loaded column without design-aid tables.

1. Determine the effective length factor K from Table 4.1.
2. Selecting an arbitrary r, obtain Kl/r and determine the corresponding allowable axial stress F_a from AISCM Tables 3-36 or 3-50 or from AISCS eqns 1.5-1 or 1.5-2.
3. Determine required area $(A = P/F_a)$, and select an appropriate rolled section.
4. Recalculate actual slenderness ratios of the selected section with respect to the x and y axes and corresponding F_a.
5. Repeat the above process until satisfactory convergence has been obtained.

When determining the slenderness ratio of a column, the radii of gyration for the x and y directions, as well as the respective unsupported lengths must be obtained. The value Kl/r must be calculated for both x and y directions, and the larger of the two values shall be used for determining the allowable compressive stress F_a.

Example 4.15. Design a W 14 column pinned at both ends to support an axial load of 610 kips. $L_x = L_y = 21$ ft, 6 in.

Solution.

$$K = 1.0$$

$$\text{Assume } \frac{Kl}{r} = 70$$

$$F_a = 16.43 \text{ ksi}$$

$$A_{\text{req}} = \frac{P}{F_a} = \frac{610 \text{ k}}{16.43 \text{ ksi}} = 37.13 \text{ in.}^2$$

Try W 14 × 120

$$A = 35.3 \text{ in.}^2, \quad r_x = 6.24 \text{ in.}, \quad r_y = 3.74 \text{ in.}$$

$$A_s L_k = L_y, r_y \text{ governs}$$

$$\frac{Kl_x}{r_x} = \frac{1.0 \times 21.5 \times 12}{6.24 \text{ in.}} = 41.35$$

$$\frac{Kl_y}{r_y} = \frac{1.0 \times 21.5 \times 12}{3.74 \text{ in.}} = 68.98 \text{ governs}$$

$$F_a = 16.53 \text{ ksi}$$

$$A_{\text{req}} = \frac{P}{F_a} = \frac{610 \text{ k}}{16.53 \text{ ksi}} = 36.90 \text{ in.}^2$$

$$36.90 \text{ in.}^2 > 35.2 \text{ in.}^2 \qquad \text{Try next heaviest section}$$

Try W 14 × 132

$$A = 38.8 \text{ in.}^2, \quad r_x = 6.28 \text{ in.}, \quad r_y = 3.76 \text{ in.}$$

$$\frac{Kl_y}{r_y} = \frac{1.0 \times 21.5 \times 12}{3.76 \text{ in.}} = 68.6 \text{ (critical)}$$

$$F_a = 16.59 \text{ ksi}$$

$$f_a = \frac{610 \text{ k}}{38.8 \text{ in.}^2} = 15.72 \text{ ksi} < 16.59 \text{ ksi}$$

Use W 14 X 132.

Example 4.16. Design a W 10 column pinned at top and bottom to support an axial load of 200 kips. $L_x = 14$ ft, 0 in., $L_y = 7$ ft, 0 in.

Solution.

$$K = 1.0$$

$$\text{Assume } \frac{Kl}{r} = 70$$

$$F_a = 16.4 \text{ ksi}$$

$$A_{\text{req}} = \frac{P}{F_a} = \frac{200 \text{ k}}{16.4 \text{ ksi}} = 12.2 \text{ in.}^2$$

Try W 10 X 45

$$A = 13.3 \text{ in.}^2, \quad r_x = 4.32 \text{ in.}, \quad r_y = 2.01 \text{ in.}$$

$$\frac{Kl_x}{r_x} = \frac{1.0 \times 14 \times 12}{4.32} = 38.9$$

$$\frac{Kl_y}{r_y} = \frac{1.0 \times 7 \times 12}{2.01} = 41.8 \text{ governs}$$

$$F_a = 19.05 \text{ ksi}$$

$$A_{\text{req}} = \frac{P}{F_a} = \frac{200 \text{ k}}{19.05 \text{ ksi}} = 10.50 \text{ in.}^2$$

$$10.50 \text{ in.}^2 < 13.3 \text{ in.}^2 \quad \text{Try next lighter section}$$

Try W 10 X 39

$$A = 11.5 \text{ in.}^2, \quad r_x = 4.27 \text{ in.}, \quad r_y = 1.98 \text{ in.}$$

$$\frac{Kl_y}{r_y} = \frac{1.0 \times 7 \times 12}{1.98} = 42.4 \text{ (critical)}$$

$$F_a = 19.0 \text{ ksi}$$

$$f_a = \frac{200 \text{ k}}{11.5 \text{ in.}^2} = 17.4 \text{ ksi} < 19.0 \text{ ksi}$$

We note that if W 10 X 33, the next lighter section, is used

$$A = 9.71 \text{ in.}^2$$

$$f_a = \frac{200 \text{ k}}{9.71 \text{ in.}^2} = 20.60 \text{ ksi} > 19.0 \text{ ksi} \quad \text{N.G.}$$

Use W 10 X 39.

Example 4.17. Design the lightest column for an axial load of 350 kips. The column is pinned at the bottom and fixed at the top, but subject to sway in the x and y directions. $L_x = 22$ ft, 0 in., $L_y = 11$ ft, 0 in.

Solution. Assume $Kl/r = 70, K = 2.0$

$$F_a = 16.43 \text{ ksi}$$

$$A_{req} = \frac{P}{F_a} = \frac{350 \text{ k}}{16.43 \text{ ksi}} = 21.30 \text{ in.}^2$$

Try W 14 X 74, W 12 X 72, and W 10 X 77

For W 14 X 74

$$A = 21.8 \text{ in.}^2, \quad r_x = 6.04 \text{ in.}, \quad r_y = 2.48 \text{ in.} \quad K_x = K_y = 2.0$$

$$\frac{K_x l_x}{r_x} = \frac{2.0 \times (22.0 \times 12)}{6.04} = 87.4$$

$$\frac{K_y l_y}{r_y} = \frac{2.0 \times (11.0 \times 12)}{2.48} = 106.5 \text{ governs}$$

$$F_a = 12.14 \text{ ksi}$$

$$A_{req} = \frac{350 \text{ k}}{12.14 \text{ ksi}} = 28.83 \text{ in.}^2 > 21.8 \text{ in.}^2 \quad \text{N.G.}$$

Try W 14 section with area between 21.8 in.^{2} and 28.83 in.^{2}

Try W 14 X 90

$$A = 26.5 \text{ in.}^2, \quad r_x = 6.14 \text{ in.}, \quad r_y = 3.70 \text{ in.}$$

$$\frac{K_x l_x}{r_x} = \frac{2.0 \times (22.0 \times 12)}{6.14} = 86.0 \text{ governs}$$

$$\frac{K_y l_y}{r_y} = \frac{2.0 \times (11.0 \times 12)}{3.70} = 71.4$$

$$F_a = 14.67 \text{ ksi}$$

$$A_{\text{req}} = \frac{350 \text{ k}}{14.67 \text{ ksi}} = 23.86 \text{ in.}^2$$

$26.5 \text{ in.}^2 > 23.86 \text{ in.}^2$ Section is adequate.

For W 12 X 72

$$A = 21.1 \text{ in.}^2, \quad r_x = 5.31 \text{ in.}, \quad r_y = 3.04 \text{ in.}$$

$$\frac{K_x l_x}{r_x} = \frac{2.0 \times (22.0 \times 12)}{5.31} = 99.4 \text{ governs}$$

$$\frac{K_y l_y}{r_y} = \frac{2.0 \times (11.0 \times 12)}{3.04} = 86.8$$

$$F_a = 13.05 \text{ ksi}$$

$$A_{\text{req}} = \frac{350 \text{ k}}{13.05 \text{ ksi}} = 26.82 \text{ in.}^2 > 21.1 \text{ in.}^2 \quad \text{N.G.}$$

Try W 12 X 87

$$A = 25.6 \text{ in.}^2, \quad r_x = 5.38 \text{ in.}, \quad r_y = 3.07 \text{ in.}$$

$$\frac{K_x l_x}{r_x} = \frac{2.0 \times (22.0 \times 12)}{5.38} = 98.1 \text{ governs}$$

$$\frac{K_y l_y}{r_y} = \frac{2.0 \times (11.0 \times 12)}{3.07} = 86.0$$

$$F_a = 13.22 \text{ ksi}$$

$$A_{\text{req}} = \frac{350 \text{ k}}{13.22 \text{ ksi}} = 26.48 \text{ in.}^2 > 25.6 \text{ in.}^2 \quad \text{N.G.}$$

Try W 12 X 96

No need to check, because section is heavier than adequate W 14 X 90.

For W 10 X 77

$$A = 22.6, \quad r_x = 4.49 \text{ in.,} \quad r_y = 2.60 \text{ in.}$$

$$\frac{K_x l_x}{r_x} = \frac{2.0 \times (22.0 \times 12)}{4.49} = 117.6 \text{ governs}$$

$$\frac{K_y l_y}{r_y} = \frac{2.0 \times (11.0 \times 12)}{2.60} = 101.5$$

$$F_a = 10.63 \text{ ksi}$$

$$A_{req} = \frac{350 \text{ k}}{10.63 \text{ ksi}} = 32.93 \text{ in.}^2 > 22.6 \text{ in.}^2 \quad \text{N.G.}$$

Due to large difference in required area and actual area, a section with cross-sectional area nearer to 30 in.^{2} is necessary.

Try W 10 X 100

No need to check, because section is heavier than adequate W 14 X 90.

Use W 14 X 90 as lightest section available.

4.5 COLUMN DESIGN USING AISCM TABLES

To design a concentrically loaded column by using the columns tables of the AISCM, the procedure below must be followed.

1. Determine the effective length factor K from AISCM Table C.1.8.1, and calculate the effective length KL in feet, for the y direction.
2. From the tables, select an appropriate section based on the effective length and the axial load P.
3. Dividing the effective length $K_x L_x$ by the value r_x/r_y of the selected section, obtain the effective length with respect to the minor axis equivalent in load-carrying capacity to the actual effective length about the major axis. The column then must be designed for the larger of the two effective lengths $K_y L_y$ or $K_x L_x/(r_x/r_y)$.

Example 4.18. Select the lightest W section for a column supporting a 300-kip load, fixed at the bottom and pinned at the top. $L_x = L_y = 17$ ft, 6 in. Refer to AISC columns tables.

Solution.

$$K = 0.80 \quad \text{(AISC Table C 1.8.1)}$$

$$KL = 0.80 \times 17.5 \text{ ft} = 14.0$$

W shapes capable of supporting load

W 8 × 67,	$P = 304$ k
W 10 × 68,	$P = 339$ k
W 12 × 65,	$P = 341$ k
W 14 × 68,	$P = 332$ k

NOTE: W 10 × 60 and W 14 × 62 may be considered satisfactory, as they are only 1% overstressed with $P = 297$ k.

Use W 12 × 65 (lightest section) or W 10 × 60 if a slight overstress is permissible.

Example 4.19. Select the lightest column of any shape of A36 steel, except angles, for the loading in example 4.16 using the AISCM tables.

Solution.

W 10 × 60	Lightest W section (slightly overstressed)
10 ϕ × 54.74 lb	Extra strong steel pipe
8 ϕ × 72.42 lb	Double-extra strong steel pipe

Use 10-in. diameter extra strong steel pipe, 54.74 lb/ft.

Example 4.20. Redesign the column shown using the AISC columns tables.

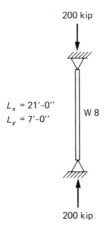

Solution.

$$K = 1.0$$

$$KL_y = 1.0 \times 7.0 \text{ ft} = 7.0 \text{ ft (weaker direction)}$$

For $(KL) = 7$ ft and loading of 200 kips, select W 8 $\times$ 40, which can carry 223 kips and has $r_x/r_y = 1.73$

$$\frac{L_x}{r_x/r_y} = \frac{21 \text{ ft}}{1.73} = 12.14 \text{ ft}$$

From the same table, for $(KL) = 12.14$ ft and $P = 200$ kips, a W 8 $\times$ 48 is needed. Use W 8 $\times$ 48.

Example 4.21. Select the lightest column from AISCM columns tables for $K_x = 2.0, K_y = 1.2, L_x = 18.0$ ft, $L_y = 12.5$ ft, $F_y = 50$ ksi, and $P = 470$ kips.

Solution.

$$K_y L_y = 1.2 \times 12.5 \text{ ft} = 15.0 \text{ ft}$$

Try W 10 $\times$ 77; $r_x/r_y = 1.73$

$$\frac{K_x L_x}{r_x/r_y} = \frac{2.0 \times 18.0 \text{ ft}}{1.73} = 20.8 \text{ ft}$$

Redesigning for $(KL) = 20.8$ ft, select W 10 $\times$ 100.

Try W 12 $\times$ 72, $r_x/r_y = 1.75$

$$\frac{K_x L_x}{r_x/r_y} = \frac{2.0 \times 18.0 \text{ ft}}{1.75} = 20.6 \text{ ft}$$

Redesigning for $(KL) = 20.6$ ft, select W 12 $\times$ 87.

Try W 14 $\times$ 82, $r_x/r_y = 2.44$

$$\frac{K_x L_x}{r_x/r_y} = \frac{2.0 \times 18.0 \text{ ft}}{2.44} = 14.8 \text{ ft} < 15.0 \text{ ft}$$

We see then that W 14 $\times$ 82 is satisfactory.

W 14 $\times$ 82 is the lightest satisfactory section.

4.6 COMBINED STRESSES

When a straight member is subjected to a bending stress, as well as a compressive stress, the member is a *column-under-bending*. As beams with significant axial stress are commonly called *beam-columns*, we will call columns with significant bending stress *column-beams*. The bending stress may be caused by (1) eccentric application of axial load, (2) applied end moments, or (3) lateral loads, such as wind forces.

A column-beam starts to deflect laterally at an axial load much smaller in magnitude than the load that can be supported by a plain column of similar properties. As the load is gradually increased, the deflection increases at a much faster rate, because the increase in moment depends not only on the increased axial load, but also on the increased eccentricity (moment arm) caused by the increased load. Thus a column-beam fails due to instability caused by excessive bending.

For a column-beam to be satisfactory for a given loading, the AISC requires that formulas 1.6-1a and 1.6-1b of the AISCM be satisfied. When f_a/F_a is less than 0.15, formula 1.6-2 may be used in lieu of formulas 1.6-1a and 1.6-1b. The three formulas are reproduced below as Eqs. 4.9, 4.10, and 4.11, respectively, for quick reference.

$$\frac{f_a}{F_a} + \frac{C_{mx} f_{bx}}{\left(1 - \dfrac{f_a}{F'_{ex}}\right) F_{bx}} + \frac{C_{my} f_{by}}{\left(1 - \dfrac{f_a}{F'_{ey}}\right) F_{by}} \leqslant 1.0 \tag{4.11}$$

$$\frac{f_a}{0.60 F_y} + \frac{f_{bx}}{F_{bx}} + \frac{f_{by}}{F_{by}} \leqslant 1.0 \tag{4.12}$$

$$\frac{f_a}{F_a} + \frac{f_{bx}}{F_{bx}} + \frac{f_{by}}{F_{by}} \leqslant 1.0 \tag{4.13}$$

To facilitate the design of column-beams, the AISC allows the use of modified versions of formulas 1.6-1a, 1.6-1b, and 1.6-2. These modified formulas, which can be found in Section 3 of the AISCM under "Combined Axial and Bending Loading," convert the moments into equivalent axial loads P', which when added to the actual axial load P give the required tabular load for which the column-beam must be designed using the columns tables. After selecting the column-beam, recalculate P', using the actual bending factors, and the column-beam must be redesigned if the original selection proves to the unsatisfactory.

Example 4.22. The top chord of a bridge truss is shown in the accompanying figure. Determine the maximum stresses at the extreme fibers of this member if it is subjected to a moment of 15 ft-k about the x axis and an axial compression of 100 kips.

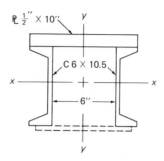

Solution.

$$A_{channel} = 3.09 \text{ in.}^2$$

$$A_{plate} = 5.0 \text{ in.}^2$$

$$I_c = 15.2 \text{ in.}^4$$

$$d_c = 6 \text{ in.}$$

$$\bar{y} = \frac{(0.5 \text{ in.} \times 10 \text{ in.}) \times 6.25 \text{ in.} + 2(3.09 \text{ in.}^2 \times 3.0 \text{ in.})}{5.0 \text{ in.}^2 + 2(3.09 \text{ in.}^2)}$$

$$= 4.45 \text{ in. from bottom}$$

$$I_x = \Sigma I_0 + Ad^2 = \frac{10(0.5)^3}{12} + 5(6.25 - 4.45)^2 + 2(15.2)$$

$$+ 2(3.09)(3 - 4.45)^2 = 59.7 \text{ in.}^4$$

Combined stress at top $= -P/A - Mc/I$ ($-$ stress indicates compression)

$$\frac{-100}{5.0 + 2(3.09)} - \frac{(15 \times 12)(6.5 - 4.45)}{59.7} = -15.13 \text{ ksi}$$

Combined stress at bottom $= -P/A + Mc/I$

$$\frac{-100}{5.0 + 2(3.09)} + \frac{(15 \times 12)(4.45)}{59.7} = 4.47 \text{ ksi}$$

Example 4.23. For the column-beam shown, determine if the member is satisfactory. Sidesway is prevented, and the member is bending about the major axis.

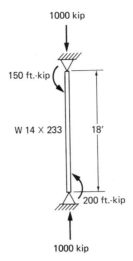

Solution. For a W 14 × 233

$$A = 68.5 \text{ in.}^2, \quad S_x = 375 \text{ in.}^3, \quad r_x = 6.63 \text{ in.}, \quad r_y = 4.10 \text{ in.}$$

$$L_c = 16.8 \text{ ft}, \quad L_u = 78.5 \text{ ft}, \quad K = 1.0$$

$$\frac{Kl}{r} = \frac{1.0 \times 18.0 \text{ ft} \times 12 \text{ in./ft}}{4.10} = 52.68$$

$$F_a = 18.11 \text{ ksi} \quad \text{(AISC Table 3-36)}$$

$$f_a = \frac{P}{A} = \frac{1000 \text{ k}}{68.5 \text{ in.}^2} = 14.60 \text{ ksi}$$

$$f_a/F_a = \frac{14.60 \text{ ksi}}{18.11 \text{ ksi}} = 0.806 > 0.15$$

AISC formulas 1.6-1a and 1.6-1b must be satisfied

$$L_c < L < L_u \quad F_b = 0.6 \times F_y = 22.0 \text{ ksi}$$

$$f_{b \text{ max}} = \frac{M_{\max}}{S_x} = \frac{200 \text{ ft-k} \times 12 \text{ in./ft}}{375 \text{ in.}^3} = 6.4 \text{ ksi}$$

$$C_m = 0.6 - 0.4\left(\frac{150 \text{ ft-k}}{200 \text{ ft-k}}\right) = 0.30 \quad \text{Use } C_m = 0.4$$

$$\text{(AISCS 1.6.1)}$$

$$F'_e = \frac{12\pi^2 E}{23(Kl_b/r_b)^2} = \frac{12\pi^2(29,000)}{23\left(\dfrac{1.0(18 \times 12)}{6.63}\right)^2} = 140.5 \text{ ksi}$$

(also AISC Table 9, Appendix A)

$$\frac{f_a}{F_a} + \frac{C_m f_{bx}}{(1-(f_a/F'_{ex}))F_{bx}} \leqslant 1.0 \qquad \text{(AISCS 1.6.1)}$$

$$\frac{14.6}{18.11} + \frac{0.4 \times 6.4}{(1-(14.6/140.5)) \times 22.0} = 0.94 < 1.0 \quad \text{ok}$$

$$\frac{f_a}{0.6 F_y} + \frac{f_{bx}}{F_{bx}} \leqslant 1.0 \qquad \text{(AISCS 1.6.1)}$$

$$\frac{14.6}{22.0} + \frac{6.4}{22.0} = 0.95 < 1.0 \quad \text{ok}$$

The W 14 X 223 column-beam is satisfactory.

Example 4.24. Design the column of 18-ft length shown below for an axial load of 300 kips.

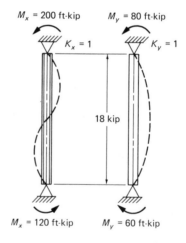

Solution. The column will be designed first by the approximate method of the AISC on p. 3-10, and then results will be refined by the use of modified formulas 1.6-1a, 1.6-1b, or 1.6-2.

$$C_m = 0.60 - 0.40 \frac{M_1}{M_2} \geqslant 0.4 \qquad \text{(AISCS 1.6.1)}$$

$$C_{mx} = 0.60 - 0.40 \frac{120}{200} = 0.36 \quad \text{Use } 0.40$$

$$C_{my} = 0.60 - 0.40 \frac{-60}{80} = 0.90$$

First trial (AISCM, p. 3-10)

For $KL = 1.0 \times 18$ ft $= 18$ ft

$$m = 2.1$$

Assume $U = 3.0$

$$P_{\text{eff}} = P_0 + M_x m + M_y mU$$

$$P_{\text{eff}} = 300 + \left[200 \times 2.1 \times \frac{0.40}{0.85} \right] + \left[80 \times 2.1 \times 3 \times \frac{0.90}{0.85} \right] = 1031 \text{ k}$$

Try W 14 $\times$ 193 as second trial

$$m = 1.7$$
$$U = 2.29$$

$$P_{\text{eff}} = 300 + \left[200 \times 1.7 \times \frac{0.40}{0.85} \right] + \left[80 \times 1.7 \times 2.29 \times \frac{0.90}{0.85} \right] = 790 \text{ k}$$

Try W 14 $\times$ 159.

Refine by use of modified equation 1.6-1a, 1.6-1b, or 1.6-2 (AISCM, p. 3-9).

$$L_c = 16.4 \text{ ft} < L = 18.0 \text{ ft} < L_u = 57.2 \text{ ft}$$

$F_{bx} = 0.60\, F_y = 22.0$ ksi (AISCS 1.5.1.4.4)

$F_{by} = 0.75\, F_y = 27.0$ ksi (AISCS 1.5.1.4.3)

$A = 46.7$ in.^{2}, $S_x = 254$ in.^{3}, $r_x = 6.38$ in., $S_y = 96.2$ in.^{3}, $r_y = 4.0$ in.

$B_x = 0.184$, $B_y = 0.485$, $a_x = 283 \times 10^6$, $a_y = 111 \times 10^6$

$$\frac{Kl_x}{r_x} = \frac{1.0 \times (18 \times 12)}{6.38} = 33.8 \quad F'_{ex} = 130.80 \text{ ksi} \quad \text{(AISCS Table 9)}$$

$$\frac{Kl_y}{r_y} = \frac{1.0 \times (18 \times 12)}{4.00} = 54.0 \quad \begin{array}{l} F_a = 17.99 \text{ ksi} \quad \text{(AISCS Table 3-36)} \\ F'_{ey} = 51.21 \text{ ksi} \quad \text{(AISCS Table 9)} \end{array}$$

$$f_{bx} = \frac{200 \text{ ft-k} \times 12 \text{ in./ft}}{254 \text{ in.}^3} = 9.45 \text{ ksi}$$

$$f_{by} = \frac{80 \text{ ft-k} \times 12 \text{ in./ft}}{96.2 \text{ in.}^3} = 9.98 \text{ ksi}$$

$$f_a = \frac{300 \text{ k}}{46.7 \text{ in.}^2} = 6.42 \text{ ksi}$$

$$\frac{f_a}{F_a} = \frac{6.42 \text{ ksi}}{17.99 \text{ ksi}} = 0.36 > 0.15$$

Use modified formula 1.6-1a (AISCM p. 3-9)

$$P + \left[B_x M_x C_{mx} \left(\frac{F_a}{F_{bx}} \right) \left(\frac{a_x}{a_x - P(Kl)^2} \right) \right] + \left[B_y M_y C_{my} \left(\frac{F_a}{F_{by}} \right) \left(\frac{a_y}{a_y - P(Kl)^2} \right) \right]$$

$$300 + \left(0.184 \times 200 \times 12 \times 0.40 \times \frac{17.99}{22} \times \frac{283 \times 10^6}{283 \times 10^6 - 300(18 \times 12)^2} \right)$$

$$+ \left(0.485 \times 80 \times 12 \times 0.90 \times \frac{17.99}{27} \times \frac{111 \times 10^6}{111 \times 10^6 - 300(18 \times 12)^2} \right)$$

Equivalent load = 771.5 k < 840 k   ok.

Verify modified formula 1.6-1b

$$P \left(\frac{F_a}{0.60 \, F_y} \right) + \left[B_x M_x \left(\frac{F_a}{F_{bx}} \right) \right] + \left[B_y M_y \left(\frac{F_a}{F_{by}} \right) \right]$$

$$300 \left(\frac{17.99}{22} \right) + \left[0.184 \times 200 \times 12 \times \frac{17.99}{22} \right] + \left[0.485 \times 80 \times 12 \times \frac{17.99}{27} \right]$$

Equivalent load = 916.7 k > 840 k   N.G.

Try W 14 × 176

$$A = 51.8 \text{ in.}^2, \quad S_x = 281 \text{ in.}^3, \quad r_x = 6.43 \text{ in.}, \quad S_y = 107 \text{ in.}^3, \quad r_y = 4.02 \text{ in.}$$

Because the lighter section satisfies all conditions except formula 1.6-1b, we need verify only the one equation

$$\frac{f_a}{0.60 \, F_y} + \frac{f_{bx}}{F_{bx}} + \frac{f_{by}}{F_{by}} \leqslant 1.0$$

Since $L_c = 16.5$ ft $< L = 18.0$ ft $< L_u = 62.6$ ft

$$F_{bx} = 0.60 \, F_y = 22 \text{ ksi}$$

$$F_{by} = 0.75 \, F_y = 27 \text{ ksi}$$

$$f_a = \frac{300 \text{ k}}{51.8 \text{ in.}^2} = 5.79 \text{ ksi}$$

$$f_{bx} = \frac{200 \text{ ft-k} \times 12 \text{ in./ft}}{281 \text{ in.}^3} = 8.54 \text{ ksi}$$

$$f_{by} = \frac{80 \text{ ft-k} \times 12 \text{ in./ft}}{107 \text{ in.}^3} = 8.97 \text{ ksi}$$

$$\frac{5.79}{22.0} + \frac{8.54}{22.0} + \frac{8.97}{27.0} = 0.98 < 1.0 \quad \text{ok}$$

Use W 14 X 176.

Example 4.25. Assuming sidesway is prevented, select a W 10 column for the situation shown below, when the bending occurs about the strong axis.

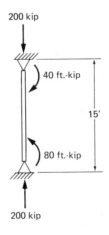

Solution.

$$K = 1.0$$

$$KL = 15.0 \text{ ft}$$

For $(KL) = 15.0$ ft and $P = 200$ kips, try W 10 X 49 $(B_x = 0.264)$

$$P + P' = 200 \text{ k} + 0.264(80 \text{ ft-k} \times 12 \text{ in./ft}) = 453.4 \text{ k}$$

For $(KL) = 15.0$ ft and $P = 453.4$ kips, try W 10 $\times$ 100 $(B_x = 0.263)$

$$P + P' = 200 \text{ k} + 0.263(80 \text{ ft-k} \times 12 \text{ in./ft}) = 452.5 \text{ k}$$

Try W 10 $\times$ 100

$$A = 29.4 \text{ in.}^2, \quad S_x = 112 \text{ in.}^3, \quad r_x = 4.60 \text{ in.}, \quad r_y = 2.65 \text{ in.}$$

$$B_x = 0.263, \quad a_x = 92.7 \times 10^6, \quad L_c = 10.9 \text{ ft}, \quad L_u = 48.2 \text{ ft}$$

$$C_m = 0.6 - 0.4(M_1/M_2) = 0.6 - 0.4\left(\frac{-40}{80}\right) = 0.8$$

$$\frac{Kl}{r} = \frac{1.0 \times 15.0 \text{ ft} \times 12 \text{ in./ft}}{2.65 \text{ in.}} = 67.9$$

$$F_a = 16.65 \text{ ksi}$$

$$F_{bx} = 22.0 \text{ ksi since } L_c < L < L_u$$

$$f_a = \frac{P}{A} = \frac{200 \text{ k}}{29.4 \text{ in.}^2} = 6.8 \text{ ksi}$$

$$\frac{f_a}{F_a} = \frac{6.8}{16.65} = 0.41 > 0.15$$

Since column table loads are for concentrically loaded columns, a modified form of AISC formulas 1.6-1a and 1.6-1b may be used (AISCM, p. 3-9).

$$P + P_x' = P + \left[B_x M_x C_{mx}\left(\frac{F_a}{F_{bx}}\right)\left(\frac{a_x}{a_x - P(Kl)^2}\right)\right]$$

$$\text{Required load} = 200 + 0.263 \times (80 \times 12) \times 0.8 \times \frac{16.65}{22.0}$$

$$\times \frac{92.7 \times 10^6}{(92.7 \times 10^6) - 200 \times (1.0 \times 180)^2}$$

$$= 364.4 \text{ k}$$

Try W 10 $\times$ 77

$$A = 22.6 \text{ in.}^2, \quad B_x = 0.263, \quad a_x = 67.9 \times 10^6, \quad r_y = 2.6$$

$$\frac{Kl}{r} = \frac{1.0 \times 180 \text{ in.}}{2.6 \text{ in.}} = 69.23$$

$$F_a = 16.51 \text{ ksi}$$

$$\text{Required load} = 200 + 0.263 \times (80 \times 12) \times 0.8 \times \frac{16.51}{22.0}$$

$$\times \frac{67.9 \times 10^6}{(67.9 \times 10^6) - 200 \times (1.0 \times 180)^2}$$

$$= 367.6 < 373 \text{ for W } 10 \times 77 \quad \text{ok}$$

$$P + P'_x = P\frac{F_a}{0.6 F_y} + B_x M_x \frac{F_a}{F_{bx}}$$

$$\text{Required load} = 200\left(\frac{16.51}{22.0}\right) + 0.263 \times (80 \times 12)\left(\frac{16.51}{22.0}\right)$$

$$= 339.6 \text{ k} < 373 \text{ k for W } 10 \times 77 \quad \text{ok}$$

Use W 10 × 77.

4.7 COLUMN BASE PLATES

Steel base plates are generally required under columns for distributing the load over a sufficient area of bearing support. Recognizing the fact that, for cases where the base plate overhangs only a small amount beyond the column dimensions, i.e., m and n dimensions are small (see Fig. 4.6), the bending stress in the base plate is more critical at the column web, halfway between the two flanges, than at the locations n or m distances away from the plate edges, the AISCM requires that the value n' be determined along with the previously required values of m and n. The values n' determines the required thickness at the column web for the portion of the base plate bounded by the column web, the two flanges, and the free edge, while n and m are used for determining the base plate thickness that projects beyond the column flanges as inverted cantilevers.

The AISC recommends the following steps for designing base plates:

1. Establish bearing values of concrete according to AISCS 1.5.5.
2. Determine required area of base plate ($A = P/F_p$).
3. Establish B and N, preferably rounded to full inches, such that m and n are approximately equal and $B \times N \geqslant A$. When either value is limited by other considerations, that dimension shall be established first and the other from $B \times N \geqslant A$.
4. Determine $m = (N - 0.95\ d)/2$, $n = (B - 0.80\ b)/2$, and n' from Table C on AISCM p. 3-100.
5. Determine actual bearing pressure on support ($f_p = P/(B \times N)$).

6. Use largest of the values of m, n, or n' to solve for the thickness t by the formula

$$t = (m, n, \text{ or } n') \times \sqrt{f_p/0.25 F_y} \qquad (4.14)$$

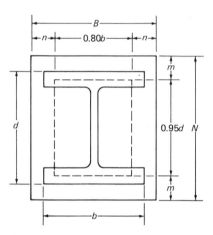

Fig. 4.6. Column base plate. These are generally used under columns to distribute the column load over a sufficient area of concrete pier or foundation.

Fig. 4.7. Pin-connected bases of trusses for wall-roof structural system for the atrium of an office building.

Example 4.26. A W 8 × 58 column supports an axial load of 300 kips. Design a base plate for the column if the supporting reinforced concrete footing has a very large pedestal and $f'c = 3000$ psi.

Solution.

$$F_p = 0.35 f'c\sqrt{A_2/A_1} \leqslant 0.7 f'c \quad \text{(AISCS 1.5.5)}$$

Use $F_p = 0.7 f'c$, since $A_2 \gg A_1$

$$A_{req} = \frac{P}{F_p} = \frac{300 \text{ k}}{2.1 \text{ ksi}} = 142.86 \text{ in.}^2$$

Choose base plate 11 × 13 in, so that $A = 11$ in. × 13 in. = 143 in.^{2}, and m and n are almost equal.

$$m = \frac{N - 0.95\, d}{2} = \frac{13 - 0.95(8.75)}{2} = 2.345 \text{ in.}$$

$$n = \frac{M - 0.8\, b}{2} = \frac{11 - 0.8(8.22)}{2} = 2.21 \text{ in.}$$

$$n' = 3.14 \quad \text{(AISC, Table 3)} \quad \text{(AISCM, p. 3-100)}$$

$$f_p = \frac{P}{B \times N} = \frac{300 \text{ k}}{11 \text{ in.} \times 13 \text{ in.}} = 2.1 \text{ ksi}$$

$$t_p = n'\sqrt{\frac{f_p}{0.25\, F_y}} = 3.14\sqrt{\frac{2.1}{0.25 \times 36}} = 1.52 \text{ in.}$$

Use $t_p = 1\frac{5}{8}$ in.

Use 11 × $1\frac{5}{8}$ in. × 1 ft, 1 in. base plate.

Example 4.27. Redesign the base plate of the previous problem for pedestal size of the same area as the base plate.

Solution.

$$F_p = 0.35\, f_c' = 0.35 \times 3000 \text{ psi} = 1.05 \text{ ksi} \quad \text{(AISCS 1.5.5)}$$

$$A_{req} = \frac{P}{F_p} = \frac{300 \text{ k}}{1.05 \text{ ksi}} = 285.7 \text{ in.}^2$$

Choose base plate 16 in. $\times$ 18 in., so that A = 16 in. $\times$ 18 in. = 288 in.^{2}

$$m = \frac{N - 0.95\, d}{2} = \frac{18 - 0.95(8.75)}{2} = 4.84 \text{ in.}$$

$$n = \frac{B - 0.8\, b}{2} = \frac{16 - 0.8(8.22)}{2} = 4.71 \text{ in.}$$

$$n' = 3.14 \quad \text{(AISC Table 3)} \quad \text{(AISCM, p. 3-100)}$$

$$f_p = \frac{P}{B \times N} = \frac{300 \text{ k}}{16 \text{ in.} \times 18 \text{ in.}} = 1.04 \text{ ksi}$$

$$t_p = m\sqrt{\frac{f_p}{0.25\, F_y}} = 4.84 \sqrt{\frac{1.04}{0.25 \times 36}} = 1.65 \text{ in.}$$

Use $t_p = 1\frac{3}{4}$ in.

Use 16 $\times$ $1\frac{3}{4}$ in. $\times$ 1 ft, 6 in. base plate.

PROBLEMS TO BE SOLVED

4.1. Determine the Euler stress and the critical load for the pin-connected column as shown, with a length of 24 ft and E = 29,000 ksi. (For an isometric view of the column see Fig. 1.10.)

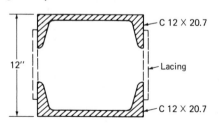

4.2. For the pin-connected column shown, determine the Euler stress and critical load if $L = 18$ ft and $E = 29,000$ ksi.

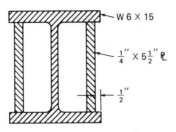

4.3. Rework Problems 4.1 and 4.2 assuming the columns are fully fixed at both ends.

4.4. For the cantilever (flagpole type) columns shown, determine the Euler stress and critical load for each column if the unsupported length is 20 ft and $E = 29,000$ ksi.

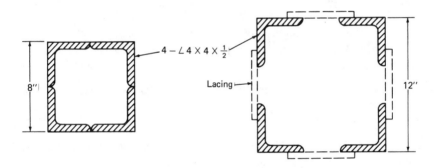

4.5. Calculate the allowable compressive axial stress F_a for (1) a W 14 × 145 column with one end fixed and the other pinned, and (2) a W 8 × 35 column with both ends pinned. Use $L = 30$ ft and AISC Section 1.5.1.3.

4.6. For the columns shown in Problems 4.1 and 4.2, calculate the allowable stress F_a for each column for both 36- and 50-ksi steels.

4.7. Using the AISC table for allowable stress for compression members of 36 ksi, determine the allowable load P a W 8 × 67 can support with a length of 12 ft and pinned ends.

4.8. Rework Problem 4.7 assuming 50-ksi steel and pinned ends.

4.9. Determine the allowable load a structural tube TS 8 × 8 × .375 can support with a length of 18 ft and pinned ends.

4.10. Determine the allowable axial load a truss member can support if the member is made from two $6 \times 6 \times \frac{1}{2}$-in. angles separated by a $\frac{1}{2}$-in. gap back-to-back. Assume the member is the truss top chord, subject to compressive stress, and the unsupported length is 10 ft.

4.11. Design a W 12 column, pinned at both ends, to support an axial load of 450 kips: $L_x = L_y = 14$ ft; $F_y = 36$ and 50 ksi.

4.12. Rework Problem 4.11 if $L_x = 16$ ft and $L_y = 8$ ft.

4.13. Design a WT 7 section for a truss top chord if it will be subject to an axial load of 120 kips. Assume the member will be laterally supported every 12 ft.

4.14. Design the lightest wide-flange column for an axial load of 225 kips. The column is fixed at the bottom, pinned at the top, and subject to sway. Assume $L_x = 20$ ft and $L_y = 10$ ft.

4.15. Determine if a W 12 × 58 with axial load $P = 200$ kips, weak axis bending $M_y = 40$ ft-kips at the top and $M_y = 20$ ft-kips at the bottom is adequate for unbraced lengths of $L_x = L_y = 18$ ft. The frame is secured against sway.

4.16. A W 14 × 202 column of loads $P = 500$ kips, $M_x = 200$ ft-kips and $M_y = 70$ ft-kips has unbraced lengths of $L_x = L_y = 14$ ft. The frame is braced against sidesway and ends of columns are restrained. Is the column safe? If the moments are due to transverse wind loads and an allowable stress factor of 1.33 is used, is the column safe?

4.17. Redesign the column of Problem 4.14, without sidesway, but with a moment of 55 ft-kips applied to the top of the column about the major $(x-x)$ axis.

4.18. A W 12 × 170 column is subject to an axial load of 700 kips and a moment of 85 ft-kips, applied about the major axis. Determine the maximum stresses at the extreme fibers of the member due to the loads.

4.19. Design the lightest wide-flange column for the loading shown. Assume the 20-kip load is applied at the face of the column flange.

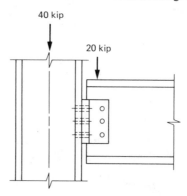

4.20 and 4.21. For the columns shown, design the lightest W 14 sections if 65-kip loads are applied at the locations shown by dots. $L = 22$ ft, $K_x = K_y = 1.0$.

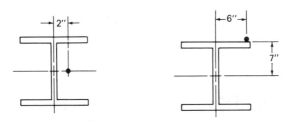

4.22. A W 12 × 65 column supports an axial load of 360 kips. Design a base plate for the column if the supporting concrete foundation is very large. $F_y = 36$ ksi and $f'c = 3000$ psi.

4.23. Design a base plate for a W 14 × 120 column supporting an axial load of 650 kips. Assume pedestal is same size as base plate and $f'c = 3500$ psi.

5
Bolts and Rivets

5.1 RIVETED AND BOLTED CONNECTIONS

Connections in structural steel are usually made with fasteners (rivets or bolts) or welds. Riveted and bolted connections will be covered in this chapter, and welds will be discussed in Chapter 6.

5.2 RIVETS

Riveting, for many years, was the sole method of connecting structural steel. In recent years, due to ease and economy of welding and high-strength bolting, the use of rivets has declined rapidly. The rivets used in construction work are usually made from a soft grade of steel that does not become brittle when heated and hammered with a riveting gun. Rivets are manufactured with one formed head and are installed in holes that are punched or drilled $\frac{1}{16}$-inch larger in diameter than the nominal diameter of the rivet. The rivet is usually heated to a cherry-red color (approximately $1800°$) before placing it in the hole, and a second head is formed on the other end by a riveting hammer or a pressure type riveter. While the second head is being formed, the heat-softened shank is forced to fill the hole completely. As the rivet cools, it shrinks, and squeezes the connected parts together, causing friction between them. However, this friction is usually neglected in calculations.

According to the AISC, rivets shall conform to the provisions of the "Specifications for Structural Rivets" ASTM A502, Grades 1 or 2. The size of rivets used in general steel construction ranges from $\frac{5}{8}$ to $1\frac{1}{2}$ in. in diameter by $\frac{1}{8}$-in. increments. The allowable stresses are tabulated in the AISCM in accordance with AISCS 1.5.2.

Rivets usually require a crew of three to four people to be put in place. Due to high labor costs and extreme noise, rivets are no longer commonly used in modern construction. Even though bolt material is more expensive than rivet steel, the advantages of bolt placement far outweigh the disadvantage of using rivets. Hence, rivets have largely been replaced by bolts.

5.3 BOLTS

Bolting of steel structures is a very rapid field-erection process. It requires less skilled labor than riveting or welding and therefore has a distinct advantage over the other connection methods. The joints obtained using high-strength bolts are superior to riveted joints in performance and economy, and bolting has become the leading method of fastening structural steel in the field.

There are two types of bolts that are most commonly used in steel construction. The first type, *unfinished bolts*, also called *common* or *ordinary* bolts, is primarily used in light structures subjected to static loads or for connecting secondary members; it is the cheapest type of connection available. Unfinished bolts must conform to the "Specifications for Low Carbon Steel Externally and Internally Threaded Fasteners," ASTM A307. AISCS 1.15.12 lists specific connection types for which A307 bolts may not be used.

The second type, *high-strength bolts*, is made from medium carbon, heat-treated, or alloy steel and has tensile strengths several times greater than those of ordinary bolts. High-strength bolts are tightened until they acquire very high tensile stresses, and the connected parts are clamped tightly together, permitting loads to be transferred primarily by friction. High-strength bolts shall conform to the "Specifications for Structural Joints Using ASTM A325 or A490 Bolts," AISCS 1.4.4.

Allowable tensile and shearing stresses for all types of bolts are tabulated in the AISCM in accordance with AISCS 1.5.2.

Connections made of high-strength bolts are categorized into three types: (1) friction type connection (F), (2) bearing type connection with threads included in shear plane (N) (see Fig. 5.1a), and (3) bearing type connection with threads excluded from shear plane (X), (see Fig. 5.1b). The two latter types of connections are similar to riveted connections and are designed by the same method,

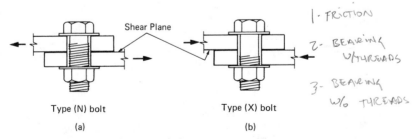

Fig. 5.1. Bearing type connections of high-strength bolts: (a) for bolts with threads included in shear plane; and (b) for bolts with threads excluded from shear plane.

the only difference being in their respective shearing and bearing capacities (see AISCM Table 1.5.2.1).

Shear connections subjected to stress reversals or severe stress fluctuations or where slippage is undesirable are required by the AISC to be of the friction type. When high-strength bolts are used in combination with welds, only those installed as friction type connections prior to welding may be considered as sharing the stresses with the weld. In friction type connections, since load is primarily transferred by friction between the connected parts, the design is based on the assumption that if and when the connection fails, the bolts will fail in shear alone, and therefore the bearing stress of the fasteners on the members need not be checked.

5.4 DESIGN OF RIVETED OR BOLTED CONNECTIONS

A connection is said to be concentrically loaded if the resultant of the applied forces passes through the centroid of the fastener group. For such cases, tests have shown that just before yielding, all fasteners in the group carry equal portions of the load. When the resultant force does not pass through the centroid, the force may be reduced to an equal force at the centroid accompanied by a moment equal in magnitude to the force times its eccentricity. In situations such as this, each fastener in the connection group is assumed to resist the axial forces uniformly and to resist the moment in proportion to its respective distance to the centroid of the connection; nevertheless, this eccentricity may be neglected in many cases (AISCS 1.15.3).

A bolted (N or X type) or riveted connection can fail four different ways. First, one of the connected members may fail in tension through one or more of the fastener holes (Fig. 5.2a). Second, if the holes are drilled too close to the

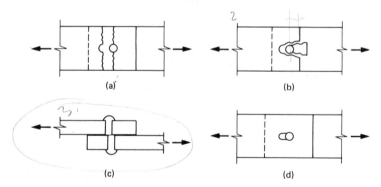

Fig. 5.2. Failure modes of connections made with bolts or rivets. (a) Tension failure of a connected part. (b) Shear failure of a connected part behind a fastener. (c) Shear failure of the fasteners. (d) Bearing failure of a connected part.

Fig. 5.3. A beam-to-column shear connection with web angles.

edge of the tension member, the steel behind the fasteners may shear off (Fig. 5.2b). Third, the fasteners may fail in shear (Fig. 5.2c). And fourth, one or more of the tension members may fail in bearing of the fasteners on the member(s) (Fig. 5.2d). To prevent failure, a connection and the connected parts must be designed to resist failure in all four of the ways mentioned.

First, to assure nonfailure of the connected parts, the members shall be designed such that the tensile stress is less than $0.60\,F_y$ on the gross area and less than $0.50\,F_u$ on the effective net area (AISCS 1.5.1.1). Therefore, the effective net area of any tensile member with bolt holes must be greater than or equal to $P/(0.50\,F_u)$; also see Chapter 1.

Second, to prevent the steel behind the fasteners from tearing off, the minimum edge distance from the center of a fastener hole to the edge, in the direction of the force, shall be not less than $2\,P/F_u\,t$ (AISCS 1.16.5.2). Here P is the force carried by one fastener, and t is the thickness of the critical connected part.

Third, to assure that the fasteners will not fail in shear, the number of fasteners should be determined to limit the maximum shear stress in the critical fastener to that listed in AISCS Table 1.5.2.1 for the particular type. To determine the minimum number of fasteners required, divide the load on the entire connection by the allowable shear stress of one fastener, $(n = P/F_v)$.

Last, to prevent a connected part from crushing due to the bearing force of the fasteners on the material, the minimum number of fasteners is determined to prevent such crushing. It has been experimentally shown that the lower bound

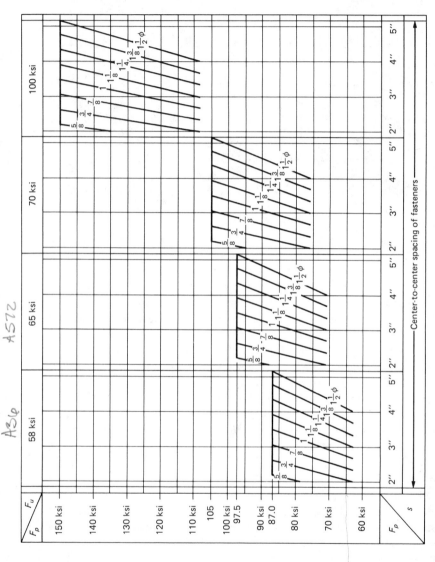

Fig. 5.4. Allowable bearing stress on structural steels whose ultimate tensile strength (F_u) equals 58, 65, 70, or 100 ksi. The allowable bearing stress depends on the fastener diameter and center-to-center spacing, as well as the connected part's ultimate strength. In the chart, the numbers $\frac{5}{8}$ through $1\frac{1}{2}$, next to each of the plots, refer to the diameter of the fastener in question.

for bearing stress F_p is given by

$$\frac{F_p}{F_u} = \frac{l_c}{d} \tag{5.1}$$

where l_c is the distance from the center of the fastener to the nearest edge of an adjacent fastener or the free edge of the connected part in the direction of stress (AISCS 1.16.5). Equation 5.1 can be rewritten as

$$\frac{F_p}{F_u} = \frac{s - d/2}{d} \tag{5.2}$$

where s is the distance between the centers of two fasteners. Using a safety factor of 2 and AISCS 1.5.1.5, we obtain

$$F_p = \frac{F_u}{2}\left(\frac{s}{d} - 0.50\right) \leqslant 1.50\,F_u \tag{5.3}$$

For convenience, Eq. 5.3 for F_p has been graphed for steels whose F_u equals 58, 65, 70, or 100 ksi (Fig. 5.4).

The analysis of F type connections is quite complex. The connection is designed similar to N or X type connections, with the appropriate allowable stresses for the shear capacity of the bolts. Though bearing should not be encountered in this type of connections, the bearing of the bolts against the connected parts must be considered in the event the friction bolts slip into bearing ("Specifications for Structural Joints Using ASTM A325 or A490 Bolts," Commentary C4). Allowable stresses for shear are given in AISCS Table 1.5.2.1. For cases of special surface treatment of the connected parts, the stresses have to be modified (generally increased) according to Table 2a in "Specifications for Structural Joints using A325 or A490 Bolts." For allowable bearing stresses, Eq. 5.3 or Fig. 5.4 may be used.

Example 5.1. Determine the maximum tensile load P that can be resisted by the connection shown. The $\frac{3}{4}$ in. × 14 in. plates are connected with $\frac{7}{8}$ in. diameter A502 Grade 1 rivets.

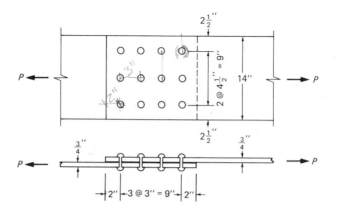

Solution. The rivets may fail in shear or bearing, or the plate in tension. Each condition must be investigated to determine the critical condition.

Shear failure in the rivet

$$P_{all} = F_v \times A_{riv} \times N_{rivets}$$

For A502 Grade 1 rivets

$$F_v = 17.5 \text{ ksi} \quad \text{(AISCS Table 1.5.2.1)}$$

$$A_{riv} = \frac{\pi d^2}{4} = \frac{\pi \left(\frac{7}{8} \text{ in.}\right)^2}{4} = 0.601 \text{ in.}^2$$

The above values can also be found in Part 4 of AISCM, "Bolts, Threaded Parts, and Rivets (Shear)."

Number of rivets $N = 12$

$$P_{all} = 17.5 \text{ ksi} \times 0.601 \text{ in.}^2 \times 12 = 126.2 \text{ k}$$

For failure in bearing

$$P_{all} = F_p \times A_{brg} \times N_{rivets}$$

$$F_p = \frac{F_u}{2}\left(\frac{s}{d} - 0.5\right) \leqslant 1.5 F_u \quad \text{between two fasteners}$$

$$F_p \leqslant \frac{58 \text{ ksi}}{2}\left(\frac{2.0 \text{ in.}}{0.875 \text{ in.}}\right) = 66.3 \text{ ksi} \quad \text{for end distance}$$

$$F_p \leqslant \frac{58 \text{ ksi}}{2} \left(\frac{3.0 \text{ in.}}{0.875 \text{ in.}} - 0.5 \right) = 84.9 \text{ ksi} \quad \text{for interior connectors}$$

$$F_p \leqslant 1.5 \times 58 \text{ ksi} = 87.0 \text{ ksi}$$

Use $F_p = 66.3$ ksi.

$$A_{brg} = d_{riv} \times t = \tfrac{7}{8} \text{ in.} \times \tfrac{3}{4} \text{ in.} = 0.656 \text{ in.}^2$$

$$P_{all} = 66.3 \text{ ksi} \times 0.656 \text{ in.}^2 \times 12 = 521.9 \text{ k}$$

Tensile capacity of member on gross area

$$F_t = 0.6 F_y = 22.0 \text{ ksi} \quad \text{(AISCS 1.5.1.1)}$$

$$A_g = \tfrac{3}{4} \text{ in.} \times 14 \text{ in.} = 10.5 \text{ in.}^2$$

$$P_{all} = F_t \times A_g = 22.0 \text{ ksi} \times 10.5 \text{ in.}^2 = 231.0 \text{ k}$$

Tensile capacity of member on effective net area (AISCS 1.14.2.2).

$$C_t = 1.0 \quad \qquad\qquad\qquad \text{(AISCS 1.14.2.2)}$$

$$F_t = 0.5 F_u = 29.0 \text{ ksi} \qquad\qquad \text{(AISCS 1.5.1.1)}$$

$$A_n = (14 \text{ in.} - 3(\tfrac{7}{8} \text{ in.} + \tfrac{1}{8} \text{ in.})) \times \tfrac{3}{4} \text{ in.} = 8.25 \text{ in.}^2$$

$$A_e = C_t \times A_n = 1.0 \times 8.25 \text{ in.}^2 = 8.25 \text{ in.}^2$$

$$A_{max} = 0.85 \times (14 \text{ in.} \times \tfrac{3}{4} \text{ in.}) = 8.93 \text{ in.}^2 \quad \text{(AISCS 1.14.2.3)}$$

$$P_{all} = F_t \times A_e = 29.0 \text{ ksi} \times 8.25 \text{ in.}^2 = 239.3 \text{ k}$$

Maximum tensile load is least value of cases calculated.

$$P_{max} = P_{all \text{ shear}} = 126.2 \text{ k}$$

Example 5.2. Determine the maximum tensile load P that can be resisted by the connection in Example 5.1 using 1 in. diameter A325 high-strength bolts (HSB), assuming standard holes and

a) a bearing connection with the threads included in the shear plane (A325-N).
b) a bearing connection with the threads excluded from the shear plane (A325-X).
c) a friction connection (A325-F).

Solution. Determine the connection for shear, bearing, and tension.

$$A_b = 0.785 \text{ in.}^2$$

$$N_b = 12$$

Bolt bearing will be same value for either friction type or bearing type connections.

$$F_p \leqslant \frac{58 \text{ ksi}}{2} \left(\frac{2.0 \text{ in.}}{1 \text{ in.}} \right) = 58.0 \text{ ksi}$$

$$F_p \leqslant \frac{58 \text{ ksi}}{2} \left(\frac{3.0 \text{ in.}}{1 \text{ in.}} - 0.5 \right) = 72.5 \text{ ksi}$$

$$F_p \leqslant 1.5 \times 58 \text{ ksi} = 87.0 \text{ ksi}$$

Use $F_p = 58.0$ ksi

$$A_{brg} = d_b \times t = 0.75 \text{ in.}^2$$

$$N_b = 12$$

$$P_{all} = F_b \times A_{brg} \times N_b = 58.0 \text{ ksi} \times 0.75 \text{ in.}^2 \times 12 = 522.0 \text{ k}$$

Tension on member will be same for all cases.

Tension on member gross area

$$P_{all} = 231.0 \text{ k} \qquad \text{Example 5.1}$$

Tension on member effective net area

$$F_t = 29.0 \text{ ksi}$$

$$A_n = (14 \text{ in.} - 3(1 \text{ in.} + \tfrac{1}{8} \text{ in.})) \times \tfrac{3}{4} \text{ in.} = 7.97 \text{ in.}^2$$

$$A_e = 1.0 \times 7.97 \text{ in.}^2 = 7.97 \text{ in.}^2$$

$$P_{all} = F_t \times A_e = 29.0 \text{ ksi} \times 7.97 \text{ in.}^2 = 231.1 \text{ k}$$

Shear capacity of connectors

$$P_{all} = F_v \times A_{bolt} \times N_b$$

a) $F_v = 21.0$ ksi (AISC, Table 1.5.2.1)

$$P_{all} = 21.0 \text{ ksi} \times 0.785 \text{ in.}^2 \times 12 = 197.8 \text{ k}$$

b) $F_v = 30.0$ ksi
$P_{all} = 30.0$ ksi $\times 0.785$ in.$^2 \times 12 = 282.6$ k

c) $F_v = 17.5$ ksi
$P_{all} = 17.5$ ksi $\times 0.785$ in.$^2 \times 12 = 164.9$ k

The bolt shear capacity is critical in the cases of a) and c), and tension on the gross area in the case of b).

Maximum tensile load P

a) $P_{max} = 197.8$ k

b) $P_{max} = 231.0$ k

c) $P_{max} = 164.9$ k

Example 5.3. A truss member in tension consists of a single angle $4 \times 4 \times \frac{1}{2}$ in. The member is connected to a $\frac{1}{2}$-in. gusset plate with $\frac{3}{4}$-in. diameter A325-N bolts as shown. Determine the maximum load the connection can carry if bolts are spaced $2\frac{1}{2}$ in. on center.

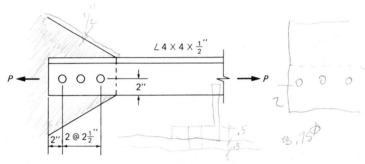

Solution. Eccentricity between the gravity axes of such members and the gage lines for their riveted or bolted end connections may be neglected (AISCS 1.15.3).

$F_v = 21.0$ ksi, $F_t = 22.0$ ksi

$A_b = 0.442$ in.^{2}

$A_g = 3.75$ in.^{2}

$A_{max} = 0.85 \times 3.75$ in.$^2 = 3.19$ in.^{2} (AISCS 1.14.2.3)

$A_e = 2.81$ in.^{2} (Example 1.5)

$P = 22.0$ ksi $\times 3.75$ in.$^2 = 82.5$ k

$P = 29.0$ ksi $\times 2.81$ in.$^2 = 81.5$ k

Bolt shear

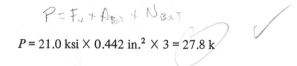

$$P = 21.0 \text{ ksi} \times 0.442 \text{ in.}^2 \times 3 = 27.8 \text{ k}$$

Bolt bearing

$$F_p = \frac{58 \text{ ksi}}{2}\left(\frac{2.5}{.75} - 0.5\right) = 82.2 \text{ ksi}$$

$$F_p = \frac{58 \text{ ksi}}{2}\left(\frac{2.0}{.75}\right) = 77.3 \text{ ksi governs}$$

$$F_p = 1.50 \times 58 \text{ ksi} = 87 \text{ ksi}$$

$$P = 77.3 \text{ ksi} \times \left(\frac{1}{2}\text{ in.} \times \frac{3}{4}\text{ in.}\right) \times 3 = 87.0 \text{ k}$$

Use least value

$$P_{\text{all}} = 27.8 \text{ k (bolt shear)}$$

Example 5.4. For the plates shown, determine the allowable value of P using 1-in. diameter A307 bolts.

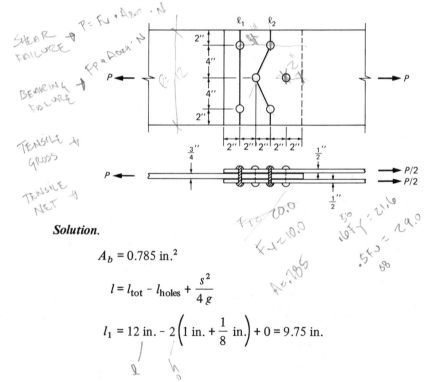

Solution.

$$A_b = 0.785 \text{ in.}^2$$

$$l = l_{\text{tot}} - l_{\text{holes}} + \frac{s^2}{4g}$$

$$l_1 = 12 \text{ in.} - 2\left(1 \text{ in.} + \frac{1}{8}\text{ in.}\right) + 0 = 9.75 \text{ in.}$$

$$l_2 = 12 \text{ in.} - 3\left(1 \text{ in.} + \frac{1}{8} \text{ in.}\right) + (2)^2/(4 \times 4) = 8.875 \text{ in. governs}$$

$$l_{max} = 0.85 \times 12 \text{ in.} = 10.2 \text{ in.}$$

a) P_{all} of plate

Gross area

$$P = F_y \times A_g = 22.0 \text{ ksi} \times (\tfrac{3}{4} \text{ in.} \times 12 \text{ in.}) = 198.0 \text{ k}$$

Effective net area

$$P = F_u \times A_e = 29.0 \text{ ksi} \times 1.0(\tfrac{3}{4} \text{ in.} \times 8.875 \text{ in.}) = 193.0 \text{ k}$$

b) P_{all} by shear

Bolts are in double shear.

$$P = F_v \times A_b \times N_b \times 2 = 10.0 \text{ ksi} \times 0.785 \text{ in.}^2 \times 6 \times 2 = 94.2 \text{ k}$$

c) P_{all} by bearing

Bearing on $\tfrac{3}{4}$-in. plate is more critical than two $\tfrac{1}{2}$-in. plates.

$$P = F_p \times d_b \times t_p \times N_b$$

$$F_p \leqslant \frac{58 \text{ ksi}}{2} \times \left(\frac{4.0}{1.0} - 0.5\right) = 72.5 \text{ ksi}$$

$$F_p \leqslant \frac{58 \text{ ksi}}{2} \times \left(\frac{2.0}{1}\right) = 58.0 \text{ ksi}$$

$$F_p \leqslant 1.5 \times 58 \text{ ksi} = 87.9 \text{ ksi}$$

Use $F_p = 58.0$ ksi

$$P = 58 \text{ ksi} \times 1.0 \text{ in.} \times 0.75 \text{ in.} \times 6 = 174 \text{ k}$$

Case b) governs.

$$P_{all} = 94.2 \text{ k}$$

Connections subjected to shear and torsion should be designed by the method explained in Example 5.5. The external effects are transferred to the centroid

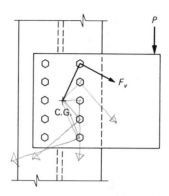

Fig. 5.5. Bolted or riveted connection subject to shear and torsion.

of the connection as two forces in the horizontal and vertical directions, respectively, and a moment. The fasteners carry the direct forces equally. The shear stresses produced in the fasteners by the moment are perpendicular to the lines connecting the fastener to the centroid, and the magnitude of the shear stress carried by each connector is proportional to the distance of the fastener from the centroid (see Fig. 5.5).

Example 5.5. Find the load on fastener A in the general situation shown in the sketch.

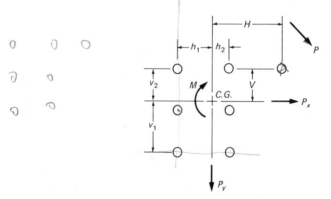

Solution. The centroid of the fastener group is first located, and the force P carried is transformed to that point as a vertical component P_y, horizontal component P_x, and an applied moment M.

It is assumed that the two components are equally divided to all the fasteners as resisting forces. The magnitude of the resisting force in each connector due to the moment is proportional to the distance of that connector to the centroid, and its direction is perpendicular to that line.

If we designate for the ith connector by R_{mi}, its resisting force, and by d_i, its distance to the centroid, its resisting moment is

$$M_r = d_i \times R_{mi}$$

The applied moment then becomes

$$\Sigma (d_i \times R_{mi}) = M$$

Also referring to the farthest connector for d_{max} and $R_{m\ max}$,

$$R_{mi} = d_i \times \frac{R_{m\ max}}{d_{max}}$$

and

$$M = \frac{\Sigma d_i^2 \times R_{m\ max}}{d_{max}}$$

From here we obtain

$$R_{m\ max} = \frac{M\, d_{max}}{\Sigma d_i^2}$$

and for any fastener,

$$R_{mp} = \frac{M\, d_p}{\Sigma d_i^2}$$

Also breaking up R_{mp} into its x and y components, we obtain

$$R_{mx} = \frac{Mv}{\Sigma d_i^2} \qquad R_{my} = \frac{Mh}{\Sigma d_i^2}$$

Here v and h are the ordinate and the abscissa of the connector under consideration with respect to the centroid.

Also,

$$\Sigma d_i^2 = \Sigma (v_1^2 + h_i^2)$$

With n being the number of fasteners, we can then calculate the components of the force in the connector as

$$R_x = \frac{P_x}{n} + \frac{Mv}{\Sigma v_i^2 + \Sigma h_i^2}$$

and

$$R_y = \frac{P_y}{n} + \frac{Mh}{\Sigma v_i^2 + \Sigma h_i^2}$$

The total resisting force in the connector is

$$R = \sqrt{R_x^2 + R_y^2}$$

Because the most critically loaded connector may not be determined by inspection prior to analysis, it may be necessary to verify all "candidates" for the most critically loaded connector.

NOTE: This problem has been solved by the elastic method in AISCM "Eccentric Loads on Fastener Groups" write up.

Ultimate strength method

A method based on the ultimate strength capacity, somewhat more liberal than the method above, but still safe, has been developed in the eighth edition of the AISCM (1978). Its use, as presented in the *Manual*, is applicable to connector groups following a definite pattern and eccentric vertical load.

Example 5.6. Determine the resultant load on the most stressed fastener in the eccentrically loaded connection shown. Use the elastic method.

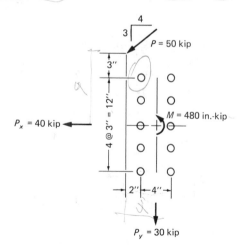

Solution. Components of the 50-kip load are 30 kips vertically down and 40 kips to the left.

Centroid of fasteners by inspection is 9 in. from the bottom and 4 in. from either side of the plate.

$$M = 40 \text{ k} \times 9 \text{ in.} + 30 \text{ k} \times 4 \text{ in.} = 480 \text{ in.-k}$$

$$\Sigma d^2 = \Sigma(h^2 + v^2) \quad \text{where } h \text{ is the horizontal distance from center of gravity to each fastener, and } v \text{ is the vertical distance from center of gravity to each fastener}$$

$$= \Sigma h^2 + \Sigma v^2 = 10(2 \text{ in.})^2 + 4(6 \text{ in.})^2 + 4(3 \text{ in.})^2$$

$$= 220 \text{ in.}^2$$

At corner (critical) fasteners,

$$V_M = \text{vertical load due to moment} = \frac{Mh}{\Sigma d^2}$$

$$= \frac{480 \text{ in.-k} \times 2 \text{ in.}}{220 \text{ in.}^2} = 4.36 \text{ k}$$

$$H_M = \text{horizontal load due to moment} = \frac{Mv}{\Sigma d^2}$$

$$= \frac{480 \text{ in.-k} \times 6 \text{ in.}}{220 \text{ in.}^2} = 13.09 \text{ k}$$

$$V_S = \text{vertical load due to direct component}$$

$$\frac{P_y}{N} = \frac{30 \text{ k}}{10} = 3 \text{ k}$$

$$H_S = \text{horizontal load due to direct component}$$

$$\frac{P_x}{N} = \frac{40 \text{ k}}{10} = 4 \text{ k}$$

Fastener A is the critical fastener, as there the loads added to each other

$$V_M = 4.36 \text{ k} \downarrow \quad H_M = 13.09 \text{ k} \quad H_S = 4.0 \text{ k} = A \quad 18.61 \text{ k}$$

$$V_S = 3.0 \text{ k} \downarrow$$

$$R = \sqrt{V^2 + H^2} = \sqrt{(4.36 + 3)^2 + (13.09 + 4)^2}$$

$$= 18.61 \text{ k}$$

Maximum resultant is on fastener A

$$R = 18.61 \text{ k}$$

Example 5.7. Determine the load P that can be carried by the bracket when $\frac{3}{4}$-in. diameter A325-F bolts are used. Use the elastic method.

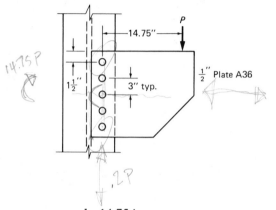

Solution.

$$l = 14.75 \text{ in.}$$

$\bar{y} = 6$ in. from bottom bolt (by inspection)

$$\Sigma d^2 = \Sigma v^2 = 2(3 \text{ in.})^2 + 2(6 \text{ in.})^2 = 90 \text{ in.}^2$$

$$M = Pl = 14.75 \, P$$

Maximum load is on top and bottom bolt

$$V_S = \text{vertical load due to shear} = \frac{P}{N} = \frac{P}{5} = 0.2 \, P$$

H_M = horizontal load due to moment

$$= \frac{M_v}{\Sigma d^2} = \frac{14.75 \, P \times 6 \text{ in.}}{90 \text{ in.}^2} = 0.983 \, P$$

$$R_{\text{tot}} = \sqrt{V^2 + H^2} = \sqrt{(.2 \, P)^2 + (.983 \, P)^2} = 1.003 \, P$$

Shear capacity of $\frac{3}{4}$-in. A325-F bolts

$$R_v = 7.7 \text{ k} \quad \text{(AISC Table I-D)}$$

$$1.003 \, P = 7.7 \text{ k}$$

$$P = \frac{7.7 \text{ k}}{1.003} = 7.68 \text{ k}$$

The maximum load that can be carried by the bracket is $P = 7.68$ k.

Check bracket plate (AISCM, p. 4-87).

$$S = 14 \text{ in.}^3 \quad \text{for } \tfrac{1}{2} \text{ in.} \times 15 \text{ in. plate}$$

$$M = 14.75 \text{ in.} \times P$$

$$F_b = 22.0 \text{ ksi} \quad \text{(AISCS 1.5.1.4.1)}$$

$$M_{all} = 22.0 \text{ ksi} \times 14 \text{ in.}^3 = 308 \text{ in.-k} \qquad M = f_b \cdot S$$

$$P = \frac{308 \text{ in.-k}}{14.75 \text{ in.}} = 20.9 \text{ k} > 7.68 \text{ k} \qquad \text{ok}$$

Example 5.8. Determine the maximum load on the fastener group shown, and verify the plate.

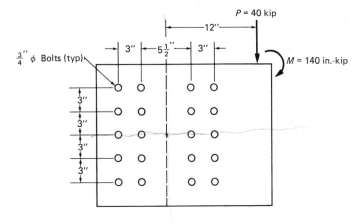

Solution.

$$M_{tot} = (40 \text{ k} \times 12 \text{ in.}) + 140 \text{ in.-k} = 620 \text{ in.-k}$$

$$l = M/P = \frac{620 \text{ in.-k}}{40 \text{ k}} = 15.5 \text{ in.}$$

$\bar{y} = 6$ in. from bottom row of bolts (by inspection)

$\bar{x} = 5.75$ in. from left vertical row of bolts (by inspection)

$$\Sigma v^2 = 8((6 \text{ in.})^2 + (3 \text{ in.})^2) = 360 \text{ in.}^2$$

$$\Sigma h^2 = 10((5.75 \text{ in.})^2 + (2.75 \text{ in.})^2) = 406.25 \text{ in.}^2$$

$$\Sigma d^2 = \Sigma v^2 + \Sigma h^2 = 360 + 406.25 = 766.25 \text{ in.}^2$$

$$V_S = P/N = \frac{40\ k}{20} = 2\ k$$

$$V_M = \frac{Mc}{J_p} = \frac{Mh}{\Sigma d^2} = \frac{620\ in.\text{-}k \times 5.75\ in.}{766.25\ in.^2} = 4.65\ k \text{ (for a corner fastener)}$$

$$V_{tot} = 2.0\ k + 4.65\ k = 6.65\ k$$

$$H_S = 0$$

$$H_M = \frac{Mc}{J_p} = \frac{Mv}{\Sigma d^2} = \frac{620\ in.\text{-}k \times 6\ in.}{766.25\ in.^2} = 4.85\ k$$

$$H_{tot} = 0 + 4.85\ k = 4.85\ k$$

$$R_{tot} = \sqrt{V^2 + H^2} = \sqrt{(6.65)^2 + (4.85)^2} = 8.23\ k$$

The maximum load on the critical fastener is 8.23 k

Plate dimensions

$$M = 140\ in.\text{-}k + 40\ k \times (12\ in. - 5.75\ in.) = 390\ in.\text{-}k$$

$$S = \frac{390\ in.\text{-}k}{22\ ksi} = 17.73\ in.^3$$

Referring to bracket plates table (AISCM, p. 4-87), a plate of $\frac{11}{16}$-in. thickness and 15-in. depth is needed.

Use 15 in. $\times$ $\frac{11}{16}$ in.

For design of connection by use of AISC tables, see example 5.18.

Example 5.9. Determine P for the connection shown. The rivets are $\frac{7}{8}$-in. diameter A502 Grade 1.

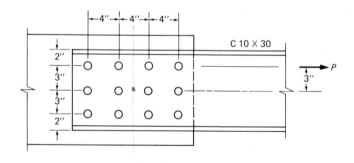

Solution. For C 10 X 30,

$$A = 8.82 \text{ in.}^2, \quad t_f = 0.436 \text{ in.}, \quad t_w = 0.673 \text{ in.}, \quad I_x = 103 \text{ in.}^4$$
$$\Sigma d^2 = \Sigma(h^2 + v^2) = 6(6 \text{ in.})^2 + 6(2 \text{ in.})^2 + 8(3 \text{ in.})^2 = 312 \text{ in.}^2$$
$$M = 3P$$

Maximum load occurs at a corner rivet.

$$V_M = \frac{Mh}{\Sigma d^2} = \frac{3P \times 6}{312} = 0.0577\,P$$

$$H_M = \frac{Mv}{\Sigma d^2} = \frac{3P \times 3}{312} = 0.0288\,P$$

$$V_S = 0$$

$$H_S = P/N = P/12 = 0.0833\,P$$

$$R = \sqrt{V^2 + H^2}$$

$$= \sqrt{(0.0577\,P)^2 + (0.0288\,P + 0.0833\,P)^2}$$

$$= 0.1261\,P$$

Capacity of $\frac{7}{8}$-in. A502-1 rivets in shear

$$R_v = 10.5 \text{ k} \quad \text{(AISC, Table I-D)}$$

Capacity of one $\frac{7}{8}$-in. A502-1 rivet in bearing on 0.673-in. flange

$$R_b = (0.673 \text{ in.} \times \tfrac{7}{8} \text{ in.}) \times 84.0 \text{ ksi} = 49.5 \text{ k}$$

(see Fig. 5.4, $s = 3$, $F_u = 58$ ksi)

Shear governs

$$P = \frac{10.5 \text{ k}}{0.1261} = 83.27 \text{ k}$$

Capacity of C 10 X 30

$$A_{\text{net}} = 8.82 \text{ in.}^2 - 3 \times 0.673 \text{ in.} \times \left(\frac{7}{8} \text{ in.} + \frac{1}{8} \text{ in.}\right) = 6.801 \text{ in.}^2$$

$$I_{\text{net}} = 103 \text{ in.}^4 - 0.673 \text{ in.} \times \left(\frac{7}{8} \text{ in.} + \frac{1}{8} \text{ in.}\right) \times (3 \text{ in.})^2 \times 2 = 90.9 \text{ in.}^4$$

$$S_{net} = \frac{90.9 \text{ in.}^4}{5 \text{ in.}} = 18.2 \text{ in.}^3$$

$$M = 3P, \quad F_b = 27.0 \text{ ksi}$$

$$\frac{\frac{P}{8.82}}{0.60 \times 36} + \frac{\frac{3P}{18.2}}{27.0} \leqslant 1.0 \quad \text{(AISCS 1.6.2)}$$

$$0.0052\,P + 0.0061\,P = 0.0113\,P \leqslant 1.0$$

$$P \leqslant 88.5 \text{ k}$$

Rivet capacity governs. Maximum load is 83.27 kips.

Example 5.10. Determine the size of A325-X high-strength bolts for this group of connectors.

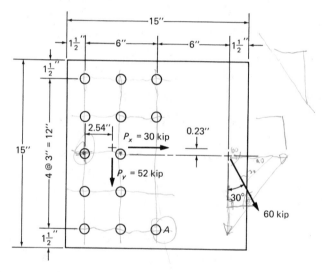

Solution. Locate the center of gravity with respect to the x and y axes.

$$\bar{x} = \frac{3 \times 6 \text{ in.} + 5 \times 3 \text{ in.}}{13} = 2.54 \text{ in.}$$

$$\bar{y} = \frac{3 \times 6 \text{ in.} + 3 \times 3 \text{ in.} - 2 \times 3 \text{ in.} - 3 \times 6 \text{ in.}}{13} = 0.23 \text{ in.}$$

$\Sigma v^2 = (5.77 \text{ in.})^2 \times 3 + (2.77 \text{ in.})^2 \times 3 + (0.23 \text{ in.})^2 \times 2$
$\qquad + (3.23 \text{ in.})^2 \times 2 + (6.23 \text{ in.})^2 \times 3 = 260.3 \text{ in.}^2$

$\Sigma h^2 = (2.54 \text{ in.})^2 \times 5 + (0.46 \text{ in.})^2 \times 5 + (3.46 \text{ in.})^2 \times 3 = 69.2 \text{ in.}^2$

$\Sigma d^2 = \Sigma v^2 + \Sigma h^2 = 260.3 \text{ in.}^2 + 69.2 \text{ in.}^2 = 329.5 \text{ in.}^2$

$P_x = 0.866 \times 60 \text{ k} = 52.0 \text{ k}$

$P_y = 0.50 \times 60 \text{ k} = 30.0 \text{ k}$

$M = 9.46 \text{ in.} \times 52.0 \text{ k} - 0.23 \text{ in.} \times 30.0 \text{ k} = 485 \text{ in.-k}$

$$V_s = \frac{52.0 \text{ k}}{13} = 4.0 \text{ k}$$

$$H_s = \frac{30.0 \text{ k}}{13} = 2.3 \text{ k}$$

By inspection, bolt A is critical.

$$V_m = \frac{485 \text{ in.-k}}{329.5 \text{ in.}^2} \times 3.46 \text{ in.} = 5.09 \text{ k}$$

$$H_m = \frac{485 \text{ in.-k}}{329.5 \text{ in.}^2} \times 6.23 \text{ in.} = 9.17 \text{ k}$$

$V_s = 4.0 \text{ kip}$

$V_m = 5.09 \text{ kip}$

$H_s = 2.30 \text{ kip}$ $H_m = 9.17 \text{ kip}$

$$R = \sqrt{(2.30 + 9.17)^2 + (5.09 + 4.0)^2} = 14.64 \text{ k}$$

A bolt of $\frac{7}{8}$ in. diameter is needed ($r_v = 18.0$ k).

Rivets and bolts under both shear and tension shall be designed such that the tensile and shearing stresses do not exceed those allowed by formulas given in Table 5.1 (AISCS Table 1.6.3).

Table 5.1. Allowable Tensile Stress (F_t) for Fasteners in Bearing Type Connections (AISCS Table 1.6.3, reprinted with permission).

Description of Fastener	Threads Not Excluded from Shear Planes	Threads Excluded from Shear Planes
Threaded parts		
A449 bolts over $1\frac{1}{2}$-in. diameter	$0.43\,F_u - 1.8\,f_v \leqslant 0.33\,F_u$	$0.43\,F_u - 1.4\,f_v \leqslant 0.33\,F_u$
A325 bolts	$55 - 1.8\,f_v \leqslant 44$	$55 - 1.4\,f_v \leqslant 44$
A490 bolts	$68 - 1.8\,f_v \leqslant 54$	$68 - 1.4\,f_v \leqslant 54$
A502 Grade 1 rivets	$30 - 1.3\,f_v \leqslant 23$	
A502 Grades 2 and 3 rivets	$38 - 1.3\,f_v \leqslant 29$	
A307 bolts	$26 - 1.8\,f_v \leqslant 20$	

Example 5.11. For the connection shown, determine P_{max} when

a) 10-$\frac{7}{8}$-in. A502 Grade 1 rivets are used.

b) 10-$\frac{7}{8}$-in. A490 bolts are used with threads not excluded from shear plane.

NOTE: Disregard effect of prying. For prying design, see Section 7.5.

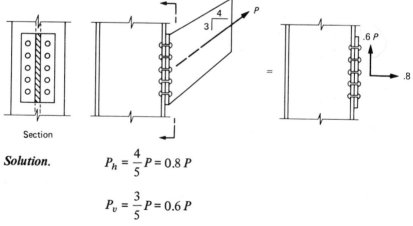

Section

Solution.

$$P_h = \frac{4}{5}P = 0.8\,P$$

$$P_v = \frac{3}{5}P = 0.6\,P$$

$$f_v = P_v/A = \frac{0.6\,P}{10 \times 0.601} = 0.100\,P$$

$$f_t = P_h/A = \frac{0.8\,P}{10 \times 0.601} = 0.133\,P$$

a) $\quad f_t = 30 - 1.3\,f_v \leqslant 23$ ksi $\quad$ (AISCS Table 1.6.3)

$\quad 0.133\,P = 30 - 1.3(0.100\,P) \leqslant 23$

$\quad P = 114.1$ k

$\quad f_v = 0.100\,P = 11.4$ ksi

$\quad f_t = 0.133\,P = 15.2$ ksi < 23 ksi

b) $\quad f_t = 68 - 1.8\,f_v \leqslant 54$ ksi $\quad$ (AISCS Table 1.6.3)

$\quad 0.133\,P = 68 - 1.8(0.100\,P) \leqslant 54$

$\quad P = 217.5$ k

$\quad f_v = 0.100\,P = 21.7$ ksi

$\quad f_t = 0.133\,P = 28.9$ ksi < 54 ksi

Example 5.12. Determine whether the connection shown is satisfactory using $\frac{7}{8}$-in. A307 bolts, each having an area of 0.6 in.^{2}

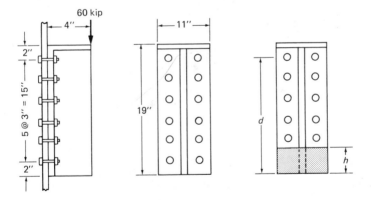

Solution. The centroid could be located by setting up a quadratic equation in terms of h, but is more conveniently obtained by trial and error. First assume the center of gravity of compression and tension areas at a distance h from the bottom, usually $h = (\frac{1}{6}$ to $\frac{1}{7})d$.

Assume $h = 3$ in.

Check to see if center of gravity is sufficiently near assumed location.

Moment of compression area $= 3$ in. $\times 11$ in. $\times 1.5$ in. $= 49.5$ in.^{3}

$$\text{Moment of tension area} = (2 \times 0.6 \times 14) + (2 \times 0.6 \times 11)$$
$$+ (2 \times 0.6 \times 8) + (2 \times 0.6 \times 5)$$
$$+ (2 \times 0.6 \times 2)$$
$$= 48 \text{ in.}^3$$

$$48.0 \text{ in.}^3 \approx 49.5 \text{ in.}^3$$

Say assumed center of gravity is ok.

Calculate moment of inertia of tension and compression areas.

$$I = \Sigma(I_0 + Ad^2)$$

For bolt areas, I_0 can be neglected.

$$I = \frac{11\,(3)^3}{12} + (33 \times 1.5^2) + (2 \times 0.6 \times 14^2) + (2 \times 0.6 \times 11^2)$$

$$+ (2 \times 0.6 \times 8^2) + (2 \times 0.6 \times 5^2) + (2 \times 0.6 \times 2^2) = 591 \text{ in.}^4$$

Shear force on each bolt

$$\frac{P}{N_b} = \frac{60 \text{ k}}{12} = 5.0 \text{ k/bolt}$$

Shear stress on each bolt

$$\frac{P}{A} = \frac{5.0 \text{ k}}{0.6 \text{ in.}^2} = 8.33 \text{ ksi}$$

Tension force on extreme top bolt

$$\frac{Mc}{I} = \frac{(60 \text{ k} \times 4 \text{ in.}) \times 14 \text{ in.}}{591 \text{ in.}^4} = 5.69 \text{ ksi}$$

For A307 bolts, $F_t = 26.0 - 1.8\,f_v \leqslant 20 \text{ ksi}$ (AISCS 1.6.3)

$$= 26.0 - 1.8\,(8.33) = 11.0 \text{ ksi} < 20.0 \text{ ksi}$$

Connection is satisfactory, because $f_v = 8.33$ ksi $< F_v = 10.0$ ksi and $f_t = 5.61$ ksi $< F_t = 11.0$ ksi.

5.5 FASTENERS FOR HORIZONTAL SHEAR IN BEAMS

In a built-up member or a plate girder subjected to transverse loading, there is a tendency for the connected elements to slide horizontally if they are not fastened together; see Fig. 2.2. If the elements are connected continuously, horizontal shearing stress will develop in the connection. This shear, measured per unit length for the entire width, is called *shear flow* and is expressed as

$$q = \frac{VQ}{I} \tag{5.4}$$

where V is the vertical shear force, I is the total moment of inertia of the member, and Q is the moment of the areas on one side of the sliding interface with respect to the centroid of the section. The spacing of fasteners is determined by dividing the shearing capacity of the fasteners to be used by the shear flow q. The maximum spacing between fasteners must also satisfy AISCS 1.18.2.3 and 1.18.3.1, as specified in AISCS 1.10.4. Where the component of a built-up compression member consists of an outside plate, the maximum spacing between bolts and rivets connecting the plate to the other member shall not exceed the thickness of the thinner plate times $127/\sqrt{F_y}$ nor 12 in. When the fasteners are staggered, the maximum spacing on each gage line shall not exceed the thickness of the thinner plate times $190/\sqrt{F_y}$ nor 18 in. (see Fig. 3.2). The spacing of fasteners connecting two rolled shapes in contact with each other shall not exceed 24 in. (AISCS 1.18.2.3).

The longitudinal spacing of bolts and rivets connecting a plate to a rolled shape or to another plate in a built-up tension member shall not exceed 24 times the thickness of the thinner plate nor 12 in. (see Fig. 3.2). When two or more shapes in contact with one another form a built-up tension member, they shall be intermittently fastened at least 24 in. on center (AISCS 1.18.3.1).

Example 5.13. Check to see if this connection is adequate with $\frac{3}{4}$-in.-diameter connectors and

a) using A502 Grade 1 rivets
b) using A325-F bolts with threads not excluded from shear plane.

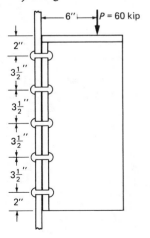
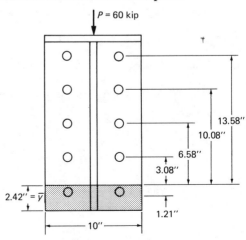

Solution. Assume neutral axis of connection $(\bar{y})$ as 2.42 in. from bottom of bracket.

Check assumption

$$\text{Moment of compression area} = 10 \text{ in.} \times 2.42 \text{ in.} \times 1.21 \text{ in.} = 29.28 \text{ in.}^3$$

$$\text{Moment of tension area} = 2 \times 0.442 \text{ in.}^2 \times (3.08 \text{ in.} + 6.58 \text{ in.}$$
$$+ 10.08 \text{ in.} + 13.58 \text{ in.})$$

$$= 29.45 \text{ in.}^3$$

Say assumption of neutral axis is ok.

$$I = \frac{10\,(2.42)^3}{3} + 2 \times 0.442 \times (3.08^2 + 6.58^2 + 10.08^2 + 13.58^2)$$

$$= 346.74 \text{ in.}^4$$

$$f_v = \frac{60}{10 \times 0.442} = 13.57 \text{ ksi}$$

$$f_{t\,\text{top}} = \frac{(60 \times 6) \times 13.58}{346.74} = 14.1 \text{ ksi}$$

a) For A502-1 rivets,

$F_v = 17.5 \text{ ksi} > 13.57 \text{ ksi}$ ok

$F_t = 30.0 - 1.3\,(13.57) = 12.4 \text{ ksi} < 14.1 \text{ ksi}$ N.G.   (AISCS Table 1.6.3)

A502-1 rivets are not adequate for loading shown.

b) For A325-F bolts,

$F_v = 17.5 \text{ ksi} > 13.57 \text{ ksi}$ ok

$F_t = 55 - 1.8\,(13.57) = 30.6 \text{ ksi} > 14.1 \text{ ksi}$ ok   (AISCS Table 1.6.3)

A325-F bolts are adequate for loading shown.

Example 5.14. Calculate the theoretical spacing of $\frac{3}{4}$-in.-diameter A325-N bolts for the beam shown. $I_{\text{tot}} = 461 \text{ in.}^4$

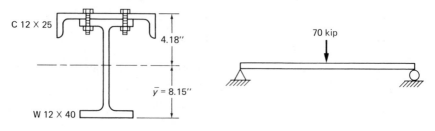

Solution.

$$A_{channel} = 7.35 \text{ in.}^2$$

$$Q = Ad = 7.35 \text{ in.}^2 \times (4.18 \text{ in.} - 0.674 \text{ in.}) = 25.77 \text{ in.}^3$$

$$V_{max} = 35 \text{ k}$$

$$q = \frac{VQ}{I} = \frac{35 \text{ k} \times 25.77 \text{ in.}^3}{461 \text{ in.}^4} = 1.96 \text{ k/in.}$$

Shear capacity of $\frac{3}{4}$-in. A325 N bolts

$$R_v = 9.3 \text{ k}$$

Spacing of bolts

$$\frac{N_b \times R_v}{q} = \frac{2 \times 9.3 \text{ k}}{1.96 \text{ k/in.}} = 9.48 \text{ in.}$$

Use two rows of bolts spaced at 9 in.

Example 5.15. For the beam shown, what is the required spacing of $\frac{3}{4}$-in. A490-N bolts at a section where the external shear is 100 kips?

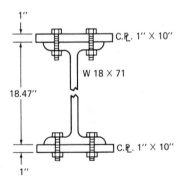

Solution. For W 18 × 71,

$$A = 20.8 \text{ in.}^2, \quad d = 18.47 \text{ in.}, \quad b_f = 7.64 \text{ in.}, \quad t_f = 0.81 \text{ in.}, \quad I = 1170 \text{ in.}^4$$

Allowable loss of material per flange is the amount of loss greater than 15% of the gross flange area (AISCS 1.10.1).

$$A_{g\, flg} = (7.64 \text{ in.} \times 0.81 \text{ in.}) + (1 \text{ in.} \times 10 \text{ in.}) = 16.19 \text{ in.}^2$$

Loss of flange area that can be neglected

$$0.15 \times 16.19 \text{ in.}^2 = 2.43 \text{ in.}^2$$

Actual loss of material

$$2 \times (\tfrac{3}{4} \text{ in.} + \tfrac{1}{8} \text{ in.}) (0.81 \text{ in.} + 1.0 \text{ in.}) = 3.17 \text{ in.}^2$$

Reduction in loss of flange material to be accounted for

$$3.17 \text{ in.}^2 - 2.43 \text{ in.}^2 = 0.74 \text{ in.}^2$$

Allowable area per flange $= 16.19 \text{ in.}^2 - 0.74 \text{ in.}^2$
$$= 15.45 \text{ in.}^2$$

$$I_{tot} = \Sigma(I_0 + Ad^2)$$

$$I_{tot} = 1170 \text{ in.}^4 + \left(\frac{10 \text{ in. } (1 \text{ in.})^3}{12}\right) \times 2 + 2(1 \text{ in.} \times 10 \text{ in.} \times (9.74 \text{ in.})^2)$$

$$= 3069 \text{ in.}^4$$

The effective moment of inertia can be calculated by deducting the moment of inertia of the hole area in excess of 15% of the gross flange area.

$$I_{\text{hole loss}} = \Sigma(Ad^2) = 2 \times 0.74 \text{ in.}^2 \times \left(10.24 \text{ in.} - \frac{1.81 \text{ in.}}{2}\right)^2$$

$$= 129 \text{ in.}^4$$

$$I_{net} = I_{tot} - I_{\text{hole loss}} = 3069 \text{ in.}^4 - 129 \text{ in.}^4$$

$$= 2940 \text{ in.}^4$$

$$Q = Ad = 1 \text{ in.} \times 10 \text{ in.} \times 9.74 \text{ in.} = 97.4 \text{ in.}^3$$

$$q = \frac{VQ}{I} = \frac{100 \text{ k} \times 97.4 \text{ in.}^3}{2940 \text{ in.}^4} = 3.31 \text{ k/in.}$$

Shear capacity of $\tfrac{3}{4}$-in. A490-N bolts

$$R_v = 12.4 \text{ k}$$

Spacing of bolts

$$\frac{N_b \times R_v}{q} = \frac{2 \times 12.4 \text{ k}}{3.31 \text{ k/in.}} = 7.49 \text{ in.}$$

Use two rows of bolts spaced at $7\tfrac{1}{4}$ in.

5.6 AISCM TABLES FOR RIVET AND BOLT DESIGNS

The tables in AISCM, pp. 4-3 through 4-11, can be used to determine areas, allowable tensile, shear, and bearing loads for rivets and bolts must commonly used in steel construction.

Tension Tables I-A and I-B list the allowable tensile stresses for various bolts, rivets, and grades of steels used to produce threaded fasteners and the allowable tensile loads for $\frac{5}{8}$- to $1\frac{1}{2}$-in. diameter fasteners of those materials.

Shear Table I-D lists the allowable shear stresses and allowable loads for $\frac{5}{8}$- to $1\frac{1}{2}$-in. diameter rivets and bolts made from the materials mentioned above. To use these tables the designer must locate the allowable load under the column for the fastener diameter in question and referenced to the row corresponding to the material of the fastener, type of connection, and the hole type.

Bearing Tables I-E for F_u equal to 58, 65, 70, and 100 ksi of the connected part's material give the allowable bearing loads, in kips, of $\frac{3}{4}$, $\frac{7}{8}$, and 1-in. diameter fasteners on material thicknesses from $\frac{1}{8}$ to 1 in., based on the fastener spacing. As the bearing values in the tables are limited by the double shear capacity of A490-X bolts, some values in the tables are not listed. To determine these values, the designer must either multiply the value corresponding to 1-in.-thick material by the actual thickness of the connected material or refer to Fig. 5.4 of this text.

"Framed Beam Connections," Tables II-A, II-B, and II-C, in AISCM, pp. 4-24 through 4-27, are provided to help determine the number of bolts required at ends of simply supported beams framing into other members when the reactions and the fastener type are known. In the general groupings of fastener types, under the column of the selected fastener diameter, the designer must locate the allowable load, in kips, that is equal to or slightly greater than the required connection capacity. Proceeding to the left from that load, the designer can determine the required number of bolts (n) and the length of the framing angles (L and L'). Along the top of the tables, the suggested thickness of these angles is tabulated according to the bolt type and diameter. If the length L (or L') exceeds the depth of the beam minus the flange thicknesses and fillets, the diameter of the bolts should be increased or the type changed to a stronger one to reduce the number of bolts required.

If and when the tabulated allowable load in Tables II-A and II-B is followed by a superscript, the allowable net shear on the framing angles must be checked. From Table II-C, the designer can determine the allowable load on two angles for a given thickness (t) and length (L), as determined from Tables II-A and II-B. If the tabulated load is greater than the beam reaction, then the angles are acceptable. Otherwise, the angle thickness should be increased. The designer should note that this table is for A36 double angles only.

When a beam is coped at one or both ends, web tear out (block shear) shall be checked (AISC Commentary 1.16.5). Using Table I-G under "Bolts and Rivets," a beam's resistance to block shear can be quickly computed. Once coefficients C_1 and C_2 are determined from the table, the resistance to block shear, in kips (R_{BS}) is equal to $(C_1 + C_2)F_u\,t$ and shall be greater than or equal to the beam reaction.

For eccentric loads on fastener groups, the tables in AISCM, pp. 4-62 and 4-70, are extremely useful for determining the maximum load on the critical fastener in the group for a given vertical load. Using l, the value of a coefficient C can be obtained from the appropriate table. If the capacity of one fastener (r_v) is known, the allowable load P can be determined by multiplying r_v times the coefficient C $(P = C \times r_v)$. Conversely, the required shear capacity of the fastener can be determined from $r_v = P/C$ when P is known and C is obtained from the tables. The designer should note that these "Eccentric Loads on Fastener Groups" tables are based on the ultimate method of analysis and that the longhand approach, using the elastic method described in AISCM, p. 4-58, will give much more conservative results.

Example 5.16. For the section shown, what is the required spacing of $\frac{3}{4}$-in. A325-N bolts if external shear = 150 kips?

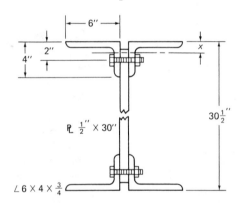

Solution. For $6 \times 4 \times \frac{3}{4}$ in. angle,

$$A = 6.94 \text{ in.}^2$$

$$I = 8.68 \text{ in.}^4$$

$$x = 1.08 \text{ in.}$$

$$I = \Sigma(I_0 + Ad^2)$$

$$= \frac{0.5 \text{ in.} \times (30)^3}{12} + 4(8.68 \text{ in.}^4) + 4(6.94 \text{ in.}^2)(15.25 \text{ in.} - 1.08 \text{ in.})^2$$

$$= 6734 \text{ in.}^4$$

$$Q = Ad = 2 \times 6.94 \text{ in.}^2 \times (15.25 \text{ in.} - 1.08 \text{ in.})$$

$$= 196.7 \text{ in.}^3$$

$$q = \frac{VQ}{I} = \frac{150 \text{ k} \times 196.7 \text{ in.}^3}{6734 \text{ in.}^4}$$

$$= 4.38 \text{ k/in.}$$

Capacity of $\frac{3}{4}$-in. A325-N bolts in double shear

$$R_v = 2 \times 9.3 \text{ k} = 18.6 \text{ k}$$

Spacing of bolts

$$\frac{R_v}{q} = \frac{18.6 \text{ k}}{4.38 \text{ k/in.}} = 4.25 \text{ in.}$$

Space bolts at $4\frac{1}{4}$ in. center to center.

Example 5.17. Rework Example 5.1 using the AISC Tables in Section 4 of the AISCM.

Solution. From Table I-D Shear (AISCM, p. 4-5)

$$\text{Area of } \tfrac{7}{8} \text{ in. rivet} = 0.6013 \text{ in.}^2$$

$$F_v = 17.5 \text{ ksi}$$

$$P_{all} \text{ single shear} = 10.5 \text{ k}$$

$$P_{all} \text{ 12 rivets} = 12 \times 10.5 \text{ k} = 126.0 \text{ k}$$

From Table I-E Bearing, $F_u = 58$ ksi, minimum fastener spacing of 3.06 in., and material thickness of 1 in. (AISCM, p. 4-6)

$$P_p = 76.1 \text{ k}$$

for $\frac{3}{4}$-in. material thickness,

$$P_p = \tfrac{3}{4} \times 76.1 \text{ k} = 57.1 \text{ k}$$

$$P_{all} \text{ 12 rivets} = 12 \times 57.1 \text{ k} = 685 \text{ k}$$

From Example 5.1,

Tensile capacity of member = 231.0 k (gross area)

Maximum capacity of connection is least value of above.

$$P_{max} = 126.0 \text{ k}$$

Example 5.18. Using the AISC tables, determine the maximum load on the fastener group shown.

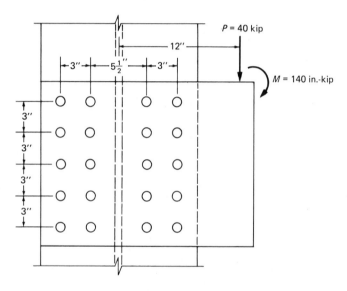

Solution. The table to be used is Table XVIII (AISCM, p. 4-70)

$$M_{tot} = 40 \text{ k} \times 12 \text{ in.} + 140 \text{ in.-k} = 620 \text{ in.-k}$$

$$\text{Equivalent } l = \frac{M}{P} = \frac{620 \text{ in.-k}}{40 \text{ k}} = 15.5 \text{ in.}$$

For $l = 12$ in. and $n = 5$,

$$C = 7.75$$

For $l = 16$ in. and $n = 5$,

$$C = 6.18$$

For $l = 15.5$ in. and $n = 5$,

$$C = 7.75 - \frac{3.5}{4}(7.75 - 6.18) = 6.376$$

$$r_v = \frac{P}{C} = \frac{40}{6.376} = 6.27 \text{ k}$$

Maximum load on critical fastener = 6.27 k

Example 5.19. Rework Example 5.7 using the eccentric loads on fastener groups tables in the AISCM.

Solution. The table to be used is Table X (AISCM, p. 4-62)

$$l = 14.75 \text{ in.}$$

For $l = 14.75$ in. and $n = 5$,

$$C = 1.40 - \frac{2.75}{4}(1.40 - 1.06) = 1.166$$

Shear capacity of $\frac{3}{4}$-in. A325-F bolts

$$R_v = 7.7 \text{ k}$$

$$P_{\text{all}} = C \times R_v = 1.166 \times 7.7 \text{ k} = 8.98 \text{ k}$$

Example 5.20. For a W 36 × 260 beam with an end reaction of 390 kips, select the number of $\frac{7}{8}$-in.-diameter A490-X bolts required for framing the beam. Assume 50-ksi steel, with $F_u = 65$ ksi for the beam and A36 steel connection angles.

Solution. For W 36 × 260,

$$t_w = 0.840 \text{ in.}$$

Using Table II-A under bolt type A490-X and bolt diameter $\frac{7}{8}$ in., select $n = 9$ corresponding to a shear force of 408 k.

From the table, suggested angle thickness is $\frac{5}{8}$ in. and length is $26\frac{1}{2}$ in.

Since the 408-k value has a superscript c, check shear capacity of $\frac{5}{8}$ in. thick angle from Table II-C (AISCS 1.5.1.2.2).

From Table II-C for bolt diameter = $\frac{7}{8}$ in., $l = 26\frac{1}{2}$ in. and angle thickness of $\frac{5}{8}$ in., allowable shear on angles is 393 k $>$ 390 k.   ok

Check bearing of nine bolts on 0.840-in. web for F_u = 65 ksi.

From Table I-E, for bolt diameter of $\frac{7}{8}$ in., bolt spacing of 3 in., and F_u = 65 ksi, allowable bearing load on 1-in.-thick material = 83.3 k.

Allowable bearing load = 9 $\times$ 0.84 in. $\times$ 83.3 k = 629.7 k $>$ 390 k   ok

Check bearing of 9 bolts on two $\frac{5}{8}$-in. angles with F_u = 58 ksi.

From Table I-E, allowable bearing load on $\frac{5}{8}$-in.-thick material = 46.4 k/bolt

Allowable bearing load = 9 $\times$ 2 $\times$ 46.4 k = 836 k $>$ 390 k

Because the beam is not coped, web tear out need not be checked.

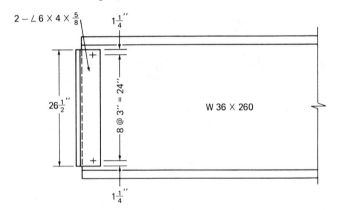

Example 5.21. Select the number of $\frac{3}{4}$-in.-diameter high-strength A325-N bolts needed for framing of a W 18 $\times$ 50 beam carrying a total load, symmetrically placed, of 102 kips. Assume A36 steel and that the beam will be coped.

Solution.

$$\text{Reaction of W 18} \times 50 = \frac{102 \text{ k}}{2} = 51 \text{ k}$$

For W 18 $\times$ 50,

$$t_w = 0.355 \text{ in.}$$

From Table II-A, for A325-N bolt, $\frac{3}{4}$-in. diameter, and an allowable shear of 55.7 k ($>$51 k reqd), select $n = 3$, angle length $l = 8\text{-}\frac{1}{2}$ in., and angle thickness = $\frac{5}{16}$ in.

Check shear capacity of two 6 X 4 X $\frac{5}{16}$ in. angles, 8-$\frac{1}{2}$ in. long.

From Table II-C, allowable load on angle if $\frac{3}{4}$ in. bolts are used is 65.9 k $>$ 51 k ok.

NOTE: If the 55.7-k value in Table II-A has no superscript, it is not required to check the shear capacity of the angle for $F_y = 36$ ksi.

Check bearing of bolts on web

From Table I-E, for $\frac{3}{4}$-in.-diameter bolt and 3-in. spacing, find 65.3 k as allowable load on 1-in. thick material. For 0.355-in. web thickness and three bolts,

Allowable load = 3 X 0.355 in. X 65.3 k/bolt = 69.5 k $>$ 51 k   ok

Check web tear out

From Table II-A, find $l_v = 1\text{-}\frac{1}{4}$ in. Since 6 in. X 4 in. angles are being used, determine $l_h = 4$ in. - $\frac{1}{2}$ in. - 1-$\frac{1}{4}$ in. = 2-$\frac{1}{4}$ in.

From Table I-G, select $C_1 = 1.50$, corresponding to $l_v = 1\text{-}\frac{1}{4}$ in. and $l_h = 2\text{-}\frac{1}{4}$ in. Also select $C_2 = 0.99$, corresponding to $n = 3$ and bolt diameter = $\frac{3}{4}$ in.

Web tear out resistance of beam = $(C_1 + C_2) F_u t$

$(1.50 + 0.99)$ X 58 ksi X 0.355 in. = 51.3 k $>$ 51 k   ok

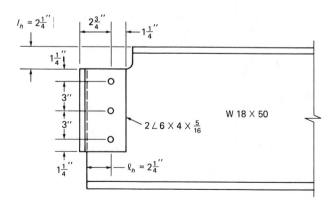

PROBLEMS TO BE SOLVED

5.1 through 5.4. Calculate the capacity of the tension members and connections shown.

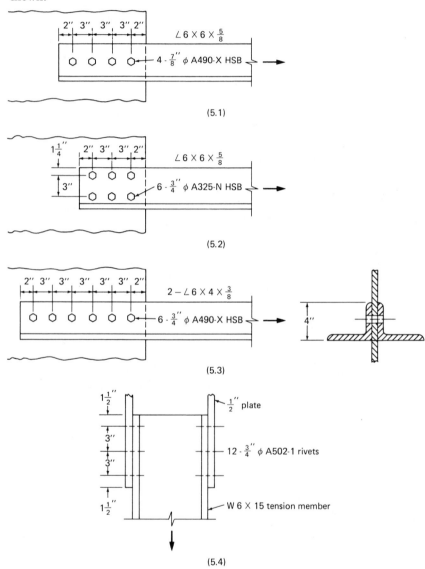

(5.1)

(5.2)

(5.3)

(5.4)

5.5. Determine the stagger s such that the loss due to the holes is equal to the loss of two holes only. Also determine the number of $\frac{7}{8}$-in. ϕ A325-F bolts required to carry the load.

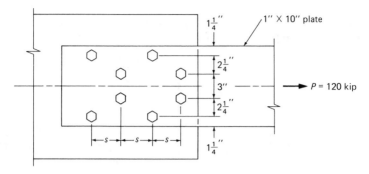

5.6. Determine P and the minimum thickness of the plate for the connection shown. Do not use AISC design aid tables.

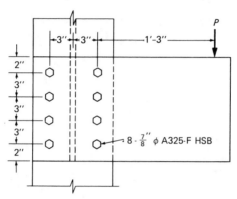

5.7. In the existing construction shown, A502-1 rivets of $\frac{7}{8}$-in. diameter were used. Inspection revealed that the rivets denoted by A are not usable. Determine the allowable load for the connection with only five good rivets.

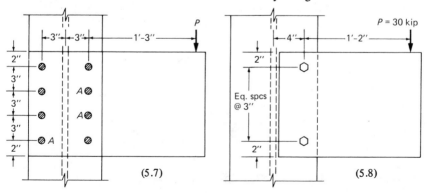

5.8. Determine the number of $\frac{3}{4}$-in. ϕ A325-X high-strength bolts required to carry the 30-k load shown.

5.9. Determine the maximum allowable load P for the connection shown. Assume six $\frac{3}{4}$-in. ϕ A 490-F bolts. Do not use AISC design tables.

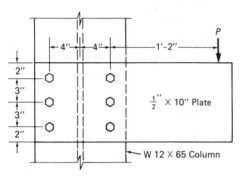

5.10 and 5.11. For the connections shown, determine the maximum allowable loads P. Assume eight 1-in. ϕ A325-X bolts.

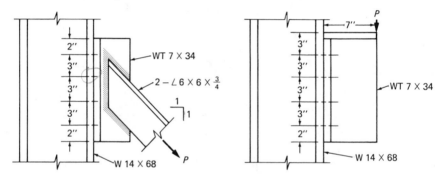

5.12 and 5.13. Calculate the maximum allowable spacing of $\frac{7}{8}$-in. ϕ A490-X bolts for the beams shown. Maximum shear force in the beams is 160 kips. $F_y = 50$ ksi.

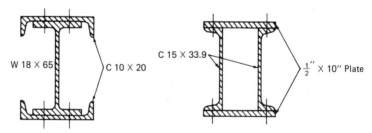

5.14. Rework Problem 5.6 using AISC design aid tables.

5.15. Rework Problem 5.8 using AISC design aid tables.

5.16. Rework Problem 5.9 using AISC design aid tables.

5.17. A W 18 × 65 beam carrying a total load of 300 kips frames into a W 14 × 68 column flange. Assuming $\frac{3}{4}$-in. ϕ A490-X bolts and F_y = 36 ksi, completely design and sketch the framed beam connection using AISC design aid tables.

5.18. Rework Problem 5.17 assuming the top flange of the W 18 × 65 beam is coped to frame into the web of a W 24 × 62 girder.

5.19. A W 24 × 76 beam supports a concentrated load of 300 kips at one of its third points. Assuming 50-ksi steel for the beam, 36-ksi steel for the framing angles, and A325-X bolts, design the end connections for both ends of the beam. The beam need not be coped. Use same diameter bolts for both ends.

5.20. Rework Problem 5.19 but assume A490-N bolts and coped ends.

6
Welded Connections

6.1 WELDED CONNECTIONS

Welding is the joining of two pieces of metal by the application of heat or through fusion. Most welds are of the fusion type, consisting of the shielding arc welding process. The surfaces of the members to be connected are heated, allowing the parts to fuse together, usually with the addition of other molten metal. When the molten metal solidifies, the bond between the members is completed.

Arc welding is performed by either an electric arc or gas process. Shielding is added to control the molten metal and improve the weld properties by preventing oxidation and is usually provided by slag around the electrode used for joining the metals.

Weld symbols used on structural drawings have been developed by the American Welding Society and are reprinted in many manuals including the AISCM. Any combination of symbols can be used to identify any set of conditions regarding the weld joint. Included in the information is weld type, size, length, and special processes.

6.2 WELD TYPES

The different types of welds are fillet, penetration (groove), plug, and slot welds, the most common being fillet and penetration welds.

Fillet welds are triangular in shape, with each leg connected along the surface of one member. When the hypotenuse is concave, the outer surface is in tension after shrinkage occurs and often cracks. The outer surfaces of convex shaped welds are in compression after shrinkage and usually do not crack. For this reason, convex fillet welds are preferred.

The fillet weld size is defined as the side length (leg) of the largest triangle that can be inscribed within the weld cross section. In most cases, the legs are of

210

Fig. 6.1. Crown Hall, Illinois Institute of Technology campus, Chicago.

equal length. Welds with legs of different lengths are less efficient than those with equal legs (see Fig. 6.3).

Groove welds are used to join the ends of members that align in the same plane. They are commonly used for column splices, butting of beams to columns, and the joining of flanges of plate girders.

If a groove weld is of greater thickness than the members joined, the weld is said to have *reinforcement*. This reinforcement adds a little strength to the weld but has the advantage of ease of application. This reinforcement, however, is not advantageous for connections subjected to fatigue. Tests have shown that stress concentrations develop in the reinforcement and contribute to early failure. For this condition, reinforcement is added and then ground flush with the joined members.

Plug and slot welds are formed in openings in one of two members and are joined to the top of the other member. Occasionally they may be used to "stitch"

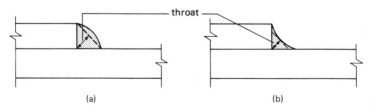

Fig. 6.2. Types of fillet welds: (a) convex fillet weld; (b) concave fillet weld.

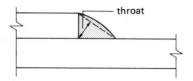

Fig. 6.3. Unequal leg fillet weld.

together cover plates to another rolled section. Another common practice is to add plug welds when the fillet weld is not of adequate length.

6.3 INSPECTION OF WELDS

Failure of welds usually originates at an existing crack or imperfection caused by shrinkage, poor workmanship, or slag inclusion. For this reason, it is advantageous to grind smooth the surface of welds and inspect for imperfections. Inspection of welds for cracks can be done in one of the following ways.

Use of dye penetrants. The surface of the weld, after being cleaned of slag, is painted with a dye dissolved in a penetrant. Due to capillary action, the dye enters the cracks. The surface is then cleaned of excess penetrant, but the dye in the cracks remains, and show up as fine lines.

Magnetic inspection. This method is based on the knowledge that discontinuities disturb the uniformity of magnetic fields. In this method a strong magnet is moved along the weld, and the field generated is studied. Any sudden variation indicates the existence of a flaw.

Magnetic particle testing. This method is also based on the property of cracks in welds to distrub magnetic fields. In this method, iron powder is scattered over the weldment, and a strong magnetic field is generated in the member. Cracks in the weld cause the particles to form a pattern that can be seen.

Radiographic inspection. An x-ray generating source placed on one side of the welded surface exposed a film placed on the other side with an image of

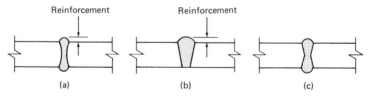

Fig. 6.4. Types of groove welds: (a) square groove weld; (b) single-V groove weld; (c) double-V groove weld.

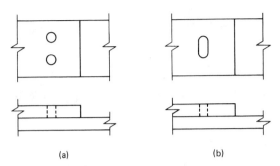

(a)   (b)

Fig. 6.5. Plug and slot welds: (a) plug weld; (b) slot weld.

the weld and any discontinuities in it. After the film is developed, it is studied to determine the quality of the weld and locate any flaws.

Ultrasonic testing. An ultrasonic transducer, which injects ultrasonic pulses into the weldment, is carried along the weld. Reflections of the pulses off the front surface and the back surface of the weld are projected on the screen of a cathode ray tube as peaks from a horizontal base line. The presence of a third peak indicates the existence of a flaw.

Although all five methods can be used in the field and in the shop for the inspection of welds, magnetic and radiographic inspection are seldom used in the field due to the greater amount of equipment and power required.

6.4 ALLOWABLE STRESSES ON WELDS

Previous AISC Specifications listed permissible weld stresses for electrodes and "matching base metal." Base metal is defined as the steel on which welding is done, and, in the case of two strength levels of steel, the weaker of the two is taken into consideration. The AISCS now refers to the current American Welding Society (AWS) recommendations for allowable weld stresses, where various structural steel grades matched with weld metal requirements are tabulated. Currently either E-60 or E-70 electrodes may be used. The number following the "E" indicates the ultimate tensile capacity of the weld metal in ksi. However, no electrodes with properties less than that of E-70 may be used with base metals with yield stress greater than 42 ksi.

AISCS Table 1.5.3 lists allowable weld stress and required strength levels for various types of welds and associated stresses. For fillet welds, the type most commonly used, allowable weld shear stress is limited to 0.30 times the nominal tensile strength of the weld metal. However, the shear stress on the base metal shall not exceed 0.40 times the metal yield stress. Thus, with high weld metal stress and 36-ksi steel, it is possible for the base metal stress to control. Various

design and base metal yield stress combinations are demonstrated in Example 6.1.

NOTE: In all the following examples, it is to be assumed that E70 electrodes are to be used unless noted otherwise.

Example 6.1. Show that for A36 steel and E60 weld electrode, the fillet weld connection is designed according to weld metal stress, and for E70 weld electrode the base metal stress is as critical as the weld metal stress.

Solution. AISC specifies that the base metal stress shall not exceed 0.40 times the metal yield stress (AISCS, Table 1.5.3).

$$F_y = 36.0 \text{ ksi}$$

$$F_v = 0.40 \times 36.0 \text{ ksi} = 14.4 \text{ ksi}$$

AISC also specifies that allowable weld shear stress on effective area is limited to 0.30 times the nominal tensile strength of weld metal. Thus, the allowable stresses are

E60 $0.30 \times 60.0 \text{ ksi} = 18.0 \text{ ksi}$

E70 $0.30 \times 70.0 \text{ ksi} = 21.0 \text{ ksi}$ (AISCS, Table 1.5.3)

For fillet welds, the effective area throat is the shortest distance from the root to the face of the weld. Thus

For E60, $\sin 45° \times 18.0 \text{ ksi} = 12.73 \text{ ksi}$

Allowable base metal stress = $14.4 \text{ ksi} > 12.73 \text{ ksi}$

Design connection for weld metal stress.

For E70, $\sin 45° \times 21.0 \text{ ksi} = 14.85 \text{ ksi}$

Allowable base metal stress = $14.4 \text{ ksi} \approx 14.85 \text{ ksi}$

Because the stresses are nearly equal, it is generally acceptable to design the weld for the weld metal stress. However, it is necessary to investigate the design stresses for uncommon weld and base metal combinations, (i.e., when the weld metal is other than the "matching" for base metal).[1]

---

[1] See American Welding Society Structural Welding Code, AWS D1.1-80, Table 4.1.1.

6.5 EFFECTIVE AND REQUIRED DIMENSIONS OF WELDS

According to AISCS 1.14.6, the area of groove and fillet welds is considered as the effective length of the weld times the effective throat thickness. The effective throat thickness for fillet welds is defined as the shortest distance from the root to the face of the weld. For welds with equal legs, the distance is equal to $\sqrt{2}/2$ or 0.707 times the length of the legs. The effective throat thickness of a complete penetration groove weld is the thickness of the thinner part joined. Reinforcement, if used, is not to be added to the throat thickness. The effective length of the weld is the width of the part joined.

The size of fillet welds shall be within the range of minimum (AISCS 1.17.2) and maximum (AISCS 1.17.3) weld sizes. The minimum weld size is determined by the thicker of the parts joined, as given in Table 6.1. Nevertheless, it need not exceed the thickness of the thinner part unless a larger size is required by calculated stress.

The maximum size of a fillet weld is determined by the edge thickness of the member along which the welding is done. Along edges of material less than $\frac{1}{4}$ in. thick, the maximum size may be equal to the thickness of the material. Along the edges of material $\frac{1}{4}$ in. thick or more, the maximum weld size is $\frac{1}{16}$ in. less than the thickness of the material, unless the weld is built up to obtain full throat thickness (AISCS 1.17.3).

Wherever possible, returns shall be included for a distance not less than twice the nominal size of the weld (AISCS 1.17.7). If returns are used, the effective length of the returns is added to the effective weld length (AISCS 1.14.6.2).

The minimum length of fillet welds shall be not less than four times the nominal size, or else the size of the weld shall be considered not to exceed one-fourth of its effective length (AISCS 1.17.4). For intermittent welds, the length shall in no case be less than $1\frac{1}{2}$ in. (AISCS 1.17.5).

For plug and slot welds, the effective shearing area is the nominal cross-sectional area of the hole or slot in the plane of the faying (connected surface)

Table 6.1. Minimum Size Fillet Weld (AISC Table 1.17.2A, reprinted with permission).

Material Thickness of Thicker Part Joined (Inches)	Minimum[a] Size of Fillet Weld (Inches)
To $\frac{1}{4}$ inclusive	$\frac{1}{8}$
Over $\frac{1}{4}$ to $\frac{1}{2}$	$\frac{3}{16}$
Over $\frac{1}{2}$ to $\frac{3}{4}$	$\frac{1}{4}$
Over $\frac{3}{4}$	$\frac{5}{16}$

[a]Leg dimension of fillet welds.

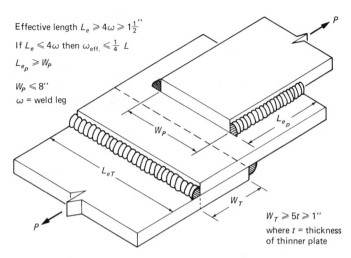

Fig. 6.6. Minimum effective length and overlap of lap joints.

member. Fillet welds in slots or holes are not to be considered as plug or slot welds and must be calculated as fillet welds (AISCS 1.17.8).

Size requirements for plug and slot welds are given in AISCS 1.17.9. For plug welds, the diameter of the hole shall not be less than the thickness of the part containing it plus $\frac{5}{16}$ in., rounded to the next greater odd $\frac{1}{16}$ in., nor shall it be greater than $2\frac{1}{4}$ times the thickness of the weld metal. The minimum center-to-center spacing is four times the diameter of the hole.

The length of a slot weld shall not exceed 10 times the thickness of the weld. The minimum width of the slot is the thickness of the part containing it plus $\frac{5}{16}$ in., rounded to the next greater odd $\frac{1}{16}$ in., nor shall it be greater than $2\frac{1}{4}$ times the thickness of the weld. The ends of the slot shall have the corners rounded to a radius not less than the thickness of the part containing it. The minimum spacing of slot welds transverse to their length shall be four times the width of the slot. Minimum center-to-center spacing in the longitudinal direction shall be twice the length of the slot.

In material $\frac{5}{8}$ in. or less in thickness, the weld thickness for plug or slot welds shall be the same as the material. In material over $\frac{5}{8}$ in. thick, the weld must be at least one-half the thickness of the material, but not less than $\frac{5}{8}$ in.

Welds may be used in combination with bolts, provided the bolts are of the friction type (AISCS 1.15.10) and are installed prior to welding.

6.6 DESIGN OF SIMPLE WELDS

Fillet welds are usually designed such that the shear on the effective weld area is less than that allowed by the AISCS 1.5.3. Once the type and size of weld to

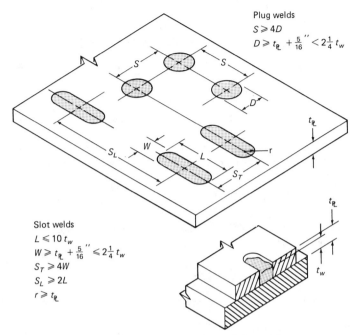

Fig. 6.7. Requirements for plug and slot welds.

be used have been determined, the required weld length is determined by

$$l_w = \frac{P}{W_c} \tag{6.1}$$

where P is the design load and W_c is the weld capacity per inch of weld. The weld length is taken to the next greater $\frac{1}{4}$ in. The allowable shearing stresses for various electrodes and base metals are given in AISC Table 1.5.3. Since the stress is calculated through the effective throat for the length of the weld, the weld strength per inch of length is determined as the allowable stress times the effective throat. Allowable stresses for welds are given in AISC Table 1.5.3 (see Section 6.4, *Allowable Stresses on Welds*).

The weld strength is usually calculated for 1 in. length of $\frac{1}{16}$-in. fillet weld. Thus the stress value of various welds is determined by multiplying the weld stress per sixteenth by the number of sixteenths. Returns should be added to fillet welds when feasible. In some cases, it is easier to continuously weld from one corner to the other. In any case, the effective weld length is the sum of all the weld lengths, including returns (AISCS 1.14.6).

Example 6.2. Determine the strength of a $\frac{5}{16}$-in. fillet weld.

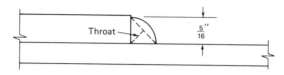

Solution. The allowable stress of E70 electrodes is 0.30 the nominal tensile strength of the weld metal (AISC, Table 1.5.3).

$$\text{Allowable weld stress} = 0.30 \ (70 \text{ ksi}) = 21.0 \text{ ksi}$$

The weld strength per inch (W_c) is the allowable stress times the effective throat. For a $\frac{5}{16}$-in. weld, the throat is

$$(\sin 45°) \times (\tfrac{5}{16} \text{ in.}) = 0.707 \times \tfrac{5}{16} \text{ in.} = 0.221 \text{ in.}$$

$$W_c = \text{Allowable weld stress} \times t_{\text{throat}} = 21.0 \text{ ksi} \times 0.221 \text{ in.} = 4.64 \text{ k/in.}$$

The effective throat thickness for equal leg welds per sixteenth inch of weld is

$$(\tfrac{1}{16} \text{ in.}) \times (0.707) = 0.0442 \text{ in.}$$

Multiplying the throat by the allowable stress gives the weld strength per inch per sixteenth. For E70 electrodes, the allowable stress is

$$0.0442 \text{ in.} \times 21.0 \text{ ksi} = 0.928 \text{ k/in./sixteenth inch}$$

From this, the weld strength can be determined by multiplying the number of sixteenths in the weld leg. As above

$$5 \text{ sixteenths} \times 0.928 \text{ k/in./sixteenth} = 4.64 \text{ k/in.}$$

Example 6.3. A $\frac{5}{16}$-in. fillet weld is used to connect two plates as shown. Determine the maximum load that can be applied to the connection.

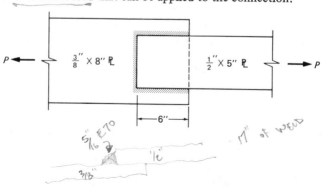

Solution. For welding

$$P_{max} = l_{weld} \times W_{capacity}$$

For connected members

$$P_{max} = A_g \times 0.6\,F_y \quad \text{or}$$

$$P_{max} = A_n \times 0.5\,F_u \quad \text{(AISCS 1.5.1)}$$

Since $A_n = A_g$ and $0.5\,F_u > 0.6\,F_y$, use

$$P_{max} = A \times 0.6\,F_y$$

Using the general relationship shown in Example 6.1

$$W_c = 5 \text{ sixteenths} \times 0.928 \text{ k/in./sixteenth} = 4.64 \text{ k/in.}$$

$$P_{max} = l_w \times W_c = (6 \text{ in.} + 5 \text{ in.} + 6 \text{ in.}) \times 4.64 \text{ k/in.} = 78.9 \text{ k}$$

For $\frac{1}{2}$ in. $\times$ 5-in. bar

$$P_{max} = A \times 0.6\,F_y = (\tfrac{1}{2} \text{ in.} \times 5 \text{ in.}) \times (22.0 \text{ ksi}) = 55.0 \text{ k}$$

For $\frac{3}{8}$-in. $\times$ 8-in. bar

$$P_{max} = A \times 0.6\,F_y = (\tfrac{3}{8} \text{ in.} \times 8 \text{ in.}) \times (22.0 \text{ ksi}) = 66.0 \text{ k}$$

The connections capacity is 55.0 kips resulting from the $\frac{1}{2}$ in. $\times$ 5 in. member allowable stress.

Example 6.4. For the members shown, determine the minimum fillet length to develop full tensile strength

a) without end returns
b) with end returns
c) continuously welded.

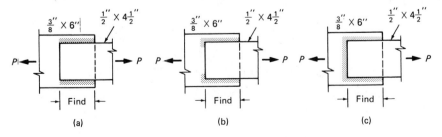

(a)   (b)   (c)

Solution. For welding

$$P_{max} = l_w \times W_c$$

For connected members

$$P_{max} = A_{net} \times 0.6\,F_y \qquad \text{(AISC 1.5.1)}$$

Determine the tensile capacity of the members and required weld length to carry the load.

$$P_{max} = A_g \times 0.6\,F_y = 2.25 \text{ in.}^2 \times 22.0 \text{ ksi} = 49.5 \text{ kips}$$
$$l_w = P_{max}/W_c$$

To minimize the length of the weld, use the maximum weld size, whose size is $\frac{1}{16}$ in. less than the thickness of the material whose edge is being welded (AISCS 1.17.3).

$$\text{Maximum weld size} = \frac{1}{2} \text{ in.} - \frac{1}{16} \text{ in.} = \frac{7}{16} \text{ in.}$$

$$W_c = 7 \text{ sixteenths} \times 0.928 \text{ k/in./sixteenth} = 6.5 \text{ k/in.}$$

$$l_w = \frac{49.5 \text{ k}}{6.5 \text{ k/in.}} = 7.6 \text{ in.}$$

a) For fillets without end returns, half of the required weld length will be placed on each longitudinal side. Each side must be longer than the perpendicular distance between them (AISCS 1.17.4).

$$l_{w_2} = \frac{7.6 \text{ in.}}{2} = 3.8 \text{ in.} < 4.5 \text{ in.}$$

Use $l_w = 4\frac{1}{2}$ in. with $\frac{7}{16}$-in. longitudinal weld.

b) End returns shall be no less than twice the nominal size of the weld (AISCS 1.17.7).

$$(2 \times l_w) + 2 \times (2 \times \tfrac{7}{16} \text{ in.}) \geqslant 7.60 \text{ in.}$$

Use $l_w = 3$ in. with $\frac{7}{16}$-in. welds with $\frac{7}{8}$-in. end returns (minimum).

c) For continuous welds, the effective length is the overall length of all full-size fillets (AISCS 1.14.6.2).

Use $\frac{7}{16}$-in. weld for a length of 2 in. on each side, with the far side continuously welded.

Example 6.5. Determine the allowable tensile capacity of the connection shown. Use a $\frac{3}{8}$-in. fillet weld with E60 electrodes.

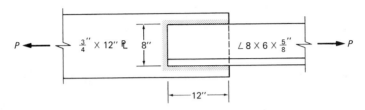

Solution. For welding

$$P_{max} = l_w \times W_c$$

For the angle

$$P_{max} = A_{max} \times F_t, \qquad F_t = 0.6\, F_y$$

For E60 welds, the capacity per sixteenth inch is

$$\tfrac{60}{70}\ (0.928 \text{ k/in./sixteenth}) = 0.795 \text{ k/in./sixteenth}$$

$$W_c = 6 \times 0.795 \text{ k/in./sixteenth} = 4.77 \text{ k/in.}$$

$$l_w = (12 \text{ in.} + 8 \text{ in.} + 12 \text{ in.}) = 32 \text{ in.}$$

$$P_{max} = 4.77 \text{ k/in.} \times 32 \text{ in.} = 152.6 \text{ k} \quad \text{WELD}$$

$$A_{pl} = 0.75 \text{ in.} \times 12 \text{ in.} = 7.5 \text{ in.}^2 \quad 9$$

$$A_{angle} = 8.36 \text{ in.}^2$$

$$P_{max} = A_{pl} \times F_t = 7.5 \text{ in.}^2 \times (22.0 \text{ ksi}) = 165 \text{ k} \quad \text{STEEL}$$

As the weld capacity is less than the steel capacity, the allowable tensile capacity is 152.6 kips.

At times, combinations of welds may be necessary to develop sufficient strength. For weld combinations, the effective capacity of each type shall be separately computed, with reference to the axis of the group, in order to determine the allowable capacity of the combination (AISCS 1.15.9).

Example 6.6. A 12×30 channel is to be connected to another member using $\frac{7}{16}$-in. E70 fillet welds. Due to clearance limitations, the two members can overlap only 4 in. as shown. Determine the distance l so the connection capacity is 170 kips.

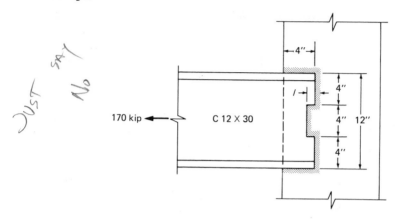

Solution.

$$l_w = P_{max}/W_c$$

$$W_c = 7 \times 0.928 \text{ k/in.}/16\text{th} = 6.5 \text{ k/in.}$$

$$l_w = \frac{170 \text{ k}}{6.5 \text{ k/in.}} = 26.15 \text{ in.}$$

The maximum length outside the slot is

$$(4 \text{ in.} + 4 \text{ in.} + 4 \text{ in.} + 4 \text{ in.}) = 16.0 \text{ in.}$$

The fillet weld length required in the slot is

$$26.15 \text{ in.} - 16.0 \text{ in.} = 10.15 \text{ in.} \quad \text{say } 10\tfrac{1}{4} \text{ in.}$$

$$\text{slot} = 4 \text{ in.} + (21) = 10\tfrac{1}{4} \text{ in.}$$

$$l = 3\tfrac{1}{8} \text{ in.} \quad \text{use } 3\tfrac{1}{4} \text{ in.}$$

Example 6.7. Redesign the channel connection in Example 6.6 as a combination fillet and slot weld to support the same applied load.

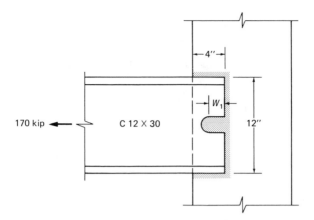

Solution. Because the web thickness is $\frac{1}{2}$ in., the maximum fillet weld size is $\frac{7}{16}$ in. (AISCS 1.17.3).

$$W_c = 7 \times 0.928 \text{ k/in.}/16\text{th} = 6.5 \text{ k/in.}$$

Allowable slot weld stress = 0.30×70.0 ksi = 21.0 ksi (AISCS, Table 1.5.3).

Assume the width of the slot weld to be 1 in.

Capacity of fillet weld

$$P_{\max} = 6.5 \text{ k/in.} \times (4 \text{ in.} + 12 \text{ in.} - 1 \text{ in.} + 4 \text{ in.}) = 123.5 \text{ k}$$

Balance to be resisted by slot weld

$$P_{bal} = 170.0 \text{ k} - 123.5 \text{ k} = 46.5 \text{ k}$$

Design of slot weld

The thickness of slot welds in material $\frac{5}{8}$ in. or less in thickness shall be equal to the thickness of the material (AISCS 1.17.9).

Min. width of slot weld = $\frac{1}{2}$ in. + $\frac{5}{16}$ in. = $\frac{13}{16}$ in.   (AISCS 1.17.9)

Max. width of slot weld = $2\frac{1}{4} \times \frac{1}{2}$ in. = $1\frac{1}{8}$ in.

Max. length of slot weld = $10 \times \frac{1}{2}$ in. = 5 in.

Effective shearing area = width $\times$ length   (AISCS 1.14.6.3)

Weld thickness = $\frac{1}{2}$ in. (thickness of material)   (AISCS 1.17.9)

Try 1-in.-wide slot weld

$$l_w = \frac{46.5 \text{ k}}{1.0 \text{ in.}} \bigg/ 21.0 \text{ ksi} = 2.21 \text{ in.} \quad \text{say } 2\frac{1}{4} \text{ in.}$$

Try $\frac{13}{16}$ in. wide slot weld

$$l_w = \frac{46.5 \text{ k}}{\dfrac{13}{16} \text{ in.}} \bigg/ 21.0 \text{ ksi} = 2.73 \text{ in.} \quad \text{say } 2\frac{3}{4} \text{ in.}$$

Try $1\frac{1}{8}$-in.-wide slot weld

$$l_w = \frac{46.5 \text{ k}}{\dfrac{18}{16} \text{ in.}} \bigg/ 21.0 \text{ ksi} = 1.97 \text{ in.} \quad \text{say } 2 \text{ in.}$$

Use any slot weld above in combination with $\frac{7}{16}$-in. fillet weld.

When calculated stress is less than the minimum size weld, intermittent fillet welds may be used. The effective length of any segment of fillet weld shall be not less than four times the weld size nor $1\frac{1}{2}$ in. (AISCS 1.17.5)

4·(WELDSIZE) < INT WELD > 1½ in

Example 6.8. A welded plate girder consisting of $1\text{-}\frac{1}{2}$ in. $\times$ 14 in. flanges and $\frac{1}{2}$ in. $\times$ 42 in. web is subject to a maximum shear of 160 kips. Determine the fillet weld size for a continuous weld and an intermittent weld.

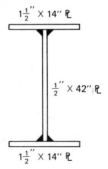

Solution. The stress in the weld is found by the shear flow at the intersection of the web and flanges. The required weld size is the shear flow divided by the allowable weld stress per inch per sixteenth inch weld leg.

$$q_v = \frac{VQ}{I}$$

$V = 160$ kips

$Q = A\bar{y} = (1.5 \text{ in.} \times 14 \text{ in.}) \times 21.75 \text{ in.} = 456.8 \text{ in.}^3$

$I = \dfrac{0.5 \text{ in.} \times (42 \text{ in.})^3}{12} + 2 \times [(1.5 \text{ in.} \times 14 \text{ in.}) \times (21.75 \text{ in.})^2]$

$\quad = 22960 \text{ in.}^4$

$q_v = \dfrac{(160 \text{ k} \times 456.8 \text{ in.}^3)}{22960 \text{ in.}^4} = 3.18 \text{ k/in.}$

$W_c = 0.928$ k/in./sixteenth weld leg

$D = \dfrac{3.18 \text{ k/in.}}{2 \times 0.928 \text{ k/in.}} = 1.72$ sixteenths $\quad$ use $\dfrac{1}{8}$ in. weld

The minimum weld size for a thickness of $\frac{3}{4}$ in. or more (thicker part joined) is $\frac{5}{16}$ in. (AISCS 1.17.2).

Use $\frac{5}{16}$-in. E70 continuous weld.

If intermittent fillet weld is to be used, assuming $\frac{3}{8}$-in. weld.

$$W_c = 6 \times 0.928 \text{ k/in./sixteenth} = 5.57 \text{ k/in.}$$

$$q_v = 3.18 \text{ k/in.} \times 12 \text{ in./ft} = 38.15 \text{ k/ft}$$

$$= \frac{38.15 \text{ k/ft}}{2 \times 5.57 \text{ k/in.}} = 3.42 \text{ in./ft} \quad \text{say } 3\frac{1}{2} \text{ in.}$$

Use $\frac{3}{8}$-in. weld $3\frac{1}{2}$ in. long, 12 in. on center.

NOTE: The use of intermittent fillet weld is acceptable for quasi-static loading. In case of fatigue loading—which is beyond the scope of our study here—intermittent fillet welds become for all practical purposes unusable.

Example 6.9. A W 24 × 104 has 1 in. × 10 in. cover plates attached to the flanges by $\frac{5}{16}$ in. E70 fillet welds as shown. Determine the length of connecting welds for a location where a 110-kip shear occurs.

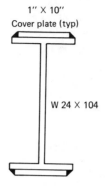

1" X 10"
Cover plate (typ)

W 24 X 104

Solution. The shear flow at the connection of cover plates is

$$q_v = \frac{VQ}{I}$$

where V = 110 kips

$$Q = A\bar{y} = (1 \text{ in.} \times 10 \text{ in.}) \times 12.53 \text{ in.} = 125.3 \text{ in.}^3$$

$$I = 3100 \text{ in.}^4 + 2 \times (1 \text{ in.} \times 10 \text{ in.}) \times (12.53 \text{ in.})^2 = 6240 \text{ in.}^4$$

$$q_v = \frac{110 \text{ k} \times 125.3 \text{ in.}^3}{6240 \text{ in.}^4} = 2.21 \text{ k/in.}$$

$$W_c = 2 \times (5 \times 0.928 \text{ k/in./sixteenth}) = 9.28 \text{ k/in.}$$

$$l_w = \frac{2.21 \text{ k/in.} \times 12 \text{ in./ft}}{9.28 \text{ k/in.}} = 2.86 \text{ in./ft} \qquad \text{say 3 in. at 12-in. spacing}$$

Use $\frac{5}{16}$-in. E70 welds 3 in. long, 12 in. on center. See note, Example 6.8.

Example 6.10. Design welded partial length cover plates for the loading shown using a W 24 × 76 section. Assume full lateral support.

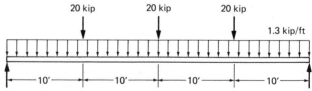

20 kip   20 kip   20 kip

1.3 kip/ft

|← 10' →|← 10' →|← 10' →|← 10' →|

Solution. Wide-flange beam properties

W 24 × 76
$$A = 22.4 \text{ in.}^2$$

$$d = 23.92 \text{ in.}$$
$$b_f = 8.99 \text{ in.}$$
$$I = 2100 \text{ in.}^4$$
$$S = 176 \text{ in.}^3$$

Shear and moment diagrams

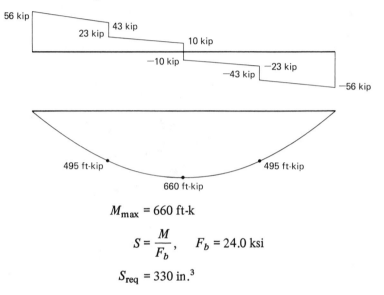

$$M_{max} = 660 \text{ ft-k}$$

$$S = \frac{M}{F_b}, \quad F_b = 24.0 \text{ ksi}$$

$$S_{req} = 330 \text{ in.}^3$$

Available capacity of W 24 × 76

$$S_{WF} = 176 \text{ in.}^3$$

Additional capacity required

$$S_{pl} = 330 \text{ in.}^3 - 176 \text{ in.}^3 = 154 \text{ in.}^3$$

Determine the moment of inertia of the entire section to determine required plate size.

$$I = I_{WF} + 2\, A_{pl}\, d^2 = 2100 \text{ in.}^4 + 2\, (bt) \left(\frac{(23.92 \text{ in.} + t)}{2} \right)^2$$

$$= 2100 \text{ in.}^4 + \frac{1}{2}\, (23.92 \text{ in.} + t)^2\, (bt)$$

$$S = \frac{I}{c} = \frac{4200 \text{ in.}^4 + (23.92 \text{ in.} + t)^2\, (bt)}{23.92 \text{ in.} + 2t} = 330 \text{ in.}^3$$

Selecting $t = 1$ in., $b = 7$ in., S becomes 329.7 in.^{3}

To determine the theoretical cutoff points, superimpose the moment capacity and moment diagrams.

Moment capacity of beam without cover plates

$$M_{WF} = \frac{S_{WF} \times F_b}{12 \text{ in./ft}} = \frac{176 \text{ in.}^3 \times 24 \text{ ksi}}{12 \text{ in./ft}} = 352 \text{ ft-k}$$

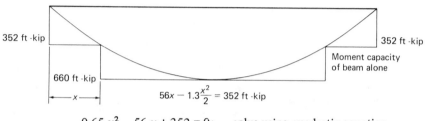

352 ft-kip   352 ft-kip

Moment capacity of beam alone

660 ft-kip

$56x - 1.3\frac{x^2}{2} = 352$ ft-kip

$0.65\, x^2 - 56\, x + 352 = 0;$ solve using quadratic equation

$$x = \frac{28 \pm \sqrt{28^2 - (.65 \times 352)}}{.65} = 6.83 \text{ ft}$$

Determine welds

Maximum shear flow at cover plates occurs at cutoff points.

$V = 56 \text{ k} - (6.83 \text{ ft} \times 1.3 \text{ k/ft}) = 47.12 \text{ k}$

$I = I_0 + Ad^2 = 2100 \text{ in.}^4 + 2(1 \text{ in.} \times 7 \text{ in.}) \left(\frac{23.93 \text{ in.}}{2} + \frac{1 \text{ in.}}{2} \right)^2 = 4274 \text{ in.}^4$

$Q = A_{pl}\, d = (1 \text{ in.} \times 7 \text{ in.}) \left(\frac{23.92 \text{ in.}}{2} + \frac{1 \text{ in.}}{2} \right) = 87.22 \text{ in.}^3$

$q = \frac{VQ}{I} = \frac{47.12 \text{ k} \times 87.22 \text{ in.}^3}{4274 \text{ in.}^4} = 0.96 \text{ k/in.}$

Use minimum intermittent fillet weld. For 1-in. cover plates, the minimum size fillet weld $= \frac{5}{16}$ in. (AISCS 1.17.2). See note, Example 6.8.

$$W_c = 0.928 \text{ k/in./sixteenth} \times 5 \text{ sixteenths} \times 2 \text{ sides} = 9.28 \text{ k/in.}$$

Force to be carried per foot $= 0.96 \text{ k/in.} \times 12 \text{ in./ft} = 11.54 \text{ k/ft}$

$$\frac{11.54 \text{ k/ft}}{9.28 \text{ k/in.}} = 1.24 \text{ in./ft}$$

Minimum weld length = 4 × weld size but not less than $1\frac{1}{2}$ in.

$4 \times \dfrac{5}{16}$ in. $= 1\dfrac{1}{4}$ in.   use $1\dfrac{1}{2}$ in. intermittent welds   (AISCS 1.17.5)

Partial length cover plates must be extended past the theoretical cutoff point for a distance capable of developing the cover plate's portion of flexural stress in the beam at the theoretical cutoff point (AISCS 1.10.4). In addition, welds connecting cover plates must be continuous for a distance dependent on plate width and weld type and must be capable of developing the flexural stresses in that length.

The AISC Commentary shows that the force to be developed by fasteners in the cover plate extension is

$$H = \frac{MQ}{I}$$

$$H = \frac{MQ}{I} = \frac{352 \text{ ft-k} \times 12 \text{ in./ft} \times 87.22 \text{ in.}^3}{4274 \text{ in.}^4} = 86.20 \text{ k}$$

Design welds in plate extension to be continuous along both edges of cover plate, but no weld across end of plate.

$$a' = 2 \times b = 14 \text{ in.}   \text{(AISCS 1.10.4.3)}$$

Weld capacity = 2 × 14 in. × (5 × 0.928 k/in.) = 129.9 k > 86.2 k

Plate extension is adequate. Use partial cover plates as shown in sketch.

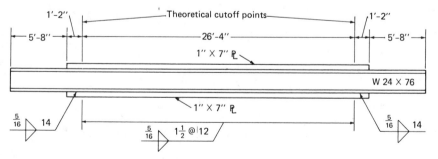

Example 6.11. For the continuous beam shown, choose a wide-flange beam to resist positive bending, and design a welded partial cover plate where required for negative bending. Assume full lateral support. In the analysis, neglect the additional stiffness created by the cover plate.

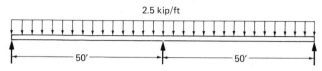

2.5 kip/ft

50′ 50′

Solution. Shear and moment diagrams

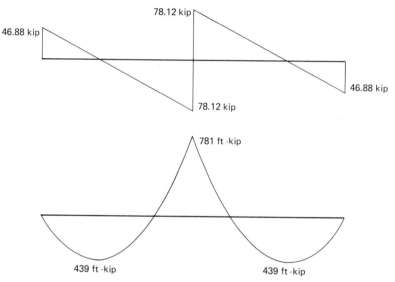

46.88 kip

78.12 kip

78.12 kip

46.88 kip

781 ft -kip

439 ft -kip 439 ft -kip

Selection of wide flange

Maximum positive moment = 439 ft-k

Here the effect of redistribution of moments has *not* been used. See Chapter 3 for discussion of redistribution (AISCS 1.5.1.4.1).

Select W 24 × 94 (M_{max} = 444 ft-k)

$$A = 27.7 \text{ in.}^2$$
$$d = 24.31 \text{ in.}$$
$$b_f = 9.065 \text{ in.}$$
$$I = 2700 \text{ in.}^4$$
$$S = 222 \text{ in.}^3$$

Total section modulus required

Maximum moment = 781 ft-k

$$S = \frac{M}{F_b} = 391 \text{ in.}^3 = \frac{I}{c}$$

$$I = 2700 \text{ in.}^4 + 2(bt)\left(\frac{(24.31 \text{ in.} + t)}{2}\right)^2$$

$$S = \frac{2700 \text{ in.}^4 + \frac{1}{2} bt (24.31 \text{ in.} + t)^2}{\frac{24.31 \text{ in.}}{2} + t} = 391 \text{ in.}^3$$

Try $t = 1\frac{3}{8}$ in.

$$b = \frac{(12.155 \text{ in.} + t) \times 391 \text{ in.}^3 - 2700 \text{ in.}^4}{\frac{1}{2} t (24.31 \text{ in.} + t)^2} = 5.71 \text{ in.} \qquad \text{say 6 in.}$$

$$I = 2700 \text{ in.}^4 + 2(6 \text{ in.} \times 1.375 \text{ in.}) \times \left(\frac{24.31 \text{ in.}}{2} + \frac{1.375 \text{ in.}}{2}\right)^2$$

$$= 5421 \text{ in.}^4$$

$$S = 400 \text{ in.}^3$$

To determine theoretical cutoff point, assume that beam has constant moment of inertia. Actual beam behavior is such that a slight difference in moments will occur with varying I.

$$1.25 x^2 - 46.88 x - 444 = 0; \quad \text{solve using quadratic equation}$$

$$x = \frac{23.44 + \sqrt{23.44^2 + (1.25 \times 444)}}{1.25} = 45.34 \text{ ft}$$

Maximum shear flow at cover plates occurs at support.

$$V = 78.1 \text{ k}$$

$$Q = A_{pl} d = (1.375 \text{ in.} \times 6 \text{ in.})\left(\frac{24.31 \text{ in.}}{2} + \frac{1.375 \text{ in.}}{2}\right) = 106.0 \text{ in.}^3$$

$$q = \frac{VQ}{I} = \frac{78.1 \text{ k} \times 106.0 \text{ in.}^3}{5421 \text{ in.}^4} = 1.53 \text{ k/in.}$$

Minimum fillet size = $\frac{5}{16}$ in. (AISCS 1.17.2).

$$W_c = 0.928 \text{ k/in./sixteenth} \times 5 \text{ sixteenths} \times 2 \text{ sides} = 9.28 \text{ k/in.}$$

Force to be carried per foot = 1.53 k/in. $\times$ 12 in./ft = 18.36 k/ft

$$\frac{18.36 \text{ k/ft}}{9.28 \text{ k/in.}} = 1.98 \text{ in.} \quad \text{say 2 in. length of intermittent weld}$$

$$\text{Minimum intermittent weld length} = 1\frac{1}{2} \text{ in.} < 2 \text{ in.} \quad \text{ok}$$

Plate extension (AISCS 1.10.4.2)

Use continuous welds along both edges of cover plate and across end of plate.

$$a' = 1.5 \times b = 9 \text{ in.}$$

$$\text{Weld capacity} = [(2 \times 9 \text{ in.}) + 6 \text{ in.}] \times (5 \times 0.928 \text{ k/in.}) = 111.4 \text{ k}$$

Required force

$$\frac{MQ}{I} = \frac{444 \text{ ft-k} \times 12 \text{ in./ft} \times 106 \text{ in.}^3}{5421 \text{ in.}^4} = 104 \text{ k} < 111.4 \text{ k} \quad \text{ok}$$

All full penetration, groove and butt welds are designed by the capacity of the base metal as given in AISC Table 1.5.3.

Example 6.12. Determine the strength of the full penetration groove welds used to connect the plates shown below in shear.

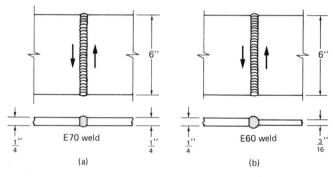

Solution. For welding

$$P_{max} = A_{weld} \times F_{weld}$$

For connected members

$$P_{max} = A_{stl} \times F_v$$

The effective area of groove welds shall be considered as the effective length of the weld times the effective throat thickness. The effective length of a groove weld shall be the width of the part joined (AISCS 1.14.6.1, Table 1.5.3).

$$F_{weld} = 0.30 \times 70.0 \text{ ksi} = 21.0 \text{ ksi}$$

a) $A_{weld} = 6.0 \text{ in.} \times \frac{1}{4} \text{ in.} = 1.5 \text{ in.}^2$

$P_{max} = 1.5 \text{ in.}^2 \times 21.0 \text{ ksi} = 31.5 \text{ k}$

$A_{stl} = 6.0 \text{ in.} \times \frac{1}{4} \text{ in.} = 1.5 \text{ in.}^2$

$P_{max} = 1.5 \text{ in.}^2 \times (0.40 \times 36.0 \text{ ksi}) = 21.6 \text{ k}$

The weld capacity is 31.5 kips. However, the connection capacity is the base metal allowable stress in shear, 21.6 kips.

b) $A_{weld} = 6.0 \text{ in.} \times \frac{3}{16} \text{ in.} = 1.125 \text{ in.}^2$

$P_{max} = 1.125 \text{ in.}^2 \times 18.0 \text{ ksi} = 20.25 \text{ k}$

$A_{stl} = 6.0 \text{ in.} \times \frac{3}{16} \text{ in.} = 1.125 \text{ in.}^2$ (least thickness member)

$P_{max} = 1.125 \text{ in.}^2 \times (0.40 \times 36.0 \text{ ksi}) = 16.2 \text{ k}$

The weld capacity is 20.25 kips. However, the connection capacity is the base metal allowable stress in shear, 16.2 kips.

6.7 REPEATED STRESSES (FATIGUE)

When a weld is subjected to repeated stresses, the centroids of the member and weld shall coincide (AISCS 1.15.2 and 1.15.3). For nonsymmetrical members, unless provision is made for eccentricity, welds of different lengths are necessary.

Example 6.13. One leg of a $6 \times 6 \times \frac{5}{8}$-in. angle is to be connected with side welds and a weld at the end of the angle to a plate behind. Design the connection to develop full capacity of the angle. Balance the fillet welds around the center of gravity of the angle, using the maximum size weld and E60 electrodes.

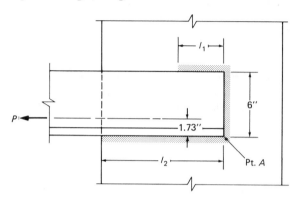

Solution. To have the shortest connection, select the maximum size weld.

$$\text{Maximum weld size} = \frac{5}{8} \text{ in.} - \frac{1}{16} \text{ in.} = \frac{9}{16} \text{ in.} \quad \text{(AISCS 1.17.3)}$$

$$P_{max} = A_{mem} \times 0.60\,F_y = 7.11 \text{ in.}^2 \times 22.0 \text{ ksi} = 156.4 \text{ k}$$

$$W_c = 9 \times 0.795 \text{ k/in./sixteenth} = 7.16 \text{ k/in.}$$

$$l_w = \frac{156.4 \text{ k}}{7.16 \text{ k/in.}} = 21.84 \text{ in.}$$

Summing moments of weld fillets about point A

$$1.73 \text{ in.} \times 21.84 \text{ in.} = 6 \text{ in.} (l_1) + 3 \text{ in.} (6 \text{ in.})$$

$$l_1 = (37.78 - 18.0)/6 = 3.3 \text{ in.} \qquad \text{say } 3\tfrac{1}{2} \text{ in.}$$

$$l_2 = 21.84 \text{ in.} - 6 \text{ in.} - 3.3 \text{ in.} = 12.54 \text{ in.} \qquad \text{say } 12\tfrac{1}{2} \text{ in.}$$

NOTE: This case is typical for a connection with repeated variations of stresses. Though desirable, it is not required to design for eccentricity between the gravity axes in statically loaded members, but is required in members subject to fatigue loading (AISCS 1.15.3).

Example 6.14. A $7 \times 4 \times \frac{3}{8}$-in. angle is to be connected to develop full capacity as shown to plate using $\frac{5}{16}$-in. welds. Determine l_1 and l_2 so the centers of gravity for the welds and angle coincide.

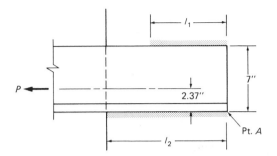

Solution.

$$P_{max} = A_{mem} \times F_t = 3.98 \text{ in.}^2 \times 22.0 \text{ ksi} = 87.56 \text{ k}$$

$$W_c = 5 \times 0.928 \text{ k/in.} = 4.64 \text{ k/in.}$$

$$l_w = 87.56 \text{ k/4.64 k/in.} = 18.87 \text{ in.}$$

Summing moments of weld fillets about point A

$$2.37 \text{ in.} \times 18.87 \text{ in.} = 7.0 \text{ in.} \times l_1$$

$$l_1 = 44.72 \text{ in.}^2 /7.0 \text{ in.} = 6.39 \text{ in.} \qquad \text{say } 6\tfrac{1}{2} \text{ in.}$$

$$l_2 = 18.87 \text{ in.} - 6\tfrac{1}{2} \text{ in.} = 12.37 \text{ in.} \qquad \text{say } 12\tfrac{1}{2} \text{ in.}$$

Lengths of welds as determined may be used. However, to satisfy AISC end return requirements, see the following example.

Example 6.15. Same as Example 6.14, but include end returns.

Solution. Side or end fillet welds terminating at ends or sides, respectively, of parts or members shall, wherever practicable, be returned continuously around the corners for a distance not less than two times the nominal size of the weld (AISCS 1.17.7).

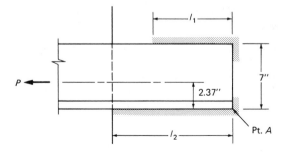

Solution.

$P = 87.56\,k$

$l_w = 18.87$ in.

Minimum end return $= 2 \times \frac{5}{16}$ in. $= \frac{5}{8}$ in.   (AISC 1.17.7)

$l_1 + l_2 + (2 \times \frac{5}{8}$ in.$) = 18.87$ in.

$l_1 + l_2 = 17.62$ in.

Summing moments about point A

2.37 in. $\times 18.87$ in. $= \frac{5}{16}$ in. $(\frac{5}{8}$ in.$) + (7$ in. $- \frac{5}{16}$ in.$)(\frac{5}{8}$ in.$) + 7(l_1)$

$l_1 = (44.72 - 0.20 - 4.18)/7 = 5.76$ in. $5\frac{3}{4}$ in.

$l_2 = 17.62$ in. $- 5\frac{3}{4}$ in. $= 11.87$ say 12 in.

6.8 ECCENTRIC LOADING

When eccentric loads are carried by welds, either shear and torsion or shear and bending may result in the weld (see Fig. 6.8). The force per unit of length is determined by

$$f = \frac{Td}{J} \tag{6.2}$$

where T is the torque, d is the distance from the center of gravity of the weld group to the point in consideration and J is its polar moment of inertia. The polar moment of inertia is given as the sum of the moment of inertia in the x and y directions, i.e., $J = I_x + I_y$. The direction of the force is perpendicular to the line connecting the point and the center of gravity of the weld group. For ease of computation, the distance d is usually reduced to its horizontal (h) and vertical (v) components.

If the horizontal and vertical components of the force applied to the centroid are, respectively, P_H and P_V, the moment M and the total length of weld l in inches, the stresses due to the direct force are

$$f_{h,P} = \frac{P_H}{l} \quad f_{v,P} = \frac{P_V}{l} \tag{6.3}$$

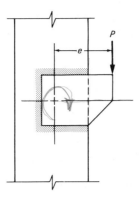

Fig. 6.8. Shear and torsion condition in welds.

Stresses due to the moment are

$$f_{h,M} = \frac{M_v}{J} \quad f_{v,M} = \frac{M_h}{J}$$ (6.4)

The resultant shear force is then computed by taking the square root of the squares of the components,

$$f = \sqrt{(f_{h,P} + f_{h,M})^2 + (f_{v,P} + f_{v,M})^2}$$ (6.5)

When computing the weld size, a trial using a 1-inch weld is used so that the weld area is easily computed. The weld size required is then determined by dividing the resultant shear force by the allowable shear stress of the weld.

Example 6.16. The weld shown is to carry an eccentric load of 5 kips acting 10 in. from the weld. Determine the minimum weld size to carry the load, using E70 electrodes.

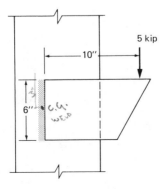

Solution. The weld shown is subject to both shear and torsion. The force caused by torsion can be computed from

$$f = \frac{Td}{J}$$

Assuming a 1-in. weld, in this case

$$J = I_{x\ \text{weld}} = \frac{1\ (6)^3}{12} = 18\ \text{in.}^4$$

$$T = 10\ \text{in.} \times 5\ \text{k} = 50\ \text{in.-k}$$

The resultant force f_r will be computed from its horizontal components f_h and f_v.

The most stressed point of the weld is the greatest distance from the weld center of gravity.

$$f_h = \frac{50\ \text{in.-k} \times 3\ \text{in.}}{18\ \text{in.}^4} = 8.33\ \text{k/in.}^2$$

$$f_v = \frac{5\ \text{k}}{6\ \text{in.}^2} = 0.83\ \text{k/in.}^2$$

$$f_r = \sqrt{(0.83)^2 + (8.33)^2} = 8.37\ \text{k/in.}^2$$

For a 1-in. weld

$$W_c = 16 \times 0.928\ \text{k/in.} = 14.85\ \text{k/in.}$$

$$W_{\text{size}} = \frac{8.37\ \text{k/in.}^2}{14.85\ \text{k/in.}} = 0.56\ \text{in.} \quad \text{use } \frac{9}{16}\ \text{in. weld}$$

Example 6.17. Determine the minimum weld size to resist the load shown.

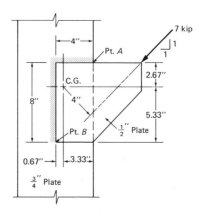

Solution. The weld shown is subject to both shear and torsion. Assuming a 1-in. weld,

$$f = \frac{Td}{J}, \quad T = 7 \text{ k} \times 4.0 \text{ in.} = 28 \text{ in.-k}$$

$$J = I_x + I_y$$

$$J = \frac{(2.67 \text{ in.})^3}{3} + \frac{(5.33 \text{ in.})^3}{3} + (4 \text{ in.}^2 \times (2.67 \text{ in.})^2) + \frac{(0.67 \text{ in.})^3}{3}$$

$$+ \frac{(3.33 \text{ in.})^3}{3} + (8 \text{ in.}^2 \times (0.67 \text{ in.})^2) = 101.33 \text{ in.}^4$$

The shearing stress of the applied force is broken down into horizontal and vertical components and combined with the torsional components.

Direct shear stress

$$f_v = \frac{0.707 \times 7 \text{ k}}{1 \text{ in.} \times (4 \text{ in.} + 8 \text{ in.})} = 0.412 \text{ k/in.}^2$$

$$f_h = \frac{0.707 \times 7 \text{ k}}{1 \text{ in.} \times (4 \text{ in.} + 8 \text{ in.})} = 0.412 \text{ k/in.}^2$$

Torsional components

Point A

$$f_v = \frac{28 \text{ in.-k} \times 3.33 \text{ in.}}{101.33 \text{ in.}^4} = 0.92 \text{ k/in.}^2$$

$$f_h = \frac{28 \text{ in.-k} \times 2.67 \text{ in.}}{101.33 \text{ in.}^4} = 0.74 \text{ k/in.}^2$$

See sketch for force orientation.

$$f_r = \sqrt{(0.412 + 0.92)^2 + (0.74 - 0.412)^2} = 1.37 \text{ k/in.}^2$$

$f_r = 1.37$ kip/in.^{2}

0.92

0.412

0.412 0.74

Weld stress—Point A

0.412 1.47

0.412

0.19

$f_r = 1.90$ kip/in.^{2}

Weld stress—Point B

Point B

$$f_v = \frac{28 \text{ in.-k} \times 0.67 \text{ in.}}{101.33 \text{ in.}^4} = 0.19 \text{ k/in.}^2$$

$$f_h = \frac{28 \text{ in.-k} \times 5.33 \text{ in.}}{101.33 \text{ in.}^4} = 1.47 \text{ k/in.}^2$$

See sketch for force orientation

$$f_r = \sqrt{(0.19 - 0.412)^2 + (0.412 + 1.47)^2} = 1.90 \text{ k/in.}^2$$

The most stressed point of the weld is Point B.

For a 1-in. weld

$$W_c = 16 \times 0.928 \text{ k/in.} = 14.85 \text{ k/in.}$$

$$W_{size} = \frac{1.90 \text{ k/in.}^2}{14.85 \text{ k/in.}} = 0.13 \text{ in.} \quad \text{use } \frac{1}{4} \text{ in. minimum weld size (AISCS 1.17.2)}$$

Example 6.18. What load P can be supported by the connection shown? The weld is $\frac{5}{16}$ in. E60 electrode.

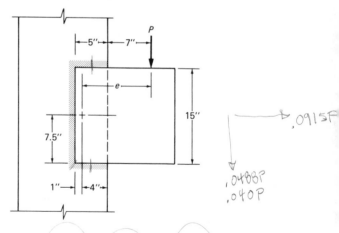

Solution.

$$f_v = \frac{Th}{J}, \quad T = Pe; \quad f_h = \frac{Td}{J}$$

Locating the C.G. of the weld group

$$\frac{(5 \text{ in.} \times 2.5 \text{ in.}) \times 2}{5 \text{ in.} \times 2 + 15 \text{ in.}} = 1 \text{ in.}$$

Assuming a 1-in. weld

$$J = (I_x + I_y)_{weld} = \left(\frac{1 (15)^3}{12}\right) + (1 \times 5)\left(\frac{15}{2}\right)^2 \times 2 + (1 \times 15)(1)^2$$

$$+ 2 \times \left[\frac{(1)^3}{3} + \frac{(4)^3}{3}\right] = 902.08 \text{ in.}^4$$

$e = 7 \text{ in.} + 4 \text{ in.} = 11 \text{ in.}$

Solving for components

torque

$$f_v = \frac{11 \text{ in.} \times P \times 4 \text{ in.}}{902.08 \text{ in.}^4} = 0.0488 P$$

$$f_h = \frac{11 \text{ in.} \times P \times 7.5 \text{ in.}}{902.08 \text{ in.}^4} = 0.0915 P$$

shear

$$f_v = \frac{P}{(5 \text{ in.} + 15 \text{ in.} + 5 \text{ in.}) \times 1 \text{ in.}} = 0.040 P$$

resultant

$$f_r = \sqrt{(0.0488 + 0.040)^2 + (0.0915)^2} \, P = 0.1275 P$$

Since f_r is determined assuming a 1-in. weld, conversion for $\frac{5}{16}$-in. weld is necessary.

$$W_c = 5 \times 0.795 \text{ k/in.} = 3.98 \text{ k/in.}$$

$$0.1275 P = 3.98 \text{ k/in.}$$

$$P = \frac{3.98 \text{ k/in.}}{0.1275 \text{ (1/in.)}} = 31.2 \text{ k}$$

For welds subjected to shear and bending, the shear is assumed to be uniform throughout the member. The weld is designed to carry the shear and maximum bending stress. The resulting shear force is the square root of the sum of the squares of the shear force and the bending force.

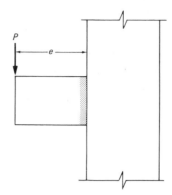

Fig. 6.9. Shear and bending condition in welds.

6.9 AISC DESIGN TABLES

The AISCM, in its "Connections" chapter (Part 4), provides tables for computing required weld sizes for connections subjected to eccentric loads. From these tables, weld dimensions, properties, and maximum allowable load can be determined.

Eccentric loads on weld groups tables in previous AISC manuals were based on elastic theory and the method of summing force vectors, as introduced in Section 6.8 of this text. The new *Manual* offers more liberal values for weld design based on ultimate strength design. The saving in welding is due to plastic behavior, where yield stress is reached in the entire weld before failure occurs.

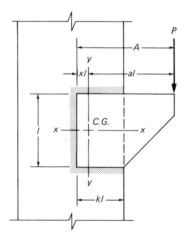

Fig. 6.10. One of several standard weld patterns that can be found in the AISCM and can be designed using tables.

Elastic analysis on weld groups gives varying factors of safety depending on the length and configuration of the weld. Ultimate strength design gives a more uniform safety factor and has been verified by experimental data. However, elastic analysis is still acceptable and is recommended for designing unusual weld patterns.

When a weld pattern is standard, i.e., a configuration tabulated in the AISCM, Part 4, the weld can be designed using the AISCM eccentric loads on weld groups tables. For welds unsymmetrical about the y-y axis, when the values of l, kl, and A are known, the center of gravity of the weld pattern can be determined from xl, where x is given in the tables based on k. Once a is known, the coefficient C is determined from the tables. Because the tables are calculated for E70 electrodes (the most commonly used), another coefficient C_1 is obtained from AISCM, p. 4-74, and used in conjunction with the tables for electrodes other than E70.

Example 6.19. Using the AISCM for eccentric loads on weld groups, determine the maximum allowable value of P for the given weld. Use a $\frac{5}{16}$-in. weld of E70 electrodes.

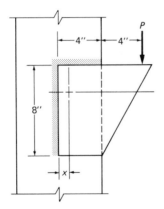

Solution.

$$P = CC_1 \, Dl$$

The weld formation is found in Table XXV. For the given conditions, $kl = 4$ in., $l = 8$ in., $l(x + a) = 8$ in., $D = 5$, $C_1 = 1.0$.

$$k = 4 \text{ in.}/l = 4 \text{ in.}/8 \text{ in.} = 0.5$$

In Table XXV the value for x under $k = 0.5$ is $x = 0.083$

$$8(0.083 + a) = 8 \text{ in.}; \quad a = 1 - 0.083 = 0.917$$

Using the chart for $k = 0.5$ and $a = 0.917$,

$$\begin{array}{cc} a & C \\ 0.90 & .409 \\ 0.917 & \qquad C = .403 \text{ (through interpolation)} \\ 1.00 & .371 \end{array}$$

$$P = .403 \, (1.0) \, 5 \, (8) = 16.1 \text{ k}$$

Example 6.20. For the welded connection below, determine the minimum size weld required to carry the applied load. Use the AISC eccentric load tables and E60 electrodes.

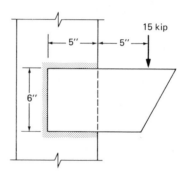

Solution.

$$D = \frac{P}{CC_1 l}$$

The weld Table XXIII is to be used.

Given values:

$$kl = 5 \text{ in.}$$
$$l = 6 \text{ in.}$$
$$l \, (x + a) = 10 \text{ in.}$$
$$P = 15 \text{ k}$$
$$C_1 = 0.857 \text{ (from table on p. 4-74 for E60 electrode)}$$
$$k = \tfrac{5}{6} = 0.83$$

Interpolating for $k = 0.83$, the value of x is determined to be 0.259.

$$l \, (x + a) = 10 \text{ in; } \quad 6 \, (0.259 + a) = 10 \text{ in.} \quad a = 1.4$$

For $k = 0.83$ and $a = 1.4$, $\quad C = 0.692$ (through interpolation)

$$D = \frac{P}{CC_1 l} = \frac{15\,k}{0.692\,(0.857)\,6} = 4.2 \quad \text{say } 5$$

Minimum weld size is $\frac{5}{16}$ in.

Example 6.21. Rework Example 6.18, solving for P using the AISC eccentric loads table. Welds are $\frac{5}{16}$ in. E60.

Solution.

$$P = CC_1 Dl$$

The weld Table XXIII is to be used.

Given values:
$kl = 5$ in.
$l = 15$ in.
$l(x + a) = 12$ in.
$D = 5$
$C_1 = 0.857$
$k = \frac{5}{15} = 0.33$

Interpolating for $k = 0.33$, the value of x is determined to be 0.066.

$l(x + a) = 12$ in.; $\quad 15(0.066 + a) = 12 \quad a = 0.734$
For $k = 0.33$ and $a = 0.734$, $\quad C = 0.630$
$P = 0.630(0.857)\,5\,(15) = 40.5\,k$

The value obtained shows an increase of 30% when compared with the elastic analysis method. Thus, a major saving can be realized for common weld groups found in the *Manual*. The elastic method is convenient to use on unusual weld groups and in all cases is conservative.

PROBLEMS TO BE SOLVED

NOTE: Unless specific mention is made, E70 weld is to be used.

6.1. Determine the strengths of $\frac{5}{16}$-in. and $\frac{1}{2}$-in. fillet welds 12 in. long for E70 and E80 electrodes.

6.2. Determine the capacities of the full penetration groove welded connections shown below.

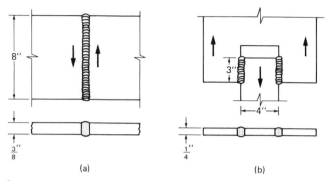

(a)   (b)

6.3. A $\frac{3}{8}$-in. fillet weld is used to connect two plates as shown. Determine the maximum load that can be applied to the connection.

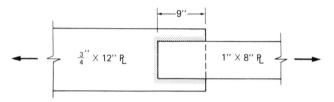

6.4. For the members shown, determine the minimum length of fillet weld required to develop full tensile strength of the plates.

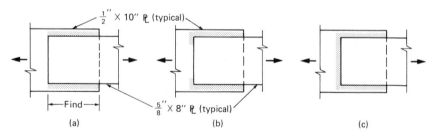

(a)   (b)   (c)

6.5. Determine the allowable tensile capacity of the connection shown. Use $\frac{5}{16}$-in. fillet weld with E60 electrodes.

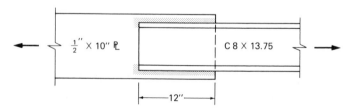

6.6. A 1-in. × 8-in. plate is to be connected to another member using $\frac{1}{2}$-in., E60 fillet welds. Due to clearance limitations, the two members can overlap only 5 in. as shown. Determine the distance l so that the connection capacity is 175 kips.

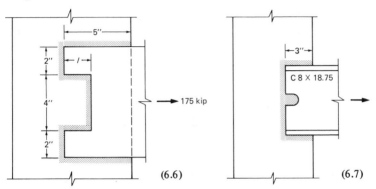

6.7. A channel is to be connected to a plate as shown. For clearance reasons, the channel cannot overlap the plate by more than 3 in. Design a combination fillet and slot weld to carry the applied load.

6.8. A welded plate girder consisting of two 2-in. × 16-in. flanges and $\frac{1}{2}$-in. × 40-in. web is subject to a maximum shear of 270 kips. Determine the fillet weld sizes for continuous welds and for intermittent welds.

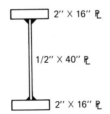

6.9. A W 21 × 101 has two 1-in. × 8-in. cover plates attached to the flanges by $\frac{3}{8}$-in. fillet welds as shown. Determine the length of connecting welds required at a location where the shear is 140 kips.

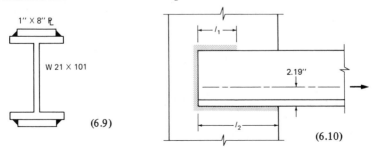

6.10. One leg of an $8 \times 8 \times \frac{1}{2}$-in. angle is to be connected with side welds and a weld at the end of the angle to a plate behind. Design the connection to develop full tensile capacity of the angle. Balance the fillet welds around the center of gravity of the angle, using $\frac{7}{16}$-in. weld and E60 electrodes.

6.11. A $6 \times 4 \times \frac{1}{2}$-in. angle is to be connected to develop full tensile capacity as shown to a plate using $\frac{3}{8}$-in. fillet welds. Determine l_1 and l_2 so that the centers of gravity for the welds and the angle coincide.

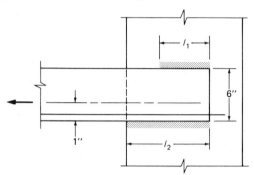

6.12. Rework Problem 6.11, but use end returns.

6.13. Design partial length cover plates for a W 21×101 rolled section for the loading shown. Assume full lateral support.

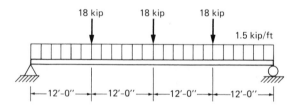

6.14. For the continuous beam shown, choose a wide-flange beam to resist the maximum positive moment, and design a welded partial cover plate where required for the negative moments. Assume full lateral support.

NOTE: To simplify the analysis of the continuous beam, neglect the variations of stiffness due to the cover plates.

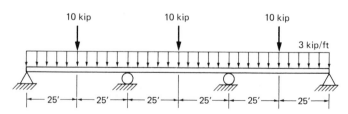

6.15. The weld shown is to carry an eccentric load of 8 kips acting 12 in. from the weld. Determine the minimum weld size required to carry the load. Do not use any tables.

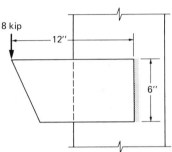

6.16. Rework Problem 6.15 using the appropriate AISC tables.

6.17. Determine the minimum weld size required to resist the load shown.

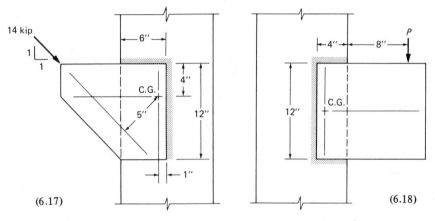

6.18. What load P can be supported by the connection shown if $\frac{3}{8}$-in. weld and E60 electrodes are used? Do not use any tables.

6.19. Rework Problem 6.18 using the appropriate AISC tables.

7
Special Connections

7.1 BEAM-COLUMN CONNECTIONS

Three types of constructions are allowed by the AISC. For each type, the strengths and kinds of connections are specified. Type 1 is rigid frame construction, whose connections are sufficiently rigid to keep the rotation of connected members virtually the same at each joint. Type 2 is simple frame construction, for which connections are made to resist shear only. Type 3 is semirigid framing, which assumes an end connection whose rigidity is somewhat between those of rigid frames and simple frames (AISCS 1.2).

The connections of simple frames are designed to carry only shear (see Fig. 7.2). They allow a certain amount of elastic rotation and some inelastic rotation to avoid overstress of the fasteners for such cases as wind loading. Simple connections are usually made of two angles framing the web of a beam or only a seat angle. For the design of such connections, refer to Chapters 5 and 6.

Type 1 rigid frames and type 3 semirigid frames have connections that are designed to carry shear and moments. Rigid connections carry full end moments and semirigid ones carry an end moment that reduces the simply supported midspan moment. However, semirigid connections are not rigid enough to completely prevent a rotation or change in angle between the connected members (see Fig. 7.3).

The AISC, in Section 1.2, states that semirigid construction will be permitted only upon evidence that the connections to be used are capable of furnishing, as a minimum, a predictable portion of full end restraints. The exact moment to be carried by the end connection is difficult to determine. Charts would be necessary relating the moment-end rotation relationship for each type of connection. A conservative design approach is to assume a fully rigid end connection for a moment less than the fully fixed one and to design the connection for that moment. The beam is then designed for the positive moment resulting from the developed negative moment at the connection (see Fig. 7.5).

250

Fig. 7.1. A beam-to-column moment connection with welded butt plate.

With the use of rigid connections, the AISC allows redistribution of moments. Compact beams and plate girders, which are rigidly framed to columns (or are continuous over supports), may be proportioned for $\frac{9}{10}$ of the negative moments produced by gravity loading, provided that the maximum positive moment is increased by $\frac{1}{10}$ of the average negative moments. The negative moment reduction may be used for proportioning the column for the combined axial and

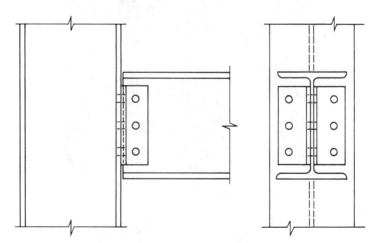

Fig. 7.2. Simple frame bolted connection. Bolts may be replaced by welds, most commonly to the beam web.

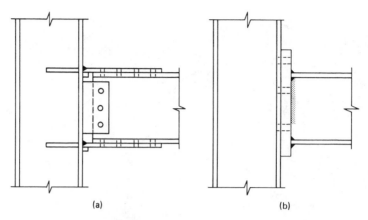

(a)   (b)

Fig. 7.3. Moment connections offering rigid and semirigid action: (a) shop-welded flange plates connected to column flange with field-bolted beam; (b) end plate shop-welded to beam and then field-bolted to column flange.

Fig. 7.4. A beam-to-column moment connection with flange angles.

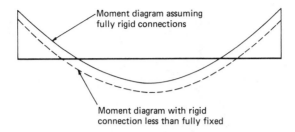

Moment diagram assuming fully rigid connections

Moment diagram with rigid connection less than fully fixed

Fig. 7.5. Redistribution of moments for semirigid connections.

bending loads, provided that the axial stress f_a does not exceed 0.15 F_a (AISCS 1.5.1.4.1) (also see "Continuous Beams" in Chapter 3).

7.2 DESIGN OF MOMENT CONNECTIONS

The requirements for rigid and semirigid connections are given in AISCS 1.15.5. Usual connection details consist of plates attached to the beam flanges and to the column flange or web and angles connecting the beam web to the column flange or web. An alternative approach, when joining the beam to the column flange, is to weld a plate to the end of the beam and then to bolt the plate to the column (see Fig. 7.3b). Stiffeners may be required to prevent the column section from either buckling or yielding due to the force applied to the column by the bending moment of the joint.

Both rigid and semirigid connections assume the web connection to carry the entire shear force and the top and bottom flange connections to carry the end moment. Fasteners are designed for the combined effects of end reaction shear and tensile or compressive forces resulting from the moment induced by the rigidity of the connection. The number of web fasteners required is determined from

$$N_b \geqslant \frac{R}{r_v} \tag{7.1}$$

where R is the end reaction and r_v is the shear capacity of one fastener. If web angles are chosen to transfer shear, they may be determined as presented in Chapter 5 and may be selected from the AISCM framed beam connections tables.

Beam end moments are assumed to be resisted totally by couples acting in the top and bottom flanges of the beam. Bolted and riveted connections are put into single shear by the tensile force at the top and compressive force at the bottom connections. Flange forces developed are determined by dividing the

Fig. 7.6. Corner detail of beam-to-column connections. Note how the large T section provides moment connection for the beam on the right while providing tension connection for the four-angle wind bracing member.

moment by the moment arm or

$$T = \frac{12\,M}{d - t_f} \tag{7.2}$$

where M is the support moment in ft kips and T is the flange force in kips. Weld length and size must similarly be designed to resist the force. For flange plate connections, the plate is designed as a tension member.

Alternatively, an end plate connection can be provided in lieu of web and flange plate connections where the end plate is shop-welded to the beam and field-bolted to the column flange. High-strength bolts are arranged near the top flange to transfer the flexural tensile force in the flange and additionally elsewhere to help resist the beam end shear. The welds connecting the end plate to the beam are usually fillet welds, though they may be full penetration type. The tension flange weld is sized to develop the full force due to bending, and the web weld is designed to carry the beam reaction. In addition, the end plate must be designed to resist the bending stress developed between connecting bolts.

The need for column web stiffeners must be investigated for the moment-

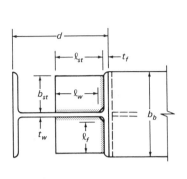
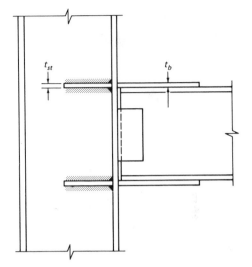

1) $b_{st} \geqslant \dfrac{b_b}{3} - \dfrac{t_w}{2}$

2) $t_{st} \geqslant \dfrac{t_b}{2}$

3) $\ell_{st} \geqslant \dfrac{d}{2}$ for one-sided loading

$\ell_{st} = d - 2\,t_f$ for both sides loaded

4) $\dfrac{b_{st}}{t_{st}} \leqslant \dfrac{95}{\sqrt{F_y}}$

5) $\ell_w = \dfrac{P_{bf} - F_{yc}t(t_b + 5\,k)}{W_{cw}}$

$\dfrac{P_{bf} - F_{yct}t(t_b + 5\,k)}{W_{cf}}$

$W_c = 0.707\,F_v t_{weld}$

Fig. 7.7. Column web stiffener requirements.

induced flange force factored according to the type of loading case. Web stiffeners are required whenever the value of A_{st} is positive (AISCS 1.15.5.2) where

$$A_{st} = \frac{P_{bf} - F_{yc}\,t(t_b + 5\,k)}{F_{yst}} \qquad (7.3)$$

The factored force $P_{bf} = 5T/3$ when T is due to live and dead load only, and $P_{bf} = 4T/3$ when T is due to live and dead loads plus wind or earthquake forces, unless local building codes dictate otherwise. The value of t_b is the thickness of the flange or flange plate transferring the concentrated force to the column. The yield stress subscripts c and st refer to the column and stiffener sections, respectively. Further, column stiffeners (or stiffener if the moment connection is to the column web) must be provided opposite the compression flange connection when

$$d_c \geq \frac{4100\, t_w^3 \sqrt{F_{yc}}}{P_{bf}} \tag{7.4}$$

where d_c is the column web depth clear of fillets (AISCS 1.15.5.3). Web stiffener(s) must be provided opposite the connection of the tension flange when

$$t_f < 0.4 \sqrt{\frac{P_{bf}}{F_{yc}}} \tag{7.5}$$

Regardless of force or formula, stiffeners, when required, must meet the requirements shown in Fig. 7.7 and AISCS 1.15.5.4. Further, the combined stiffener area, when required by Eq. 7.3, must not be less than the value A_{st} computed. This method is developed in AISCM, pp. 4-98 through 4-117 and is explained step-by-step in the following examples.[1]

Example 7.1. Design bolted flange plate moment connections (shop-welded, field-bolted[2]) for the beam, columns, and loading as shown. Use friction type $\frac{3}{4}$-in. A325-F bolts and beam allowable bending stress $F_b = 22.0$ ksi.

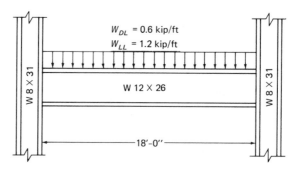

Solution. For W 8 × 31

$$d = 8.00 \text{ in.,} \quad t_w = 0.285 \text{ in.,} \quad b_f = 7.995 \text{ in.,} \quad t_f = 0.435 \text{ in.,} \quad k = \tfrac{15}{16} \text{ in.}$$

---

[1] U.S. Steel has published a design aid, *Beam-to-Column Flange Connections–Restrained Members*, which includes tables to facilitate stiffener design.
[2] Welded to column, bolted to beam. See AISCM p. 4-104.

For W 12 × 26

$d = 12.22$ in., $t_w = 0.230$ in., $b_f = 6.490$ in., $t_f = 0.380$ in., $k = \frac{7}{8}$ in.

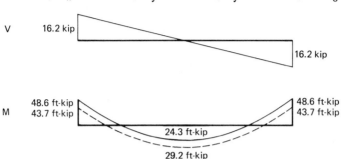

Redistributing moments for rigidly framed members,

$$M^- = 0.9 \times 48.6 \text{ ft-k} = 43.7 \text{ ft-k}$$

$$M^+ = (0.1 \times 48.6 \text{ ft-k}) + 24.3 \text{ ft-k} = 29.2 \text{ ft-k}$$

Check beam flange area reduction for two rows of $\frac{3}{4}$-in. connecting bolts

$$A_f = 0.38 \text{ in.} \times 6.49 \text{ in.} = 2.47 \text{ in.}^2$$

$$A_h = 2\,(.75 + .125) \times 0.38 = 0.665 \text{ in.}^2$$

$$\frac{A_h}{A_f} = \frac{0.665 \text{ in.}^2}{2.47 \text{ in.}^2} = 0.27 \times 100\% = 27\% \text{ flange area loss}$$

15% flange loss permitted. Reduce flange area only by excess (AISCS 1.10.1).

$$27\% - 15\% = 12\% \text{ excess flange loss}$$

$$A_n = (1.0 - 0.12) \times 2.47 \text{ in.}^2 = 2.17 \text{ in.}^2$$

$$I_n = 204 - \left[2 \times (.12 \times 2.47) \left(\frac{12.22 - 0.38}{2} \right)^2 \right] = 183.2 \text{ in.}^4$$

$$S_n = \frac{I_n}{c} = \frac{183.2 \text{ in.}^4}{6.11 \text{ in.}} = 30.0 \text{ in.}^3$$

$$S_{\text{req}} = \frac{M}{F_b} = \frac{43.7 \text{ ft-k} \times 12 \text{ in./ft}}{22 \text{ ksi}} = 23.8 \text{ in.}^3 < 30.0 \text{ in.}^3 \quad \text{ok}$$

Web shear (single shear plate)

Shear capacity of $\frac{3}{4}$-in. A325-F in single shear $r_v = 7.7$ k

Number of bolts required

$$N_b = \frac{R}{r_v} = \frac{16.2 \text{ k}}{7.7 \text{ k}} = 2.1 \qquad \text{three bolts required}$$

Shear plate

Try 9-in. shear plate 3 in. center-to-center of bolts, with $1\frac{1}{2}$ in. at ends.

$$l_n = 9 \text{ in.} - 3(.75 \text{ in.} + .125 \text{ in.}) = 6.375 \text{ in.}$$

$$F_v = 0.30 \, F_u = 17.4 \text{ ksi} \qquad \text{(AISCS 1.5.1.2.2)}$$

$$t_{pl} \geqslant \frac{R}{l_n F_v} = \frac{16.2 \text{ k}}{6.375 \text{ in.} \times 17.4 \text{ ksi}} = 0.146 \text{ in.}$$

Try $\frac{1}{4}$-in. plate

For F_u = 58 ksi, $1\frac{1}{2}$ in. edge distance, 3-in. bolt spacing, and $t = \frac{1}{4}$ in.

Bolt capacity = 10.9 k/bolt (edge distance controls)   (AISC Part 4, Table I-E)

10.9 k/bolt $> r_v$ = 7.7 k   ok

Check beam web bearing, t_w = 0.23 in.

Bolt capacity = 0.23 in. $\times$ 65.3 k (1-in. value) = 15.0 k/bolt

15.0 k $> r_v$ = 7.7 k   ok

Plate fillet weld to column flange (E70XX weld)

$$D > \frac{16.2 \text{ k}}{2(0.928 \text{ k/in.}) \times 9 \text{ in.}} = 0.97$$

Use $\dfrac{1}{8}$ in. minimum weld, $D = 2$ (AISC Table 1.17.2A)

Design of flange plates

Flange force due to moment

$$T = \frac{12 \, M}{d - t_f} = \frac{12 \text{ in./ft} \times 48.6 \text{ ft-k}}{12.22 \text{ in.} - 0.38 \text{ in.}} = 49.3 \text{ k}$$

$$A_n \geqslant \frac{T}{0.5 \, F_u} = \frac{49.3 \, \text{k}}{29 \, \text{ksi}} = 1.7 \, \text{in.}^2$$

$$A_g \geqslant \frac{T}{0.6 \, F_y} = \frac{49.3 \, \text{k}}{22 \, \text{ksi}} = 2.24 \, \text{in.}^2$$

Try $\frac{3}{8} \times 6\text{-}\frac{1}{2}$-in. plate

$$A_g = 0.375 \, \text{in.} \times 6.5 \, \text{in.} = 2.44 \, \text{in.}^2 > 2.24 \, \text{in.}^2 \qquad\qquad \text{ok}$$

$$A_n = 2.44 \, \text{in.}^2 - 2(.75 + .125) \times .375 \, \text{in.} = 1.78 \, \text{in.}^2 > 1.7 \, \text{in.}^2 \qquad \text{ok}$$

Number of bolts required

$$N_b = \frac{T}{r_v} = \frac{49.3 \, \text{k}}{7.7 \, \text{k}} = 6.4$$

Use minimum four bolts in each of two rows $\left(3\frac{1}{2} \, \text{in. gage}\right)$

Check for column web stiffeners

$$A_{st} \geqslant \frac{P_{bf} - F_{yc} \, t_{wc} \, (t_b + 5 \, k)}{F_{yst}} \qquad \text{(AISCS 1.15.5.2)}$$

$$P_{bf} = \frac{5}{3} T = 82.2 \, \text{k}$$

$$t_{wc} = 0.285 \, \text{in.}, \quad k = \frac{15}{16} \, \text{in.}$$

$$t_b = t_{pl} = 0.375 \, \text{in.}$$

$$A_{st} \geqslant \frac{82.2 - 36 \times 0.285 \left(.375 + 5 \times \dfrac{15}{16}\right)}{36} = 0.84 \, \text{in.}^2$$

Column web stiffeners are required

Design of stiffeners (AISCS 1.15.5.4)

$$b_{st} + \frac{t_w}{2} \geqslant \frac{b_b}{3}$$

$$b_{st} \geqslant \frac{b_b}{3} - \frac{t_w}{2} = \frac{6.5 \, \text{in.}}{3} - \frac{0.285 \, \text{in.}}{2} = 2.02 \, \text{in.}$$

$$t_{st} \geqslant \frac{t_b}{2} = \frac{0.375 \text{ in.}}{2} = 0.19 \text{ in.}$$

$$l_{st} \geqslant \frac{d}{2} = \frac{8.0 \text{ in.}}{2} = 4.0 \text{ in.}$$

Try $2\text{-}\frac{1}{2}$ in. $\times \frac{1}{4}$ in. $\times$ 0 ft, 4 in.

$$A_{st} = 2(.25 \times 2.5) = 1.25 \text{ in.}^2 > 0.84 \text{ in.}^2 \quad \text{ok}$$

$$\frac{2.5 \text{ in.}}{0.25 \text{ in.}} \leqslant 15.83 \quad \text{ok} \quad \text{(AISCS 1.9.1.2)}$$

Weld length

$$f_{v \text{ web}} = \frac{\text{force to be carried}}{\text{capacity of 1 in. weld}}$$

$$l_{\text{weld}} = \frac{P_{bf} - F_{yc} t_{wc} (t_b + 5 k)}{W_{cw}}$$

Use $\frac{3}{16}$ in. weld

$$W_{cw} = 0.928 \text{ k/in.} \times 3 = 2.78 \text{ k/in.}$$

$$l_{\text{weld}} = \frac{82.2 - 36 \times 0.285 \left(0.375 + 5 \times \dfrac{15}{16}\right)}{2.78} = 10.9 \text{ in.}$$

Weld will be placed on four sides (two sides of each stiffener)

$$\frac{10.9 \text{ in.}}{4} = 2.73 \text{ in.} \quad 2\frac{3}{4} \text{ in. minimum weld length}$$

Flange weld length will be same as web weld.

Assuming $\frac{3}{4}$-in. notch in plate

$$2\text{-}\frac{3}{4} \text{ in.} + \frac{3}{4} \text{ in.} = 3\text{-}\frac{1}{2} \text{ in.}$$

Use minimum stiffener width of $3\text{-}\frac{1}{2}$ in, welding along entire length.

Use end moment connection as shown.

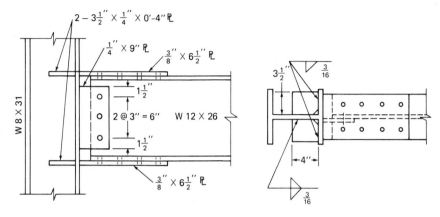

Example 7.2. For the previous example, design semirigid moment connection, such that the negative and positive moments are equal. Use $\frac{3}{4}$-in. 325-N bolts to allow slippage of connection, obtaining semirigid action.

Solution.

$$R = 16.2 \text{ k}$$

$$M^- = M^+ = \frac{wL^2}{16} = 36.45 \text{ ft-k}$$

The web shear connection will not change. Design flange plates for the reduced moment value.

$$T = \frac{12 M}{d - t_f} = \frac{12 \text{ in./ft} \times 36.45 \text{ ft-k}}{12.22 \text{ in.} - 0.38 \text{ in.}} = 36.9 \text{ k}$$

$$A_n \geqslant \frac{T}{0.5 F_y} = \frac{36.9 \text{ k}}{29 \text{ ksi}} = 1.27 \text{ in.}^2$$

$$A_g \geqslant \frac{T}{0.6 F_y} = \frac{36.9 \text{ k}}{22 \text{ ksi}} = 1.68 \text{ in.}^2$$

Try $\frac{5}{16}$-in. $\times$ $6\frac{1}{2}$-in. plate

$$A_g = 0.31 \text{ in.} \times 6.5 \text{ in.} = 2.02 \text{ in.}^2 > 1.68 \text{ in.}^2 \quad \text{ok}$$

$$A_n = 2.02 \text{ in.}^2 - 2(0.75 \text{ in.} + 0.125 \text{ in.}) \times 0.31 \text{ in.}$$

$$= 1.48 \text{ in.}^2 > 1.27 \text{ in.}^2 \quad \text{ok}$$

Number of bolts required

$$r_v = 9.3 \text{ k}$$

$$N_b = \frac{R}{r_v} = \frac{36.9 \text{ k}}{9.3 \text{ k}} = 3.97$$

Use two bolts in each of two rows ($3\text{-}\frac{1}{2}$-in. gage)

Check for column web stiffeners

$$A_{st} \geqslant \frac{P_{bf} - F_{yc} t_{wc}(t_b + 5 \text{ k})}{F_{y\,st}}$$

$$P_{bt} = \frac{5}{3}T = 61.5 \text{ k}$$

$$A_{st} \geqslant \frac{61.5 - 36 \times 0.285\left(0.313 + 5 \times \dfrac{15}{16}\right)}{36} = 0.28 \text{ in.}^2$$

Column web stiffeners are required.

$$b_{st} \geqslant \frac{b_b}{3} - \frac{t_w}{2} = 2.02 \text{ in.}$$

$$t_{st} \geqslant \frac{t_b}{2} = 0.16 \text{ in.}$$

$$l_{st} \geqslant \frac{d}{2} = 4.0 \text{ in.}$$

Try 2 in. $\times \frac{3}{16}$ in. $\times$ 0 ft, 4 in.

$$A_{st} = 2(0.188 \text{ in.} \times 2.0 \text{ in.}) = 0.75 \text{ in.}^2 > 0.28 \text{ in.}^2 \qquad \text{ok}$$

$$\frac{2.0}{0.188} \leqslant 15.83 \qquad \text{ok}$$

Weld length

$$\text{Use } \frac{3}{16} \text{ in. weld}$$

$W_c = 0.928$ k/in. $\times$ 3 = 2.78 k/in.

$$l = \frac{61.5 - (36 \times 0.285)\left(0.313 + 5 \times \dfrac{15}{16}\right)}{2.78} = 3.67 \text{ in.}$$

$\dfrac{3.67 \text{ in.}}{4} = 0.92$ in.   1-in. minimum weld length on four sides

Assuming $\frac{3}{4}$-in. notch in plate

$$1 \text{ in.} + \tfrac{3}{4} \text{ in.} = 1\text{-}\tfrac{3}{4} \text{ in.}$$

Weld entire length of plate along column flange

Use end moment connection as shown.

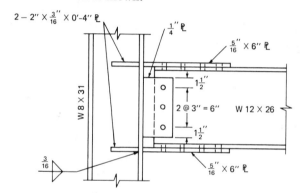

Example 7.3. A wide-flange beam spanning 35 ft supports a uniform load of 1.3 k/ft. Design the beam and semirigid connections assuming the maximum positive moment and maximum negative moment will be equal. Use web angles and tee connections to a WF 14 $\times$ 90 column in the weak axis with $\frac{3}{4}$-in. A325-X bolts. Assume full lateral support.

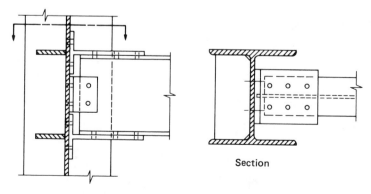

Section

Solution. For W 14 × 90 column

$$d = 14.02 \text{ in.,} \quad b_f = 14.52 \text{ in.,} \quad t_f = 0.71 \text{ in.,} \quad t_w = 0.44 \text{ in.,} \quad k = 1\text{-}\tfrac{3}{8} \text{ in.}$$

Maximum moments

$$M_{\max} = \frac{wL^2}{16} = \frac{1.3 \text{ k/ft} \times (35.0 \text{ ft})^2}{16} = 99.5 \text{ ft-k}$$

$$S_{\text{req}} = \frac{M}{F_b} = \frac{12 \text{ in./ft} \times 99.5 \text{ ft-k}}{24.0 \text{ ksi}} = 49.8 \text{ in.}^3 \qquad \text{(AISCM, Part 2)}$$

Use W 18 × 35 beam, $S_x = 57.6$ in.^{3}

$$d = 17.70 \text{ in.,} \quad b_f = 6.00 \text{ in.,} \quad t_f = 0.425 \text{ in.,} \quad t_w = 0.30 \text{ in.}$$

At the connection, design the web angles to carry full shear and the end moment by the top and bottom connections.

$$R = \frac{wL}{2} = \frac{1.3 \text{ k/ft} \times 35 \text{ ft}}{2} = 22.8 \text{ k}$$

Referring to Framed Beam Connections, a total number of two bolt lines of $\frac{3}{4}$-in. A325-X are required (beam web bolts in double shear).

$$R_{\max} = 53.0 \text{ k} \qquad \text{(AISC, Table II-A)}$$

$$R_{\max} = 33.7 \text{ k} \ (\tfrac{1}{4} \text{ in. } \pm \text{ angle}) \qquad \text{(AISC, Table II-C)}$$

$$22.8 \text{ k} < 33.7 \text{ k} \qquad \text{ok}$$

Framing angle size

Minimum angle leg along column for $5\frac{1}{2}$-in. bolt gage of column

$$\frac{5.5 \text{ in.}}{2} + 1.25 \text{ in.} - 0.15 \text{ in.} = 3.85 \text{ in.} \qquad \text{say 4 in.}$$

Use two 4 in. × 4 in. × $\frac{1}{4}$ in. beam web framing angles $5\frac{1}{2}$ in. long.

Flange tee connections

Flange force due to moment

$$T = \frac{12 M}{d - t_f} = \frac{12 \text{ in./ft} \times 99.5 \text{ ft-k}}{17.70 \text{ in.} - 0.425 \text{ in.}} = 69.1 \text{ k}$$

See Example 7.9 for design of tee section, as it also involves prying effect.

Use WT 10.5×25, $8\frac{1}{2}$ in. long

Check for column web stiffener

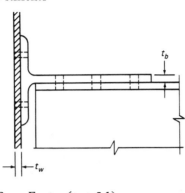

$$A_{st} \geqslant \frac{P_{bf} - F_{yc} t_{wc}(t_b + 5k)}{F_{yst}}$$

$$P_{bf} = \frac{5}{3}T = 115.2 \text{ k}$$

$$t_{wc} = 0.44 \text{ in.}$$

$$t_b = t_{wWT} = 0.380 \text{ in.}$$

$$k = 0$$

$$A_{st} \geqslant \frac{115.2 - 36 \times 0.44 \times (0.380 + 0)}{36} = 3.03 \text{ in.}^2$$

Column web stiffener is required

Design of stiffener

$$b_{st} \geqslant \frac{b_b}{3} - \frac{t_w}{2} = \frac{8.5 \text{ in.}}{3} - \frac{0.44 \text{ in.}}{2} = 2.61 \text{ in.}$$

$$t_{st} \geqslant \frac{t_b}{2} = \frac{0.40 \text{ in.}}{2} = 0.20 \text{ in.}$$

Try 4 in. $\times \frac{5}{16}$ in. $\times$ 1 ft, $0\frac{5}{8}$ in.

$$A_{st} = 0.3125 \text{ in.} \times 12.625 \text{ in.} = 3.94 \text{ in.}^2 > 3.03 \text{ in.}^2 \quad \text{ok}$$

$$\frac{12.625 \text{ in.}}{0.3125 \text{ in.}} = 40.4 > 15.83 \qquad \text{N.G.}$$

$$t_{min} = \frac{12.625 \text{ in.}}{15.83} = 0.80 \text{ in.}$$

Try 4 in. $\times \frac{7}{8}$ in. $\times$ 1 ft, 0 $\frac{5}{8}$ in.

Weld length

$$\text{Try } \frac{5}{16} \text{ in. weld}$$

$$W_{cw} = 0.928 \times 5 = 4.64 \text{ k/in.}$$

$$l_{\text{weld}} = \frac{P_{bf} - F_{yc} t_{wc} t_b}{W_{cw}} = \frac{115.2 - (36 \times 0.44 \times 0.395)}{4.64} = 23.5 \text{ in.}$$

Due to excessive length, try $\frac{3}{8}$ in. weld size

$$W_{cw} = 0.928 \text{ k/in.} \times 6 = 5.57 \text{ k/in.}$$

$$l_{\text{weld}} = \frac{115.2 - (36 \times 0.44 \times 0.395)}{5.57} = 19.6 \text{ in.}$$

Along column flanges

$$\frac{19.6 \text{ in.}}{4} = 4.9 \text{ in.} \qquad \text{Use 5 in.}$$

Along column web

$$\frac{19.6 \text{ in.}}{2} = 9.8 \text{ in.} \qquad \text{Use 10-in. minimum}$$

Assuming 1-in. notch in plate

$$5 \text{ in.} + 1 \text{ in.} = 6 \text{ in.}$$

Use 6 in. $\times \frac{7}{8}$ in. $\times$ 1 ft, 0 $\frac{5}{8}$ in. stiffener with continuous $\frac{3}{8}$-in. weld and 1-in.-45° notch at corners.

Example 7.4. Design an end plate moment connection to resist the maximum negative moment force and maximum end reaction without beam web reinforcement for a W 12 × 65 beam to W 12 × 79 column. Use A490-N bolts, F_y = 36.0 ksi.

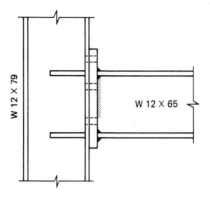

Solution.

$$F_b = 24.0 \text{ ksi}, \quad F_v = 14.4 \text{ ksi}$$

For W 12 × 79

$$d = 12.38 \text{ in.}, \quad b_f = 12.08 \text{ in.}, \quad t_f = 0.735 \text{ in.}, \quad t_w = 0.470 \text{ in.}$$

For W 12 × 65

$$d = 12.12 \text{ in.}, \quad b_f = 12.00 \text{ in.}, \quad t_f = 0.605 \text{ in.}, \quad t_w = 0.390 \text{ in.}$$

Maximum beam moment

$$M = F_b \times S = 24 \text{ ksi} \times 87.9 \text{ in.}^3 = 2110 \text{ in.-k} = 175.8 \text{ ft-k}$$

Maximum shear force

The effective area in resisting shear is the overall depth times the web thickness.

$$A_w = 0.390 \text{ in.} \times 12.12 \text{ in.} = 4.73 \text{ in.}^2$$

$$V = F_v \times A_w = 14.4 \text{ ksi} \times 4.73 \text{ in.}^2 = 68.1 \text{ k}$$

Flange force due to moment

$$T = \frac{12\,M}{d - t_f} = \frac{12 \text{ in./ft} \times 175.8 \text{ ft-k}}{12.12 \text{ in.} - 0.605 \text{ in.}} = 183.2 \text{ k}$$

Tensile flange force is resisted by four bolts in tension, at tension flange

Required nominal area per bolt to resist tension

$$F_{t\,\text{bolt}} = 54.0 \text{ ksi} \qquad \text{(AISCS, Table 1.5.2.1)}$$

$$A_b = \frac{T}{2nF_t}$$

n = two bolts per transverse line

$$A_b = \frac{183.2 \text{ k}}{2 \times 2 \times 54.0 \text{ ksi}} = 0.848 \text{ in.}^2$$

Minimum bolt size = $1\text{-}\frac{1}{8}$ in. ϕ, $A_b = 0.9940$ in.^{2}

Try minimum six bolts to resist reaction, four at tension flange and two at compression flange.

Number of bolts required to carry shear

$$r_v = 27.8 \text{ k} \qquad \text{(AISC, Table I-D)}$$

$$N_b = \frac{V}{r_v} = \frac{68.1 \text{ k}}{27.8 \text{ k}} = 2.45 < 6 \qquad \text{ok}$$

Beam flange welds must also resist tension force.

For E70XX weld, capacity per sixteenth is

$$21 \text{ ksi} \times 0.707/16 = 0.928 \text{ k/in./sixteenth}$$

$$D_{\text{weld}} \geqslant \frac{T}{0.928 \times 2\, b_f} = \frac{183.2 \text{ k}}{0.928 \times 2(12.00)} = 8.23$$

Use $\dfrac{9}{16}$ in. weld

Beam web weld must resist shear force.

Use minimum weld size for welding economy ($\frac{5}{16}$ in.)

$$L_{\text{weld}} \geqslant \frac{68.1 \text{ k}}{5 \times 0.928 \text{ k/in.}} = 14.7 \text{ in.}$$

Use $7\frac{1}{2}$ in. weld length each side of beam web as shown.

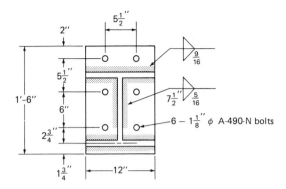

End plate design—assume width same as beam flange.

The end plate must be designed to resist bending between plate tension bolts.

$$P_e = P_f - (d_b/4) - w_t$$

$$P_f = \frac{5.5 \text{ in. } - 0.605 \text{ in.}}{2} = 2.45 \text{ in. } > d_b + \frac{1}{2} \text{ in.} \quad \text{ok}$$

$$d_b = 1.125 \text{ in.}$$

$$w_t = \frac{9}{16} \text{ in. } \times 0.707 = 0.40 \text{ in.}$$

$$P_e = 2.45 \text{ in. } - (1.125 \text{ in.}/4) - 0.40 \text{ in. } = 1.77 \text{ in.}$$

$$M_e = \alpha_m T P_e/4$$

The value of α_m is determined with the aid of AISC tables (AISCM, p. 4-113).

$$\alpha_m = C_a C_b (A_f/A_w)^{1/3} (P_e/d_b)^{1/4}$$

$$C_a = 1.14$$

$$C_b = (b_f/b_s)^{1/2} = 1$$

$$A_f/A_w = 1.706$$

$$\alpha_m = 1.14 \,(1.0)\,(1.706)^{1/3}\,(1.77/1.125)^{1/4} = 1.53$$

$$M_e = 1.53 \times 183.2 \text{ k} \times 1.77 \text{ in.}/4 = 123.7 \text{ in.-k}$$

Required plate width

$$b_s = 12 \text{ in. } \leqslant 1.15\, b_f = 1.15\,(12.00) = 13.8 \text{ in.} \quad \text{ok}$$

Required plate thickness

$$t_s \geq \left(\frac{6 M_e}{b_s F_b}\right)^{1/2} = \left(\frac{6 \times 123.7 \text{ in.-k}}{12 \text{ in.} \times 27.0 \text{ ksi}}\right)^{1/2} = 1.51 \text{ in.}$$

Try $1\frac{5}{8}$-in. plate thickness

Plate shear

$$f_v = \frac{T}{2 b_s t_s} = \frac{183.2 \text{ k}}{2 \times 12 \text{ in.} \times 1.625 \text{ in.}} = 4.70 \text{ in.} < 14.4 \text{ ksi} \quad \text{ok}$$

Use $1\frac{5}{8}$-in. end plate moment connection as previously shown.

7.3 MOMENT-RESISTING COLUMN BASE PLATES

Columns subject to both an axial and compressive force and an applied moment produce a combination of axial compressive stress and bending stress on the base

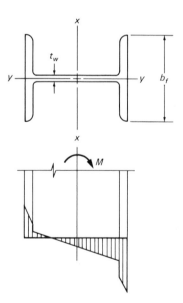

Fig. 7.8. Distribution of reaction (force per unit length) in column due to combined axial and flexural stress.

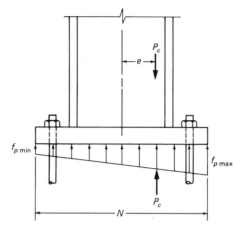

Fig. 7.9. Bearing stress of base plate when $e \leqslant N/6$.

plate. With the moment transposed to an eccentric load equal to the axial load P_c, the stress diagram is determined by superposition.

Consider a uniaxial bending condition along the Y-Y plane producing a combined axial and bending stress (see Fig. 7.8). This stress multiplied by the width of the flange and the thickness of the web yields a force distribution that is greatest at the column flanges. To ensure that the flanges, which are of a larger cross-sectional area, carry the greater stress, the column is set with eccentricity lying within the plane of the web.

When the eccentricity is less than $\frac{1}{6} N$, where N is the base plate dimension parallel to the column web, there is no uplift of the plate at the support. For this condition, the stress in the base plate is

$$f_c = \frac{P_c}{A} \pm \frac{P_c e}{S} \tag{7.6}$$

where A is the base plate area, and S is the section modulus of the base plate, equal to $BN^2/6$.

Uplift of the base plate will occur if the eccentricity is greater than $\frac{1}{6} N$. The uplift is resisted by the anchor hold-down bolts. At the extreme edge of the bearing plate, the bearing stress is maximum and decreases linearly across the plate for a distance Y (see Fig. 7.10). An approximate method of determining Y is to assume the center of gravity to be fixed at a point coinciding with the concentrated compressive force of the column flange.

When a more exact method of determining the uplift is desired, an approach similar to the design of a reinforced concrete section is employed. Assuming

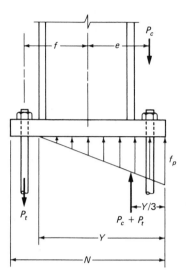

Fig. 7.10. Uplift of base plate when $e > N/6$.

plate dimensions that can be estimated by the above method, the value for Y can be determined by the cubic equation

$$Y^3 + K_1 Y^2 + K_2 Y + K_3 = 0 \qquad (7.7)$$

where $K_1 = 3(e - N/2)$, $K_2 = (6nA_s/B)(f + e)$, and $K_3 = -K_2(N/2 + f)$. Once Y is found, the tensile force on the hold-down bolts (P_t) is

$$P_t = -P_c \frac{N/2 - Y/3 - e}{N/2 - Y/3 + f} \qquad (7.8)$$

and the bearing stress is equal to $2(P_c + P_t)/YB$.[3]

Example 7.5. Design a base plate for a W 14 × 74 column to resist an axial force of 350 kips. Then test the base for the same axial load and a 90-ft-kip wind-induced bending moment applied in the major axis. Assume the base plate is set on the full area of concrete support with a compressive strength $f_c' = 3000$ psi.

---

[3]For derivation of Eqs. 7.7 and 7.8, refer to O. W. Blodgett, *Design of Welded Structures*. The James F. Lincoln Arc Welding Foundation, 1966, pp. 3.3-6 through 3.3-19.

Solution.

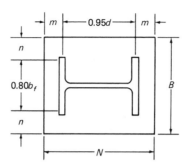

For a W 14 × 74

$$d = 14.17 \text{ in.}, \qquad b_f = 10.07 \text{ in.}$$

$$F_p = 0.35 f'c = 0.35 \times 3000 \text{ psi} = 1.05 \text{ ksi}$$

$$A_{\text{req}} = \frac{P}{F_p} = \frac{350.0 \text{ k}}{1.05 \text{ ksi}} = 333.33 \text{ in.}^2$$

Assume $N = 21$ in.

$$B = \frac{A}{N} = \frac{333.33 \text{ in.}^2}{21 \text{ in.}} = 15.87 \text{ in.}$$

Use $B = 16$ in.

$$f_{p \text{ actual}} = \frac{350 \text{ k}}{21 \text{ in.} \times 16 \text{ in.}} = 1.04 \text{ ksi}$$

$$m = \frac{N - 0.95 d}{2} = \frac{21 - 0.95(14.17)}{2} = 3.76 \text{ in.}$$

$$n = \frac{B - 0.80 b}{2} = \frac{16 - 0.80(10.07)}{2} = 3.97 \text{ in.}$$

$$n' = 4.43 \text{ in.} \qquad \text{(AISC Table C)} \qquad \text{(AISCM, p. 3-100)}$$

$$t = 4.43 \text{ in.} \sqrt{\frac{1.04 \text{ ksi}}{0.25 \times 36 \text{ ksi}}} = 1.51 \text{ in.}$$

Use base plate 16 in. × $1\frac{1}{2}$ in. × 1 ft, 9 in.

Check concrete bearing for applied moment.

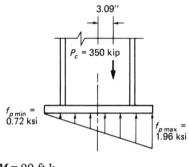

$$M = 90 \text{ ft-k}$$

$$e = \frac{M}{P_c} = \frac{90 \text{ ft-k} \times 12 \text{ in./ft}}{350 \text{ k}} = 3.09 \text{ in.}$$

$$N/6 = 3.50 \text{ in.} > 3.09 \text{ in.}$$

No uplift occurs

$$A = 21 \text{ in.} \times 16 \text{ in.} = 336 \text{ in.}^2$$

$$S = \frac{BN^2}{6} = \frac{16 \text{ in.} \times (21 \text{ in.})^2}{6} = 1176 \text{ in.}^3$$

$$f_p = \frac{P_c}{A} \pm \frac{P_c e}{S} = \frac{350 \text{ k}}{336 \text{ in.}^2} \pm \frac{350 \text{ k} \times 3.09 \text{ in.}}{1176 \text{ in.}^3} = 1.04 \pm 0.92$$

$$f_p = 1.96 \text{ ksi}, \quad 0.12 \text{ ksi}$$

$$1.96 \text{ ksi} > 1.05 \text{ ksi} \times 1.33 \text{ (allowable overstress)} = 1.40 \qquad \text{(AISCS 1.5.6)}$$

Base plate will not carry applied moment.

Example 7.6. Design a base plate to carry the axial force and bending moment in the preceding example without uplift.

Solution.

$$P_c = 350 \text{ k}$$

$$M = 90 \text{ ft-k} \times 12 \text{ in./ft} = 1080 \text{ in.-k}$$

$$F_c = \frac{P_c}{A} \pm \frac{M}{S} < 1.05 \text{ ksi} \times 1.33 = 1.40 \text{ ksi}$$

Try $N = 24$ in., $B = 19$ in.

$$A = 24 \text{ in.} \times 19 \text{ in.} = 456 \text{ in.}^2$$

$$S = \frac{BN^2}{6} = \frac{19 \text{ in.} \times (24 \text{ in.})^2}{6} = 1824 \text{ in.}^3$$

$$f_c = \frac{350 \text{ k}}{456 \text{ in.}^2} \pm \frac{1080 \text{ in.-k}}{1824 \text{ in.}^3} = 0.77 \text{ ksi} \pm 0.59 \text{ ksi} = 1.36 \text{ ksi}, \quad 0.18 \text{ ksi}$$

$$f_{c \text{ max}} = 1.36 \text{ ksi} < 1.05 \text{ ksi} \times 1.33 = 1.40 \quad \text{ok}$$

$$m = \frac{N - 0.95\, d}{2} = \frac{24 - 0.95(14.17)}{2} = 5.27 \text{ in.}$$

$$n = \frac{B - 0.80\, b}{2} = \frac{19 - 0.80(10.07)}{2} = 5.47 \text{ in.}$$

$$n' = 4.43 \text{ in.}$$

$$t = 5.47 \text{ in.} \sqrt{\frac{1.36 \text{ ksi}}{0.25 \times 36 \text{ ksi} \times 1.33}} = 1.84 \text{ in.}$$

Use base plate 19 in. $\times$ $1\frac{7}{8}$ in. $\times$ 2 ft, 0 in.

Example 7.7. Using the approximate method, design a base plate for a W 14 $\times$ 74 column to resist a combined axial force of 350 kips and bending moment due to gravity loads in the major axis of 250 ft-kips. Assume the footing of 3000-psi concrete to be quite large, making $F_p = 0.70 f'c$. Three hold-down bolts on each side of the column flange are 9 in. from the column center line. Use A36 steel and A307 bolts. Then verify solution with a more precise approach.

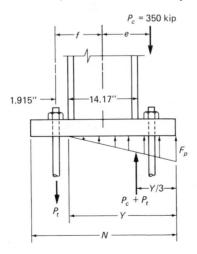

Solution. For a W 14 × 74

$$d = 14.17 \text{ in.}, \quad b_f = 10.07 \text{ in.}, \quad t_f = 0.785 \text{ in.}$$

$$F_p = 0.70 \times 3000 \text{ psi} = 2100 \text{ psi} = 2.1 \text{ ksi}$$

Assume compressive reaction from base is applied at the center of the flange, and its value is

$$P = P_c + P_t$$

$$e = \frac{M}{P_c} = \frac{250 \text{ ft-k} \times 12 \text{ in./ft}}{350 \text{ k}} = 8.57 \text{ in.}$$

Summing moments with respect to compression flange

$$P_t = \frac{350 \text{ k} \times \left(8.57 \text{ in.} - \dfrac{14.17 \text{ in.}}{2} + \dfrac{0.785 \text{ in.}}{2}\right)}{\left(\dfrac{14.17 \text{ in.}}{2} - \dfrac{0.785 \text{ in.}}{2} + 9.0 \text{ in.}\right)} = 41.88 \text{ k}$$

Total volume of compressive stress

$$\frac{1}{2} \times F_p \times Y \times B = P_c + P_t$$

$$Y = \frac{2 \times (41.88 \text{ k} + 350 \text{ k})}{2.1 \text{ ksi} \times B}$$

Try $B = 21$ in.

$$Y = \frac{2 \times 391.88 \text{ k}}{2.1 \text{ ksi} \times 21 \text{ in.}} = 17.77 \text{ in.}$$

Distance from center of flange to end of plate is $Y/3$.

$$\frac{Y}{3} = 5.923 \text{ in.}$$

Length of base plate

$$L = 2 \times 5.923 \text{ in.} + 14.17 \text{ in.} - 0.785 \text{ in.} = 25.23 \text{ in.} \quad \text{say 26 in.}$$

Recalculating $Y/3$

$$\frac{Y}{3} = \frac{26 \text{ in.} - 14.17 \text{ in.} + 0.785 \text{ in.}}{2} = 6.31 \text{ in.}$$

$$m = \frac{N - 0.95\, d}{2} = \frac{26 \text{ in.} - 0.95 \times 14.17 \text{ in.}}{2} = 6.27 \text{ in.}$$

$$n = \frac{B - 0.80\, b_f}{2} = \frac{21 \text{ in.} - 0.80 \times 10.07 \text{ in.}}{2} = 6.47 \text{ in.}$$

$$n' = 4.43 \text{ in.}$$

$$f_p = \frac{2 \times (P_c + P_t)}{3 \times Y/3 \times B} = \frac{2 \times (350 \text{ k} + 41.88 \text{ k})}{3 \times 6.31 \text{ in.} \times 21 \text{ in.}} = 1.97 \text{ ksi}$$

Use bearing stress as uniformly distributed value for a conservative calculation of thickness.

$$t = 6.47 \text{ in.} \times \sqrt{\frac{1.97 \text{ ksi}}{0.25 \times 36 \text{ ksi}}} = 3.02 \text{ in.} \quad \text{use } 3\frac{1}{4} \text{ in.}$$

Checking stress in anchor bolts

three 1-in. ϕ A307 bolts have capacity of

$$T = 3 \times 0.7854 \text{ in.}^2 \times 20 \text{ ksi} = 47.1 \text{ k} > 41.88 \text{ k} \quad \text{ok}$$

Verify using "Reinforced Concrete Beam" approach. Y is the solution of the cubic equation.

$$Y^3 + K_1 Y^2 + K_2 Y + K_3 = 0, \text{ where}$$

$$K_1 = 3\left(e - \frac{N}{2}\right)$$

$$K_2 = \frac{6\, nA_s}{B}\, (f + e)$$

$$K_3 = -K_2\left(\frac{N}{2} + f\right)$$

$$f = 1.915 \text{ in.} + \frac{14.17 \text{ in.}}{2} = 9.0 \text{ in.}$$

$$e = 8.57 \text{ in.}$$

$$B = 21 \text{ in.}$$

$$N = 26 \text{ in.}$$

$$n = 10$$

$$A_s = 3 \times 0.785 \text{ in.}^2 = 2.355 \text{ in.}^2$$

$$f_p = \frac{2\,(P_c + P_t)}{Y \times B}$$

$$P_t = -P_c \frac{\dfrac{N}{2} - \dfrac{Y}{3} - e}{\dfrac{N}{2} - \dfrac{Y}{3} + f}$$

$$K_1 = 3 \times \left(8.57 \text{ in.} - \frac{26.0 \text{ in.}}{2}\right) = -13.29 \text{ in.}$$

$$K_2 = \frac{6 \times 10 \times 2.355 \text{ in.}^2}{21 \text{ in.}} \times (9.0 \text{ in.} + 8.57 \text{ in.}) = 118.22 \text{ in.}^2$$

$$K_3 = -118.22 \text{ in.}^2 \times \left(\frac{26.0 \text{ in.}}{2} + 9.0 \text{ in.}\right) = -2600 \text{ in.}^3$$

$$Y^3 - 13.29 \text{ in.}\ Y^2 + 118.22 \text{ in.}^2\ Y - 2600 \text{ in.}^3 = 0$$

By trial and error

$$Y = 16.03 \text{ in.}$$

$$\frac{Y}{3} = 5.34 \text{ in.}$$

$$P_t = -350 \text{ k} \times \frac{\dfrac{26 \text{ in.}}{2} - \dfrac{16.03 \text{ in.}}{3} - 8.57 \text{ in.}}{\dfrac{26 \text{ in.}}{2} - \dfrac{16.03 \text{ in.}}{3} + 9.0 \text{ in.}} = 19.1 \text{ k}$$

$$f_p = \frac{2\,(19.1 \text{ k} + 350 \text{ k})}{16.03 \text{ in.} \times 21 \text{ in.}} = 2.19 \text{ ksi} \quad \text{say 4\% overstress} \quad \text{ok}$$

We see that the compressive stresses are greater, but there is little tension in the bolts compared with the previous approximation.

7.4 FIELD SPLICES OF BEAMS AND PLATE GIRDERS

Field splices may become necessary to connect beams or plate girders when long spans are desired. When the length required of a shape is not available or if available, but shipment or erection is difficult, splicing may be the solution.

A splice must develop the force due to the design load, but not less than 50% of the effective strength of the member (AISCS 1.15.7). This is usually done with a web splice designed to carry shear and a portion of the moment. If additional moment capacity is necessary, flange splices are used. An approximate alternative is to design web splices to carry the total shear and flange splices to carry the total moment.

Fig. 7.11. A column splice of two different rolled sections made with splice plates.

For groove welded splices in plate girders and beams, AISCS 1.10.8 states that the weld must develop the full strength of the smaller spliced section.

Bolted splice plates are usually fastened to the outside surface of the flanges. Plates may be used on both the outside and the inside of the flanges to place the fasteners in double shear. The splice should be of the same or similar cross section as the flange, with the number of bolts determined by the axial shear resulting from the design moment.

Web splices are usually symmetrically placed over the height between fillets (T) on beams and between welds on plate girders. They transfer the whole shear and a part of the moment. The amount of moment carried by the web is proportional to the total bending moment

$$M_w = M \frac{I_w}{I_{section}} \tag{7.9}$$

with the remainder of the moment carried by the flange splices. Once M_w and V are known, the calculation of bolts is the same as that for eccentric loading on fastener groups (see Chapter 5).

Example 7.8. Design a bolted beam splice for a W 30 × 108 at a point where the shear is 100 kips and the moment is 300 ft-kips. Use A325-X bolts.

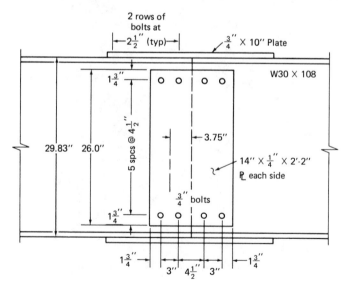

Solution. Preliminary design

Web splice

Design the web splice (plates on both sides of the web) to carry the entire shear and a portion of the moment.

Use plate height to toe of fillets

$$29\tfrac{7}{8} \text{ in.} - 2(1\tfrac{9}{16} \text{ in.}) = 26\tfrac{3}{4} \text{ in.} \quad \text{use plate} = 26 \text{ in.}$$

Try web splice with bolt spacing as shown.

For A325-X $\tfrac{3}{4}$-in. ϕ bolts

$$\text{Single shear} = 13.3 \text{ k} \quad \text{double shear} = 26.5 \text{ k}$$

Bolt shear capacity in web =

$$12 \text{ bolts} \times 26.5 \text{ k} = 318 \text{ k} > 100 \text{ k} \quad \text{ok}$$

Shear capacity of web

Try two plates 14 in $\times$ $\tfrac{1}{4}$ in. $\times$ 2 ft, 2 in.

$$\frac{1}{4} \text{ in.} \times \left(26 \text{ in.} - 6 \times \frac{7}{8} \text{ in.}\right) \times 2 = 10.37 \text{ in.}^2$$

$$f_v = \frac{100 \text{ k}}{10.37 \text{ in.}^2} = 9.64 \text{ ksi} < 14.4 \text{ ksi} \quad \text{ok}$$

Flange splice

$$\text{Force in flange} = \frac{M}{d} = \frac{300 \text{ ft-k} \times 12 \text{ in./ft}}{29.83 \text{ in.} - 0.76 \text{ in.}} = 123.8 \text{ k}$$

Gross plate area required $= P/F_t = 5.63$ in.^{2}

Effective net area required $= P/F_t = 4.27$ in.^{2}

Use plate width = 10 in.

Net plate width = 10 in. $- 2(\tfrac{3}{4}$ in. $+ \tfrac{1}{8}$ in.$) = 8.25$ in.

$A = 8.25$ in. $\times t = 4.27$ in.^{2} $\quad t = 0.52$ in. $\quad$ say $\tfrac{3}{4}$ in.

Final design

Considering that the bending moment at the splice carried by web and cover plates is proportioned to their moments of inertia

$$I_{web\ plates} = \frac{1}{4} \times \frac{(26)^3}{12} \times 2 - ((2.25)^2 + (6.75)^2 + (11.25)^2) \times 2 \times \frac{1}{4} \times \frac{7}{8}$$

$$= 655 \text{ in.}^4$$

$$I_{cover\ plates} = \left(10 - 2 \times \frac{7}{8}\right) \times 0.75 \times \left(\frac{29.83}{2} + 0.75\right)^2 \times 2$$

$$= 3037 \text{ in.}^4$$

Moment carried by web

$$\frac{300 \text{ ft-k} \times 655 \text{ in.}^4}{665 \text{ in.}^4 + 3037 \text{ in.}^4} = 53.2 \text{ ft-k}$$

Stress in flange

$$f_b = \frac{300 \text{ ft-k} \times 12 \text{ in./ft} \times \left(\frac{29.83 \text{ in.}}{2} + 0.75 \text{ in.}\right)}{3692 \text{ in.}^4} = 15.27 \text{ ksi}$$

$$F = 15.27 \text{ ksi} \times 8.25 \text{ in.} \times 0.75 \text{ in.} = 94.5 \text{ k}$$

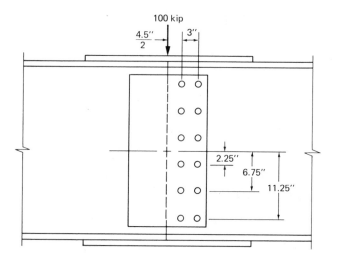

Check web bolts

Moment on fully loaded bolt

$$(2.25 \text{ in.} + 1.50 \text{ in.}) \times 100 \text{ k} = 375 \text{ in.-k} \quad \text{from eccentricity}$$
$$53.2 \text{ ft-k} \times 12 \text{ in./ft} = \underline{638 \text{ in.-k}} \quad \text{from moment}$$
$$1013 \text{ in.-k}$$

Force in extreme bolt

$$\Sigma d^2 = (1.5)^2 \times 12 + [(2.25)^2 + (6.75)^2 + (11.25)^2] \times 4$$

$$= 27.0 \text{ in.}^2 + 709 \text{ in.}^2 = 736 \text{ in.}^2$$

$$F = \sqrt{\left(\frac{100}{12} + \frac{1013}{736} \times 1.5\right)^2 + \left(\frac{1013}{736} \times 11.25\right)^2} = 18.65 \text{ k} < 26.5 \text{ k} \quad \text{ok}$$

Bolts in cover plates

$$\frac{94.5 \text{ k}}{13.3 \text{ k}} = 7.1 \text{ bolts} \quad \text{use eight bolts}$$

Bolts must be staggered as shown so that loss is equal to two holes.

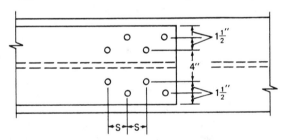

$$10 \text{ in.} - 2 \times \frac{7}{8} \text{ in.} = 10 \text{ in.} - 4 \times \frac{7}{8} \text{ in.} + 2 \times \frac{s^2}{4g}$$

$$= 10 \text{ in.} - 4 \times \frac{7}{8} \text{ in.} + \frac{2 \times s^2}{4 \times 1.5}$$

$$s = 2.29 \text{ in.} \quad \text{say } 2\frac{3}{8} \text{ in. minimum}$$

Check 3 in. minimum distance between bolts.

$$s = \sqrt{3^2 - 1.5^2} = 2.60 \text{ in.}$$

Use $s = 2\frac{5}{8}$ in.

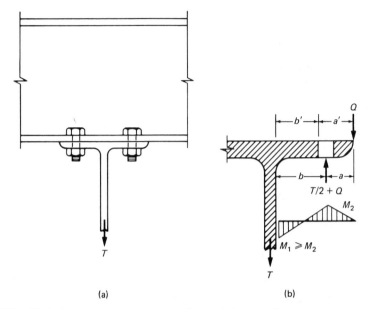

(a)   (b)

Fig. 7.12. Typical hanger type connection: (a) applied tensile force to stem causes combined tension and bending in the flanges; (b) moment diagram for flange leg.

7.5 HANGER TYPE CONNECTIONS

A tensile force applied to tee-section webs or the free legs of double angles fastened to another member, as shown in Fig. 7.12, causes tensile and bending stresses in the bolt and prying action to develop the hanger steel. Moments develop in the leg of the hanger due to the applied tensile force and the resulting prying force. Increased flange stiffness is realized by decreasing the distance b from the face of the stem or free leg of the angle. Therefore, it is recommended to keep b as small as erection methods permit.

Exact modeling of the combined tension and prying forces is difficult to perform. Therefore, it is accepted practice to use a safety factor greater than or equal to 2 for bolts in tension and tee flanges in bending.

A design procedure can be found on p. 4-88 of the AISCM, which takes prying, and also the situation of shear and prying, into account. The following example presents the design requirements recommended by the AISC for hanger type connections.

Example 7.9. Design a tee section for the semirigid moment connection in Example 7.3. The tension flange force transferred by the tee is 69.1 kips. Fasteners are to be $\frac{3}{4}$ in. A325-X bolts.

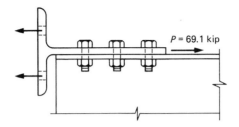

P = 69.1 kip

Solution. Number of tee flange bolts required

$$r_T = 19.4\,k \qquad \text{(AISCM, Table I-A)}$$

$$N_b = \frac{T}{r_T} = \frac{69.1\,k}{19.4\,k} = 3.56 \qquad \text{four bolts minimum}$$

$$T_b = \frac{69.1\,k}{4} = 17.28\,k$$

Number of tee web bolts required (single shear)

$$r_v = 13.3\,k$$

$$N_b = \frac{69.1\,k}{13.3\,k} = 5.20 \qquad \text{six bolts minimum}$$

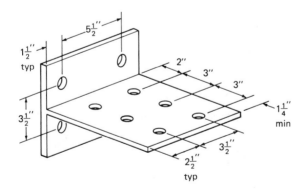

Try two flange bolts in each of two rows as shown.

$$\text{length tributary to bolt} = p = \frac{8.5\,\text{in.}}{2} = 4.25\,\text{in.}$$

Try three web bolts in each of two rows with $3\text{-}\frac{1}{2}$-in. gage.

Minimum tee length

$$d \geqslant 2.0 \text{ in.} + (2 \times 3.0 \text{ in.}) + 1.25 \text{ in.} = 9.25 \text{ in.}$$

Try WT 10.5 × 25 8-$\frac{1}{2}$ in. long

$$d = 10.415 \text{ in.}, \qquad b_f = 6.530 \text{ in.}, \qquad t_f = 0.535 \text{ in.}, \qquad t_w = 0.380 \text{ in.}$$

Check tee web for tension capacity

$$A_n \geqslant \frac{T}{0.5\,F_u} = \frac{69.1 \text{ k}}{29 \text{ ksi}} = 2.38 \text{ in.}^2$$

$$A_g \geqslant \frac{T}{0.6\,F_y} = \frac{69.1 \text{ k}}{22 \text{ ksi}} = 3.14 \text{ in.}^2$$

$A_{g\ provided} = 0.380 \text{ in.} \times 8.5 \text{ in.} = 3.23 \text{ in.}^2 > 3.14 \text{ in.}^2$ ok

$A_n = 3.23 \text{ in.}^2 - 2(.75 \text{ in.} + .125 \text{ in.}) \times 0.380 \text{ in.} = 2.57 \text{ in.}^2 > 2.38 \text{ in.}^2$ ok

Check tee flange for tension and prying action.

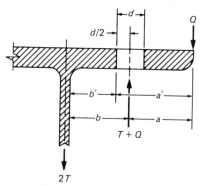

$$b = \frac{3.50 \text{ in.} - .535 \text{ in.}}{2} = 1.48 \text{ in.}$$

$$b' = b - d/2 = 1.11 \text{ in.}$$

$$a = \frac{6.530 \text{ in.} - 3.50 \text{ in.}}{2} = 1.52 \text{ in.} < 1.25\,b \quad \text{ok}$$

$$a' = a + d/2 = 1.89 \text{ in.}$$

$$d' = 0.75 \text{ in.} + 0.06 \text{ in.} = 0.81 \text{ in.}$$

$$\delta = 1 - d'/p = 1 - 0.81/4.25 = 0.81 \quad \text{(AISCM Eq. 1, p. 4-89)}$$

$$M = p\, t_f^2 F_y/8 = 4.25 \text{ in.} \times (0.535 \text{ in.})^2 \times 36 \text{ ksi}/8$$

$$= 5.47 \text{ in.-k} \quad \text{(AISCM Eq. 2)}$$

$$\alpha = (Tb'/M - 1)/\delta = 21.2 > 1.0 \quad \text{use } \alpha = 1.0 \quad \text{(AISCM Eq. 3)}$$

$$B_c = T_b \left[1 + \left(\frac{\delta\alpha}{1 + \delta\alpha}\, \frac{b'}{a'} \right) \right]$$

$$= 17.28 \text{ k} \left[1 + \frac{0.81 \times 1.0}{1 + 0.81}\, \frac{1.11}{1.89} \right] = 21.79 \text{ k} \quad \text{(AISCM Eq. 4)}$$

$$t_{f\,\min} = \left[\frac{\delta\, B_c a' b'}{p\, F_y (a' + \delta\alpha(a' + b'))} \right]^{1/2}$$

$$= \left[\frac{0.81 \times 21.79 \times 1.89 \times 1.11}{4.25 \times 36 \times (1.89 + 0.81(1.89 + 1.11))} \right]^{1/2}$$

$$= 0.237 \text{ in.} \quad \text{(AISCM Eq. 5)}$$

$$Q = B_c - T_b = 21.79 \text{ k} - 17.28 \text{ k} = 4.51 \text{ k} \quad \text{(AISCM Eq. 6)}$$

$$t_{f\,\text{provided}} = 0.535 \text{ in.} < 0.237 \text{ in.} \quad \text{ok}$$

Use WT 10.5 × 25, 8-$\frac{1}{2}$ in. long.

PROBLEMS TO BE SOLVED

7.1. Design the rigid connection shown to carry and end moment of 85 ft-kips and an end shear of 35 kips due to dead and live loads only. Use $\frac{3}{4}$-in. A325-N bearing type bolts. The connection plates are welded to the column and bolted to the beam.

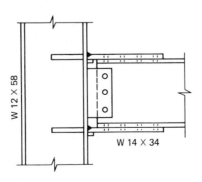

7.2. Design bolted flange plate moment connections for the beam and columns as shown. Design the connections to carry

a) Full end moment with $\frac{3}{4}$-in. A325-F bolts
b) 75% of full end moment with $\frac{3}{4}$-in. A325-N bolts (semirigid)
c) 50% of full end moment with $\frac{3}{4}$-in. A325-X bolts (semirigid).

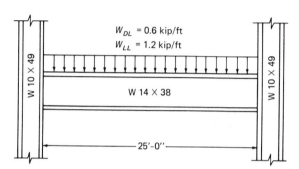

7.3. Rework Problem 7.2a for an end plate moment connection.

7.4. A W 12 × 40 beam frames into the web (weak axis) of a W 14 × 68 column. Design a semirigid moment connection of tee-sections and shear angles to transfer a moment of 50 ft-kips and shear of 23 kips. Assume full lateral support and $\frac{7}{8}$-in. A325-N bolts.

7.5. Design a base plate for a W 12 × 65 column to resist an axial force of 275 kips on an 18-in. × 18-in. support of 3000 psi (f_c') concrete. Choose the plate dimensions such that m and n are relatively equal. Then test the base plate for a 65-ft-kip bending moment due to wind along the major axis. Redesign the base plate if necessary.

7.6. Rework Problem 7.5 for 30 × 30-in. base of f_c' = 2500-psi concrete.

7.7. Using the approximate method, design a base plate for a W 12 × 65 column to resist a combined axial force of 200 kips and bending moment along the major axis of 130 ft-kips. Assume the footing of 4000-psi concrete to be the same size as the base plate. Use three 1-in. diameter anchor bolts on each side of the column flanges, 8 in. from the column centerline. Verify that hold-down anchor bolt stress does not exceed 12,000 psi.

7.8. Check Problem 7.5 using the method that assumes the base plate to act similar to a reinforced concrete beam. Assume $n = 9$.

7.9. Design a base plate for a W 14 × 90 column with 350 kips axial force, 250-ft-kip moment along the strong axis, and 150-ft-kip moment along the weak axis. Assume F_b = 0.7 and f_c' = 2.5 ksi.

7.10. Design a bolted beam splice for a W 27 × 84 at a point where the shear is 150 kips and the moment is 150 ft-kips. Assume $\frac{3}{4}$-in. A-325-N bolts.

7.11. Design a bolted beam splice to join a W 30 × 99 to a W 30 × 173 where the shear is 100 kips and the moment is 160 ft-kips. Use $\frac{3}{4}$-in. A 325-N bolts. Sketch detail showing bolt spacings and necessary shims.

7.12. A hanger is supporting a 60-kip force from a W 27 × 146 beam as shown. Verify whether a WT 12 × 47 × 8$\frac{1}{2}$ in. is adequate. Use $\frac{3}{4}$-in. A490-N bolts on 5$\frac{1}{2}$-in. gage.

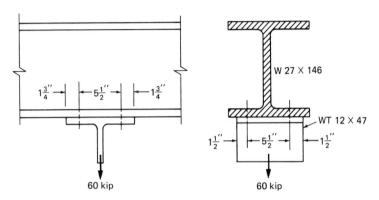

7.13. Select a double-angle hanger to support the loading shown in Problem 7.12. Use $\frac{7}{8}$-in. A325-X bolts.

8
Torsion

8.1 TORSION

Members that have loads applied away from their shear center are subject to torsion.[1] These loads, when multiplied by their distances from the shear center of the member, cause deformations due to rotation that may also include warping. The rotation of the member is measured by the angle of the twist.

8.2 TORSION OF AXISYMMETRICAL MEMBERS

For pure torsion in the elastic range, a plane section perpendicular to the member's axis is assumed to remain plane after the torque has been applied. In this plane, shearing stress and strain vary linearly from the neutral axis. This behavior and distribution of stress resulting solely from pure torsion is also known as *St. Venant torsion.* Any shearing stress produced by bending would be added to the St. Venant shear stresses.

In a solid circular cylinder subjected to torsion (Fig. 8.2), the maximum shearing stress occurs at the perimeter. The stress at any point is obtained from

$$f_v = \frac{Tr}{J} \tag{8.1}$$

and the angle of rotation is determined by

$$\phi = \frac{Tl}{GJ} \tag{8.2}$$

---

[1]For sections with a center of symmetry, the shear center coincides with the center of symmetry. Refer to Section 8.5 for discussion of shear center.

Fig. 8.1. Sears Tower, Chicago.

where T is the torque (in.-lb), r is the distance from the center of the member to the point under consideration (in.), l is the length of the cylinder (in.), G is the shear modulus $E/2(1 + \mu)$ (psi), μ is Poisson's ratio, which for steel is between .25 and .30, and J is the polar moment of inertia (in.^{4}), which for a solid round section is expressed by

$$J = \frac{\pi d^4}{32} \qquad (8.3)$$

When it is required to limit the torsional moment (torque), so that the maximum shear stress does not exceed its allowable value, Eq. 8.1 is transposed to

$$T = \frac{F_v J}{c} \qquad (8.4)$$

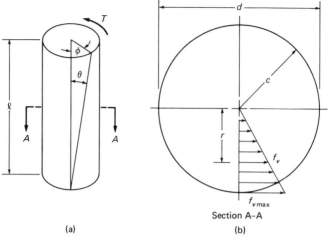

Section A-A

(a)   (b)

Fig. 8.2. Torsion on circular cross section.

where F_v is the allowable shear stress of the material, and c is the distance from the center of the member to the outer fibers.

Example 8.1. Determine the maximum allowable torque for a $1\frac{1}{2}$-in.-diameter steel shaft with an allowable stress $F_v = 14.5$ ksi.

Solution. The maximum torque is determined by the equation

$$T = F_v \frac{J}{c}$$

$$F_v = 14.5 \text{ ksi}$$

$$J = \frac{\pi d^4}{32} = 0.497 \text{ in.}^4$$

$$c = \frac{1.5 \text{ in.}}{2} = 0.75 \text{ in.}$$

$$T = \frac{14.5 \text{ ksi} \times 0.497 \text{ in.}^4}{0.75 \text{ in.}} = 9.6 \text{ in.-kips}$$

Converting to ft-kips

$$\frac{9.6 \text{ in.-kips}}{12 \text{ in./ft}} = 0.80 \text{ ft-kips}$$

Example 8.2. Determine the maximum torsion for a hollow shaft with an outside diameter (d_o) of 9 in. and an inside diameter (d_i) of 8 in. Allowable stress is 8000 psi.

Solution. In a hollow, circular cylinder, J is given by $\pi/32\,(d_o^4 - d_i^4)$ where d_o and d_i are the outer and inner diameters, respectively.

$$T = F_v \frac{J}{c}$$

$$F_v = 8000 \text{ psi}$$

$$J = \frac{\pi(d_o^4 - d_i^4)}{32} = 242 \text{ in.}^4$$

$$c = 4.5 \text{ in.}$$

$$T = \frac{8000 \text{ psi} \times 242 \text{ in.}^4}{4.5 \text{ in.}} = 430{,}200 \text{ in.-lb}$$

Converting to ft-kips

$$\frac{430{,}200 \text{ in.-lb}}{12 \text{ in./ft} \times 1000 \text{ lb/k}} = 35.85 \text{ ft-kips}$$

Example 8.3. Determine the maximum torque a standard-weight 6-in.-diameter pipe can carry when allowable shear stress is 14.5 ksi.

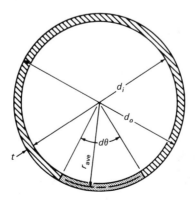

Solution.

a) Calculating J from AISC tables.

We know that $J_p = I_x + I_y$. In our case, J_p 2 $\times$ 28.1 in.^{4} = 56.2 in.^{4}

b) Inertia by integration.

Consider the pipe as a thin-walled element, such that its thickness is small with respect to the diameter. Hence, $d_o = d_i = d_{ave}$ = 6.345 in.

$$J = \int r^2 \, dA \quad \text{in our case}$$

$$dA = r_{ave} \, d\theta \, t$$

$$J = \int_0^{2\pi} tr_{ave}^3 \, d\theta = 2\pi r^3 t$$

$$= 2\pi \left(\frac{6.345 \text{ in.}}{2}\right)^3 0.28 \text{ in.} = 56.17 \text{ in.}^4$$

c) Consider $J = J_o - J_i$

$$J = J_o - J_i = \frac{\pi}{32} [(6.625 \text{ in.})^4 - (6.065 \text{ in.})^4] = 56.28 \text{ in.}^4$$

We see that all three methods result in nearly identical values.

$$T = \frac{F_v J}{c} = \frac{14.5 \text{ ksi} \times 56.2 \text{ in.}^4}{3.313 \text{ in.}} = 245.97 \text{ in.-k} = 20.50 \text{ ft-k}$$

Example 8.4. What size solid steel circular bar is required to carry a torsional moment of 1 ft-kip if the allowable stress is 14.5 ksi?

Solution.

$$c = F_v \frac{J}{T}$$

$$J = \frac{\pi(2c)^4}{32} = \frac{\pi c^4}{2}$$

$$T = 1 \text{ ft-k} \times 12 \text{ in./ft} \times 1000 \text{ lb/k} = 12,000 \text{ in.-lb}$$

$$c = \frac{14{,}500 \text{ psi} \left(\dfrac{\pi c^4}{2}\right)}{12{,}000 \text{ in.-lb}}$$

$$c^3 = \frac{(2)\, 12{,}000 \text{ in.-lb}}{14{,}500 \text{ psi} \times \pi} = 0.527 \text{ in.}^3$$

$$c = \sqrt[3]{0.527 \text{ in.}^3} = 0.808 \text{ in.}$$

Use $1\text{-}\frac{5}{8}$-in.-diameter solid bar.

Example 8.5. A 6-ft-kip torque is applied to a 4-in. standard-weight pipe. Determine the stress at the inside face and outside face of the pipe and the angle of twist if the tube is 6 ft long. $E_s = 29 \times 10^6$ psi, and $\mu = 0.25$.

Solution. Member properties

$$d_o = 4.5 \text{ in.} \qquad d_i = 4.026 \text{ in.}$$

$$f_v = \frac{Tr}{J}, \qquad \text{where } r = 2.013 \text{ in. for the shear stress at inside face}$$
$$r = 2.25 \text{ in. for the shear stress at outside face}$$

$$\phi = \frac{Tl}{GJ}$$

$$J = \frac{\pi(d_o^4 - d_i^4)}{32} = 14.47 \text{ in.}^4$$

$$l = 6 \text{ ft} \times 12 \text{ in./ft} = 72 \text{ in.}$$

Also, from AISC tables, $J = 2 \times 7.23 \text{ in.}^4 = 14.46 \text{ in.}^4$

$$G = \frac{E}{2\,(1 + \mu)} = 11.6 \times 10^6 \text{ psi}$$

$$T = 6.0 \text{ ft-k} \times 12 \text{ in./ft} \times 1000 \text{ lb/k} = 72{,}000 \text{ in.-lb}$$

Stress at the inside face

$$f_v = \frac{72{,}000 \text{ in.-lb} \times 2.013 \text{ in.}}{14.47 \text{ in.}^4} = 10{,}000 \text{ psi}$$

Stress at the outside face

$$f_v = \frac{72,000 \text{ in.-lb} \times 2.25 \text{ in.}}{14.47 \text{ in.}^4} = 11,200 \text{ psi}$$

Angle of twist

$$\phi = \frac{72,000 \text{ in.-lb} \times 72 \text{ in.}}{11.6 \times 10^6 \text{ psi} \times 14.47 \text{ in.}^4} = 0.031 \text{ radians} = 1.77 \text{ degrees}$$

8.3 TORSION OF SOLID RECTANGULAR SECTIONS

When a torsional moment is applied to a noncircular section, the cross section rotates and deforms nonuniformly. As a result, a plane section perpendicular to the member's axis does not remain plane. This additional deformation is referred to as *warping*. In the case of solid rectangular sections, twisting will occur about the axis. No shear strain will occur at the corners, and consequently shear stresses at the corners are equal to zero.

In Fig. 8.3, the shear stress distribution indicates that the maximum stress occurs at the midpoint of the long side.

This nonuniform stress distribution is in direct contrast to the stress distribution of a circular section (Fig. 8.2b).

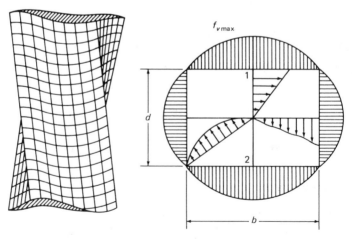

Fig. 8.3. Torsion on rectangular bars.

Table 8.1. Torsional Factors for Solid Rectangular Bars (Blodgett, *Design of Welded Structures*, reprinted with permission).

$\dfrac{b}{d}$ =	1.00	1.50	1.75	2.00	2.50	3.00	4.00	6	8	10	∞
α	.208	.231	.239	.246	.258	.267	.282	.299	.307	.313	.333
β	.141	.196	.214	.229	.249	.263	.281	.299	.307	.313	.333

To determine the stress and rotation of a rectangular section, the use of two characteristic elements α and β becomes necessary.

The α factor is used in computing the torsional resistance for the stress formula. A solid rectangular section has a torsional constant of

$$J^* = \alpha b d^3 \tag{8.5}$$

where b is the larger and d is the smaller dimension. α is obtained from Table 8.1. The maximum shear stress due to a torque is then

$$f_v = \frac{Td}{J^*} \tag{8.6}$$

and is located at the middle of the long side. The twist angle of the member is determined by

$$\phi = \frac{TL}{GJ^*} \tag{8.7}$$

where J^* is now given by

$$J^* = \beta b d^3 \tag{8.8}$$

It can be noted from Table 8.1 that α and β converge to the value $\frac{1}{3}$ as the ratio b/d becomes large. When the ratio is 4.0, the two coefficients are nearly the same, and the torsional resistance becomes the same for use in the stress and rotation formulas.

Example 8.6. Show that for a rectangular section under torsion, the shear stress at the corners is zero.

Solution. Consider a corner element as shown. Assuming that a shearing stress exists at the corner, two components can be shown parallel to the edges of the bar. However, as equal shear stress always occurs in pairs acting on mutually perpendicular planes, these components would have to be met by shear stresses lying in the plane of the outside surfaces. Because the surfaces are free surfaces, no stresses exist. Therefore, the shear stress at the corners must be zero.

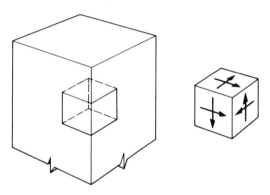

Example 8.7. Determine the maximum torque that can be applied to a 2-in. × 4-in. cantilevered steel bar. $F_v = 14.5$ ksi.

Solution.

$$F_v = \frac{Td}{J^*}, \quad T = \frac{F_v J^*}{d}, \quad J^* = \alpha bd^3$$

$d = 2$ in.

$b = 4$ in.

$b/d = 2$ From Table 8.1, $\alpha = 0.246$

$J^* = 0.246 \times 4 \text{ in.} \times (2 \text{ in.})^3 = 7.87 \text{ in.}^4$

$$T = \frac{14.5 \text{ ksi} \times 7.87 \text{ in.}^4}{2 \text{ in.}} = 57.06 \text{ in.-kip} \times \frac{1 \text{ ft}}{12 \text{ in.}} = 4.75 \text{ ft-kip}$$

Example 8.8. A 6.25-ft-kip torque is applied to a rectangular member. Determine the maximum stress when acting on a cantilevered 3-in. × 5-in. rectangular bar 4 ft long. Also find the rotation due to the applied torque if the shear modulus G is 11.6×10^6 psi.

Solution.

$$f_v = \frac{Td}{J^*}$$

$$\phi = \frac{Tl}{GJ^*}$$

$b = 5$ in.

$d = 3$ in.

$b/d = \dfrac{5}{3} = 1.67$ Interpolation must be used in Table 8.1 for b/d between 1.5 and 1.75.

$\alpha = 0.236$ J^* (stress) $= 0.236 \times 5$ in. $\times (3 \text{ in.})^3 = 31.86$ in.^{4}

$\beta = 0.208$ J^* (twist angle) $= 0.208 \times 5$ in. $\times (3 \text{ in.})^3 = 28.08$ in.^{4}

$l = 4$ ft $\times 12$ in./ft $= 48$ in.

$T = 6.25$ ft-kip $\times 12,000 = 75,000$ in.-lb

$$f_v = \frac{75,000 \text{ in.-lb} \times 3 \text{ in.}}{31.86 \text{ in.}^4} = 7062 \text{ psi}$$

$$\phi = \frac{75,000 \text{ in.-lb} \times 48 \text{ in.}}{11.6 \times 10^6 \text{ psi} \times 28.08 \text{ in.}^4} = 0.011 \text{ radians} = 0.64 \text{ degrees}$$

Example 8.9. A $1\text{-}\frac{1}{2}$-in. $\times$ 3-in. bar 6 ft long is subject to a 1000 ft-lb torsional moment. Determine if the bar is safe with an allowable shear of 14.5 ksi and an angle of rotation limited to 1.5 degrees. $G = 11.6 \times 10^6$ psi.

Solution.

$b/d = 2$; $\alpha = 0.246$, $\beta = 0.229$

$$f_v = \frac{Td}{J^*} ; \quad J^* \text{ (stress)} = 0.246 \times 3 \text{ in.} \times (1.5 \text{ in.})^3 = 2.49 \text{ in.}^4$$

$$f_v = \frac{12 \text{ in.-kip} \times 1.5 \text{ in.}}{2.49 \text{ in.}^4} = 7.23 \text{ ksi} < 14.5 \text{ ksi} \quad \text{ok}$$

$$\phi = \frac{Tl}{J^*}; \quad J^* \text{ (rotation)} = 0.229 \times 3 \text{ in.} \times (1.5 \text{ in.})^3 = 2.32 \text{ in.}^4$$

$$\phi = \frac{12 \text{ in.-kip} \times (6 \text{ ft} \times 12 \text{ in./ft})}{11.6 \times 10^3 \text{ ksi} \times 2.32 \text{ in.}^4} = 0.032 \text{ radians} = 1.84 \text{ degrees} \quad \text{N.G.}$$

The bar will rotate $0.34°$ greater than allowable. Therefore, the member is unsatisfactory due to excessive rotation.

Example 8.10. A 4-ft, simply supported solid steel bar is restrained against torsion at the ends. The bar will support a sign weighing 1.8 kips, which act at quarter points 12 in. from the bar's shear center. Design the bar to carry the applied load and torsion. $F_y = 36$ ksi.

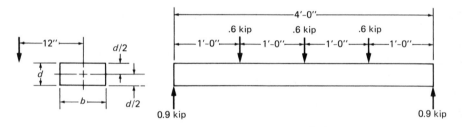

Solution. Rectangular sections bent about their weak axis develop greater bending and torsional strength due to their resistance to lateral-torsional buckling (see Commentary 1.5.1.4.3). Therefore, assume the bar to be b in. wide and $d = b/2$ in. high.

Design the section for flexure, and check the increase in shear stress due to torsion.

$$\frac{M}{F_b} = S = \frac{bd^2}{6}$$

$$M_{max} = (0.9 \text{ k} \times 2 \text{ ft}) - (0.6 \text{ k} \times 1 \text{ ft}) = 1.2 \text{ ft-kip} = 14{,}400 \text{ in.-lb}$$

$$F_b = 0.75 \, F_y = 27 \text{ ksi} \quad \text{(AISCS 1.5.1.4.3)}$$

$$S = \frac{b}{6} \left(\frac{b}{2}\right)^2 = \frac{b^3}{24}$$

$$\frac{b^3}{24} = \frac{14{,}400 \text{ in.-lb}}{27{,}000 \text{ psi}}$$

$$b^3 = 12.8 \text{ in.}^3$$

$$b = 2.34 \text{ in.} \quad \text{say } 2\frac{1}{2} \text{ in.}$$

$$d = b/2 = 1\frac{1}{4} \text{ in.}$$

Check shear stress

$$F_v = 0.40\, F_y = 14.5 \text{ ksi}$$

$$J^* = 0.246 \times 2.50 \text{ in.} \times (1.25 \text{ in.})^3 = 1.20 \text{ in.}^4$$

$$T = 0.9 \text{ k} \times 12 \text{ in.} = 10.8 \text{ in.-k}$$

$$f_{v \text{ torque}} = \frac{Tc}{J^*} = \frac{10.8 \text{ in.-k} \times 1.25 \text{ in.}}{1.20 \text{ in.}^4} = 11.25 \text{ ksi}$$

$$f_{v \text{ bending}} = \frac{3}{2} \times \frac{0.9 \text{ k}}{2.50 \times 1.25 \text{ in.}} = 0.43 \text{ ksi}$$

$$f_{v \text{ torque}} + f_{v \text{ bending}} = 11.25 + 0.43 = 11.68 \text{ ksi} < F_v$$

8.4 TORSION OF OPEN SECTIONS

Open structural steel members can often be considered as consisting of thin rectangular elements, where

$$J^* = \Sigma \frac{bd^3}{3} \qquad\qquad (8.9)$$

and b is the largest dimension of each element, and d is the smallest. For members with varying thicknesses, the average thickness is used as d. For commonly used rolled shapes, the AISCM has listed as J the torsional constant of each section. A slight variance can be observed by calculating J^* in a direct manner and comparing to the values for J listed in the properties table. These variations can be attributed to the effect of fillets, the rounded corners, and the ratio b/d not being infinite, which is implied by the factor 3 in the denominator of Eq. 8.9.

The open sections in Fig. 8.4 show their shear flow and the locations of maximum shear. It can be seen that each shape behaves as a series of rectangular

---

[2]Some authors recommend the use of $J^* = \eta \Sigma \frac{1}{3} bd^3$, where η is a coefficient slightly larger than 1 depending on the type of section (i.e., η is equal to 1.0 for angles, 1.12 for channels, 1.15 for T sections and 1.2 for I and wide flange sections).

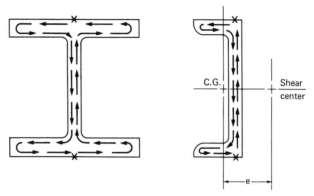

Fig. 8.4. Shear flow and maximum stress in open structural shapes in torsion.

elements with their shear stresses flowing continuously. An approximate value of the maximum shear occurring at the points indicated can be calculated from

$$f_v = \frac{Tt}{J^*} \qquad (8.10)$$

Welds that may be used for a section subjected to torsional loading also must be designed to carry the torsional stress. This stress is superimposed on the shear stress in the web. The weld is designed for the combined stresses.

Example 8.11. Determine the torsional constant J^* for a W 12 X 58, and compare the result with the value given in the AISCM.

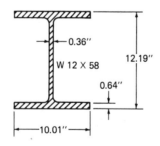

Solution. The value J^* can be determined from the formula

$$J^* = \Sigma \tfrac{1}{3} bd^3 \qquad \text{where } b = \text{width of each element}$$
$$d = \text{thickness}$$

$$J^* = \tfrac{1}{3} \, [2 \times (10.01 \text{ in.} \times (0.640 \text{ in.})^3) + (10.91 \text{ in.} \times (0.360 \text{ in.})^3)]$$

$$= 1.92 \text{ in.}^4$$

The value J from the AISCM is 2.10. The variation can be explained by the approximation of the wide flange by rectangular elements.

Example 8.12. An externally applied force of 850 lb acts 18 in. to the right of the web centerline. Design the most economic wide flange for a 10-ft cantilever. $F_v = 13.5$ ksi, and $F_b = 20$ ksi. Neglect the combination of shear stresses due to torsion and bending stresses.

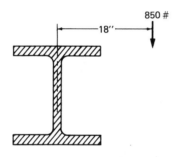

Solution.

$$\frac{t_f}{J} \leqslant \frac{F_v}{T}$$

The allowable stresses have been reduced to allow for the combined effect of the shear stress due to torsion and bending stresses, resulting in a larger shear stress. This can be seen by the use of Mohr's circle.

Design the member first for torsion, then verify for bending and shear.

The member must be found by trial and error. Use the torsional constant in the AISCM.

$$\frac{t_f}{J} \leqslant \frac{13,500 \text{ psi}}{850 \text{ lb} \times 18 \text{ in.}} = .882$$

Try W 16 $\times$ 31, $t_f = 0.440$ $J = 0.46$

$$\frac{0.440}{0.46} = 0.957 > 0.882 \quad \text{N.G.}$$

Try W 10 $\times$ 30, $t_f = 0.51$ $J = 0.62$

$$\frac{0.51}{0.62} = 0.822 < 0.882 \quad \text{ok}$$

Check for bending and shear

$$M = 0.850 \text{ k} \times 10 \text{ ft} \times 12 \text{ in./ft} = 102 \text{ in.-k}$$

$$S = 32.4 \text{ in.}^3$$

$$f_b = \frac{102}{32.4} = 3.15 \text{ ksi} < 20 \text{ ksi}$$

$$f_{v \text{ web}} = \frac{0.85}{10.47 \times 0.30} + \frac{0.85 \times 18 \times 0.30}{0.62} = 7.67 < 13.5 \text{ ksi} \quad \text{ok}$$

$$f_{v \text{ flg}} = \frac{0.85 \times 18 \times 0.51}{0.62} = 12.59 \text{ ksi} < 13.5 \text{ ksi} \quad \text{ok}$$

For the reader familiar with Mohr's circle, the combination is indicated as follows.

For an element at the top flange near the point of fixity, we have $f_x = 3.15$ and $f_y = 0$.

Considering Mohr's circle of center A,

$$OA = \frac{f_x + f_y}{2} = \frac{3.15 + 0}{2} = 1.58 \text{ ksi}$$

$$AB = \frac{f_x - f_y}{2} = \frac{3.15 - 0}{2} = 1.58 \text{ ksi}$$

$$BD = f_v = 12.59 \text{ ksi}$$

$$AD = \sqrt{AB^2 + BD^2} = 12.69 \text{ ksi} = AC = f_{v \text{ max}}$$

$$OE = f_{b \text{ max}} = OA + AE = \frac{f_x + f_y}{2} + f_{v \text{ max}} = 1.58 + 12.69 = 14.27 \text{ ksi}$$

The orientation is an angle $EAD/2$.

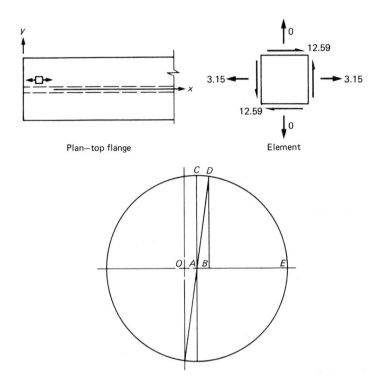

Plan—top flange   Element

Example 8.13. A wide-flange beam spanning 24 ft supports a 1-kip-per-foot brick wall, which acts 5 in. off center. Design the beam to withstand torsion and bending. Assume the beam is simply supported, but restrained against torsion at the supports. F_v = 14.5 ksi.

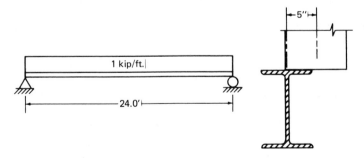

Solution. Determine the beam to carry the flexure (laterally unsupported); then check to see if it will be safe for the torsion.

$$S = \frac{M}{F_b}$$

$$F_v = \frac{Tc}{J}$$

$$M = \frac{wL^2}{8} = \frac{1 \text{ k/ft} \times (24 \text{ ft})^2}{8} = 72 \text{ ft-k}$$

Designing for flexure alone, the lightest available section is W 10 × 45. (AISC, Part 2).

$$S_x = 49.1 \text{ in.}^3$$

$$f_b = M/S_x = \frac{72 \text{ ft-k} \times 12 \text{ in./ft}}{49.1 \text{ in.}^3} = 17.60 \text{ ksi}$$

Check for torsion

$$T = \frac{5}{12} \text{ ft} \times \frac{24 \text{ ft}}{2} \times 1.0 \text{ k/ft} = 5.0 \text{ ft-k} = 60 \text{ in.-k}$$

$$J = 1.51 \text{ in.}^4$$

$$d = t_f = 0.62 \text{ in.}$$

$$f_v = \frac{60 \text{ in.-k} \times 0.62 \text{ in.}}{1.51 \text{ in.}^4} = 24.63 \text{ ksi} > 14.5 \text{ ksi} \qquad \text{N.G.}$$

Calculating the ratio d/J,

$$\frac{d}{J} = \frac{F_v}{T}$$

$$\frac{d}{J} = \frac{14.5 \text{ ksi}}{60 \text{ in.-k}} = 0.24 \text{ in.}^{-3}$$

Try W 12 × 72, $d = t_f = 0.670$ in., $J = 2.93$ in.^{4}

$$\frac{0.670 \text{ in.}}{2.93 \text{ in.}^4} = 0.229 \text{ in.}^{-3} < 0.24 \text{ in.}^{-3} \qquad \text{ok}$$

Trying for a more economical section,

$$W 14 \times 68, \qquad d = t_f = 0.720 \text{ in.}, \qquad J = 3.02 \text{ in.}^4$$

$$\frac{0.720 \text{ in.}}{3.02 \text{ in.}^4} = 0.238 \text{ in.}^{-3} < 0.24 \text{ in.}^{-3} \qquad \text{ok}$$

We see also by inspection (AISC, Part 2, "Allowable Moments in Beams") that the bending stress remains under control.

Use W 14 × 68.

Example 8.14. Determine the required weld for the section shown at a location where the shear force is 10 kips and the torque is 300 in.-kips. Use A36 steel and $\frac{5}{16}$-in. E70 intermittent weld.

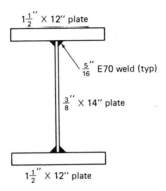

Solution. The force to be carried by the weld is produced at the interface of the flange and web. The force is the superposition of the shear flow due to vertical loading and the shear flow due to the torsional moment.

Shear stress due to direct shear force

$$f_v = \frac{V}{bd} = \frac{10}{\frac{3}{8} \times \left(14 + 2\left(1\frac{1}{2}\right)\right)} = 1.57 \text{ ksi}$$

$$J^* = \frac{1}{3} \times \left[2 \times 12 \times \left(1\frac{1}{2}\right)^3 + 14 \times \left(\frac{3}{8}\right)^3\right] = 27.25 \text{ in.}^4$$

Shear stress due to torsion

$$f_v = \frac{Tt_w}{J^*} = \frac{300 \times 0.375}{27.25} = 4.13 \text{ ksi}$$

Distribution due to direct shear — 1.57 ksi

Distribution due to torque — 4.13 ksi

Shear stress distribution in web

The shear flow to be carried by the weld is the reaction at right.

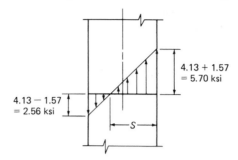

Force carried by weld,

$$q = 0.54 \times \frac{0.375}{2} + (2.6 - 0.54) \times \frac{0.375}{2} \times \frac{2}{3} = 0.36 \text{ k/in.}$$

For E70 $\frac{5}{16}$ in. weld,

$$w_l = 5 \times 0.928 \text{ k/in./sixteenth} = 4.64 \text{ k/in.}$$

For 1 foot length,

$$q = 0.36 \text{ k/in.} \times 12 \text{ in./ft} = 4.32 \text{ k/ft}$$

Length of weld required[3]

$$L = \frac{4.32 \text{ k/ft}}{4.64 \text{ k/in./ft}} = 0.93 \text{ in.} \quad \text{use } 1\frac{1}{2} \text{ in. weld at 12 in. centers (min.).}$$

Example 8.15. Determine the required intermittent weld length per foot of section for the plate section shown. Use E70 minimum size fillet weld. $V = 200$ kips; $T = 40$ ft-k.

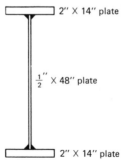

---

[3]Welds in engineering practice are generally continuous. The design of intermittent welds is provided as an exercise.

Solution.

$$V_{max} = 200 \text{ kips;} \quad T = 40 \text{ ft-kips}$$

Verify section, and design welds.

$$J = \frac{1}{3}\left[(14 \text{ in.} \times (2 \text{ in.})^3) \times 2 + \left(48 \text{ in.} \times \left(\frac{1}{2} \text{ in.}\right)^3\right)\right] = 76.7 \text{ in.}^4$$

$$f_{v\,V} = \frac{200 \text{ k}}{\frac{1}{2} \text{ in.} \times 52 \text{ in.}} = 7.69 \text{ ksi}$$

$$f_{v\,T\,fl} = \frac{40 \text{ ft-k} \times 12 \text{ in./ft} \times 2 \text{ in.}}{76.7 \text{ in.}^4} = 12.52 \text{ ksi}$$

$$f_{v\,T\,web} = \frac{40 \text{ ft-k} \times 12 \text{ in./ft} \times 0.5 \text{ in.}}{76.7 \text{ in.}^4} = 3.13 \text{ ksi}$$

$$f_{v\,web\,tot} = 7.69 \text{ ksi} + 3.13 \text{ ksi} = 10.82 \text{ ksi}$$

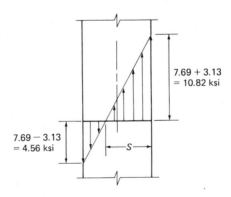

$$s = 0.5 \text{ in.} \times \frac{10.82}{4.56 + 10.82} = 0.352 \text{ in.}$$

Total shear flow

$$q = \frac{0.352 \text{ in.} \times 10.82 \text{ ksi}}{2} = 1.90 \text{ k/in.}$$

Minimum weld size = $\frac{5}{16}$ in.   (AISCS 17.2)

$$w_l = 5 \times 0.928 \text{ k/in.} = 4.64 \text{ k/in.}$$

Length of weld required per foot

$$q_{t/ft} = 2.18 \text{ k/in.} \times 12 \text{ in./ft} = 26.16 \text{ k/ft}$$

$$L = \frac{26.16 \text{ k/ft}}{4.64 \text{ k/in./ft}} = 5.6 \text{ in.} \quad \text{say 6 in.}$$

Use $\frac{5}{16}$-in. E70 weld 6 in. long, spaced 12 in. on center.

8.5 SHEAR CENTER

A prismatic member has a singular shear axis through which a force can be applied from any angle and no twisting moment will result. For any given cross section, the point on the axis is called the *shear center* of the section. If applied forces pass through the shear center, the member is subject only to flexural stresses. Forces applied away from the shear center develop a torsional moment that twists around the shear center.

Consider a section with an axis of symmetry transversely loaded in the plane of the axis, i.e., a wide flange loaded in the plane of the web (Fig. 8.5a). It can be seen that the external shear force and the internal shear forces due to the shear flow are in equilibrium. Consider now a channel loaded at the centroid (Fig. 8.5b). The forces are in equilibrium, but an unbalanced moment has developed due to the shear flow in the flanges. If the external force is applied at some distance away from the centroid, the moment can be brought to zero. The point of load application is the shear center of the figure. It can be proved

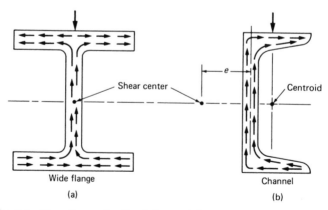

Fig. 8.5. Shear flow under direct shear and shear centers of common sections.

that the distance between the shear center and the center of the web is given by

$$e = \frac{\Sigma I_x x}{\Sigma I_x} \qquad (8.11)$$

Values of e for channel sections can be found in the tables that show properties for designing. Values for J may be obtained from *Structural Shapes*, published by Bethlehem Steel.

Example 8.16. Calculate the location of the shear center for the shape shown below.

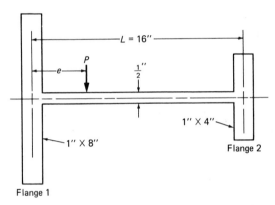

Solution. To locate the shear center, consider a force P applied to the section such that both flanges will deflect equally. Then

$$\frac{(L - e)P}{I_1} = \frac{eP}{I_2}$$

Solving for e,

$$e = \frac{LI_2}{I_1 + I_2}$$

Using the expression given

$$e = \frac{16 \text{ in.} \times 1 \text{ in.} \times \dfrac{(4 \text{ in.})^3}{12}}{1 \text{ in.} \times \dfrac{(8 \text{ in.})^3}{12} + 1 \text{ in.} \times \dfrac{(4 \text{ in.})^3}{12}} = 1.78 \text{ in.}$$

Example 8.17. Determine the approximate location of the shear center for a C 10 × 15.3. Compare the approximate location with the location given in the AISCM.

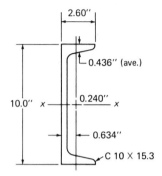

Solution. For symmetrical sections, the axis of symmetry is one shear axis. The other axis is found by

$$e = \frac{\Sigma I_x x}{\Sigma I_x}$$

where x is the distance from the Y axis to the centroid of the elements. Consider the Y axis passing through the center of the web. Assuming the flanges as separate parts,

$$e = \frac{I_{web} x + 2I_{flg} x}{I_{web} + I_{flg}} = \frac{\left(\dfrac{t_w d^3}{12} \times 0\right) + 2(t_f b)\left(\dfrac{d}{2} - \dfrac{t_f}{2}\right)^2 \left(\dfrac{b}{2} - \dfrac{t_w}{2}\right)}{I_x}$$

$t_f = 0.436$ in., $\quad t_w = 0.24$ in., $\quad b = 2.60$ in., $\quad d = 10.0$ in., $\quad I_x = 67.4$ in.^{4}

$$e = \frac{2(0.436 \text{ in.} \times 2.60 \text{ in.})\left(\dfrac{10 \text{ in.}}{2} - \dfrac{0.436 \text{ in.}}{2}\right)^2 \times \left(\dfrac{2.60 \text{ in.}}{2} - \dfrac{0.24 \text{ in.}}{2}\right)}{67.4 \text{ in.}^4}$$

$= 0.908$ in.

The value for shear center location in the AISCM is measured from the back of the web. To determine e_0,

$$e_0 = 0.908 \text{ in.} - \frac{0.24 \text{ in.}}{2} = 0.788 \text{ in.}$$

The tabular value is $e_0 = 0.796$ in. The minor variation can be explained by the approximation of the flanges, which are tapered by rectangular elements.

Example 8.18. Determine if a cantilevered C 12 X 30 10 ft long will carry the applied force shown if F_v = 14.5 ksi and F_b = 24.0 ksi.

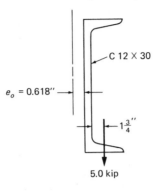

Solution.

$$T = 5.0 \text{ k} \times (0.618 \text{ in.} + 1.75 \text{ in.}) = 11.84 \text{ in.-k}$$

$$M = 5.0 \text{ k} \times 10 \text{ ft} \times 12 \text{ in./ft} = 600 \text{ in.-k}$$

$$V = 5.0 \text{ k}$$

$$J = \frac{1}{3}\left[(12.0 - 0.501) \times (0.510)^3 + 2 \times \left(3.17 - \frac{0.510}{2}\right) \times (0.501)^3\right]$$

$$= 0.753 \text{ in.}^4$$

$$S = 27.0 \text{ in.}^3$$

$$f_{v \text{ flg}} = \frac{11.84 \text{ in.-k} \times 0.501 \text{ in.}}{0.753 \text{ in.}^4} = 7.88 \text{ ksi} < 14.5 \text{ ksi}$$

$$f_{v \text{ web}} = \frac{11.84 \text{ in.-k} \times 0.510 \text{ in.}}{0.753 \text{ in.}^4} + \frac{5.0 \text{ k}}{12.0 \text{ in.} \times 0.510 \text{ in.}} = 8.02 \text{ k} + 0.82 \text{ k}$$

$$= 8.84 \text{ ksi} < 14.5 \text{ ksi}$$

$$f_b = \frac{600 \text{ in.-k}}{27 \text{ in.}^3} = 22.22 \text{ ksi} < 24 \text{ ksi} \quad \text{ok}$$

Investigating the angle of twist,

$$\phi = \frac{Tl}{GJ} = \frac{11.84 \text{ in.-k} \times 10 \text{ ft} \times 12 \text{ in./ft}}{11.6 \times 10^3 \times 0.753 \text{ in.}^4} = 0.1627 \text{ radians} = 9.32°$$

Even though stress values are within allowable limits, the beam may be considered inadequate due to excessive deformation.

Example 8.19. Determine the shear center for the box beam shown. The wide flange is a W 14 × 90 with a 1-in. plate connected to the flanges as shown.

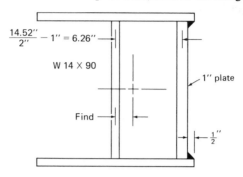

Solution. The shear center will be located somewhere along the X axis passing through the center of the members. Assume the Y axis passing through the centroid of the wide flange.

$$I_w = 999 \text{ in.}^4; \quad d \text{ (depth of plate)} = 14.02 - 2(0.710) = 12.60 \text{ in.}$$

$$e = \frac{\Sigma I_x x}{\Sigma I_x} = \frac{(999 \times 0) + \left(\dfrac{1 \times 12.60^3}{12} \times \left(\dfrac{14.52}{2} - 1 \right) \right)}{999 + \left(\dfrac{1 \times 12.60^3}{12} \right)} = 0.895 \text{ in.}$$

8.6 TORSION OF CLOSED SECTIONS

A member subject to a torsional moment is considered to be a closed section if the cross section is a continuous thin-walled tube of any shape. With these conditions, a closed section can be analyzed using Bredt's formula.

For any thin-walled closed section, the shear flow is assumed to be constant throughout the plane. Describing this flow of shear through a continuous tube section, Bredt developed the formula

$$q = \frac{T}{2A} \tag{8.12}$$

where A is the area enclosed by the center line of the section's contour. The shearing stress at any point can then be found by

$$f_v = \frac{q}{t} = \frac{T}{2At} \tag{8.13}$$

where t is the thickness of the section.

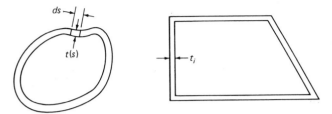

Fig. 8.6. Closed thin-walled sections.

The twist angle of a thin-walled closed section is given by the same expression (8.2)

$$\phi = \frac{TL}{GJ} \tag{8.14}$$

where

$$J = \frac{4A^2}{\int \frac{ds}{t}} \tag{8.15}$$

ds is an element of length along the midthickness of the contour wall, and t is the wall thickness of that element.

Example 8.20. Determine the maximum torsion a thin wall shaft will carry if the outside shaft diameter is 9 in. and the inside diameter is 8 in. Allowable stress is 8000 psi. Compare the answer to Example 8.2.

Solution.

$$T = 2 AtF_v$$

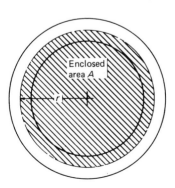

$$A = \pi \frac{d^2}{4} = 56.7 \text{ in.}^2 \ (d = d_{ave})$$

$t = 0.5$ in.

$T = 2 \times (56.7 \text{ in.}^2) \times 0.5 \text{ in.} \times (8000 \text{ psi}) = 453,600 \text{ in.-lb}$

Converting to ft-kips,

$$\frac{453,600 \text{ in.-lb}}{12 \text{ in./ft} \times 1000 \text{ lb/k}} = 37.8 \text{ ft-k}$$

The maximum torsion determined through exact use of theory in Example 8.2 was 35.84 ft-kips. We see that, although the wall is somewhat thick, the result obtained by the approximate Bredt's formula is satisfactory.

Example 8.21. Determine the maximum torque and angle of rotation for

a) a closed round tube with an outside diameter of 8 in. and inside diameter of 7-$\frac{1}{2}$ in.
b) the same thin-walled tube that has been slit along the wall.

$F_v = 14.5$ ksi, $G = 11.7 \times 10^3$ ksi, and $L = 10$ ft.

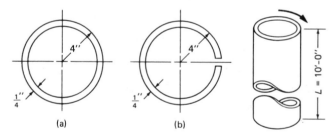

(a)   (b)

Solution.

a) Using thin-walled closed section theory,

$$T = 2 A t F_v$$

$$A = \frac{\pi d^2}{4} = 47.17 \text{ in.}^2 \left(d = d_{ave} = 7\frac{3}{4} \text{ in.}\right)$$

$t = 0.25$ in.

$T = 2 \times 47.17 \text{ in.}^2 \times 0.25 \text{ in.} \times 14.5 \text{ ksi} = 342 \text{ in.-kip} = 28.5 \text{ ft-kip}$

$$\phi = \frac{Tl}{GJ}$$

$l = 10 \text{ ft} \times 12 \text{ in./ft} = 120 \text{ in.}$

$$J = \frac{4A^2}{\int \frac{ds}{t}} = \frac{4 \times (47.12)^2}{\left(\frac{\pi \times 7.75}{0.25}\right)} = 91.19 \text{ in.}^4$$

$$\phi = \frac{342 \text{ in.-kip} \times 120 \text{ in.}}{11.7 \times 10^3 \text{ ksi} \times 91.19 \text{ in.}^4} = 0.0384 \text{ radians} = 2.20°$$

b) Consider the slit section as a thin rectangular section in torsion.

$$T = \frac{F_v J^*}{d}$$

$J^* = \alpha \, bd^3 ; \quad \text{for thin wall, } \alpha = 0.33$

$J^* = 0.33 \, (\pi \times 7.75 \text{ in.}) \times (0.25 \text{ in.})^3 = 0.1255 \text{ in.}^4$

$$T = \frac{14.5 \text{ ksi} \times 0.1255 \text{ in.}^4}{0.25 \text{ in.}} = 7.28 \text{ in.-kip} = 0.607 \text{ ft-kip}$$

$$\phi = \frac{TL}{GJ^*}$$

$L = 120 \text{ in.}$

$$\phi = \frac{7.28 \text{ in.-kip} \times 120 \text{ in.}}{11.7 \times 10^3 \text{ ksi} \times 0.1255 \text{ in.}^4} = 0.595 \text{ radians}$$

The advantage, by having closed sections over open sections, can readily be seen.

$$\text{Ratio of maximum torques} = \frac{28.50 \text{ ft-kip}}{0.607 \text{ ft-kip}} = 47$$

$$\text{Ratio of corresponding twist angles} = \frac{0.0384 \text{ rad}}{0.595 \text{ rad}} = 0.0645$$

Example 8.22. For two C 10 × 15.3 channels with a length of 8 ft, determine the torsional capacity and the angle of twist for that torque

a) if connected as shown but do not act as a closed section.
b) if connected toe-to-toe, providing a closed section.
c) if connected back-to-back, providing an open section.

$F_v = 14.5 \text{ ksi and } G = 11.7 \times 10^3 \text{ ksi.}$

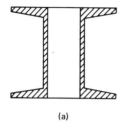

(a)   (b)   (c)

Solution.

a) $T = \dfrac{F_v J^*}{t_f}$

$J^* = \dfrac{1}{3}\left[(10 - 0.436) \times (0.24)^3 + 2\left(2.60 - \dfrac{0.24}{2}\right) \times (0.436)^3\right] \times 2$

$= 0.362$ in.^{4}

$t_f = 0.436$ in.

$T = \dfrac{14.5 \text{ ksi} \times 0.362 \text{ in.}^4}{0.436 \text{ in.}} = 12.04$ in.-k $= 1.0$ ft-k

$\phi = \dfrac{Tl}{GJ^*}$

$l = 8$ ft $\times$ 12 in./ft $= 96$ in.

$\phi = \dfrac{12.04 \text{ in.-kip} \times 96 \text{ in.}}{11.7 \times 10^3 \text{ ksi} \times 0.362 \text{ in.}^4} = 0.273$ rad $= 15.64°$

b) Treating the section as a thin-walled closed section,

$T = 2 F_v A t$

$A = (5.2 \text{ in.} - 0.24 \text{ in.}) \times (10 \text{ in.} - 0.436 \text{ in.}) = 47.44$ in.^{2}

$t = 0.24$ in.

$T = 2 \times 14.5 \text{ ksi} \times 47.44 \text{ in.}^2 \times 0.24 \text{ in.} = 330.2$ in.-k $= 27.5$ ft-k

$\phi = \dfrac{Tl}{GJ}$

$l = 96$ in.

$$J = \frac{4A^2}{2 \times \left(\dfrac{b}{t_w} + \dfrac{d}{t_f}\right)} = \frac{4 \times (47.44)^2}{2 \times \left[\dfrac{5.20 - 0.24}{0.24} + \dfrac{10.0 - 0.436}{0.436}\right]} = 105.65 \text{ in.}^4$$

$$\phi = \frac{330.2 \text{ in.-k} \times 96 \text{ in.}}{11.7 \times 10^3 \text{ ksi} \times 105.65 \text{ in.}^4} = 0.0256 \text{ rad} = 1.469°$$

If we compare the rotation that would result from the torque of (a),

$$\phi_1 = 1.469° \frac{12.05}{330.2} = 0.054°$$

c) $$T = \frac{F_v J^*}{t_f}$$

$$J^* = 2\left(\frac{1}{3} \times 2\, b_f \times t_f^3\right) + \left(\frac{1}{3} \times (d - t_f) \times (2\, t_w)^3\right)$$

$$= 0.287 + 0.353 = 0.640 \text{ in.}^4$$

$$T = \frac{14.5 \text{ ksi} \times 0.640 \text{ in.}^4}{0.436 \text{ in.}} = 21.3 \text{ in.-k} = 1.77 \text{ ft-k}$$

$$\phi = \frac{Tl}{GJ^*}$$

$$l = 96 \text{ in.}$$

$$\phi = \frac{21.3 \text{ in.-k} \times 96 \text{ in.}}{11.7 \times 10^3 \text{ ksi} \times 0.640 \text{ in.}^4} = 0.273 \text{ rad} = 15.63°$$

If we compare the rotation that would result from the torque of (a),

$$\phi_1 = 15.63° \frac{12.06}{21.3} = 8.85°$$

Example 8.23. For the two 20-ft, simply supported beams below, determine

a) the allowable torque for $F_v = 14.5$ ksi.
b) the angle of twist of an applied torque of 150-in.-kip; $G = 11.7 \times 10^3$ ksi.
c) the required length of $\frac{5}{16}$-in. E70 welds per foot; $V = 10$ kips and $T = 80$ in.-kip.

---

[4] Refer to Blodgett, *Design of Welded Structures*, 2.10-4.

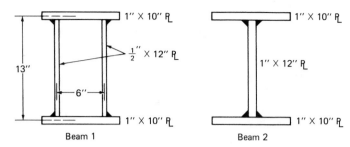

Beam 1   Beam 2

Solution.

a) Beam 1

$$T = F_v \, 2 \, (At)$$

$$A = 6 \text{ in.} \times 13 \text{ in.} = 78.0 \text{ in.}^2$$

$$t = \tfrac{1}{2} \text{ in.}$$

$$T = 14.5 \text{ ksi} \times 2 \times 78 \text{ in.}^2 \times \tfrac{1}{2} \text{ in.} = 1131 \text{ in.-kip} = 94.3 \text{ ft-kip}$$

Beam 2

$$T = F_v \, \frac{J^*}{t_f}$$

$$J^* = 0.33 \, [(10 \text{ in.} \times (1 \text{ in.})^3 \times 2) + (12 \text{ in.} \times (1 \text{ in.})^3)] = 10.67 \text{ in.}^4$$

$$t_f = 1.0 \text{ in.}$$

$$T = \frac{14.5 \text{ ksi} \times 10.67 \text{ in.}^4}{1.0 \text{ in.}} = 154.7 \text{ in.-kip} = 12.9 \text{ ft-kip}$$

b) Beam 1

$$\phi = \frac{TL}{GJ}$$

$$L = 20 \text{ ft} \times 12 \text{ in./ft} = 240 \text{ in.}$$

$$J = \frac{4 \, b^2 d^2}{2 \times \left(\dfrac{b}{t_w} + \dfrac{d}{t_f}\right)} = \frac{4 \, (13 \text{ in.})^2 \, (6 \text{ in.})^2}{2\left(\dfrac{13 \text{ in.}}{0.5 \text{ in.}} + \dfrac{6 \text{ in.}}{1 \text{ in.}}\right)} = 380.25 \text{ in.}^4$$

$$\phi = \frac{150 \text{ in.-kip} \times 240 \text{ in.}}{11.7 \times 10^3 \text{ ksi} \times 380.25 \text{ in.}^4} = 0.0081 \text{ rad} = 0.464 \text{ deg}$$

Beam 2

$$\phi = \frac{TL}{GJ^*}$$

$L = 240$ in.

$J^* = 10.67$ in.^{4}

$$\phi = \frac{150 \text{ in.-kip} \times 240 \text{ in.}}{11.7 \times 10^3 \text{ ksi} \times 10.67 \text{ in.}^4} = 0.288 \text{ rad} = 16.52 \text{ deg}$$

c) Beam 1

Consider web

$$\text{Shear } f_v = \frac{10 \text{ k}}{2(0.5 \text{ in.}) \times 14 \text{ in.}} = 0.7 \text{ k/in.}^2$$

Shear flow $q_v = 0.7$ k/in.$^2 \times 0.5$ in. $= 0.35$ k/in.

$$\text{Torsion } f_v = \frac{80.0 \text{ in.-kip}}{2 \times 78.0 \text{ in.}^2 \times 0.5 \text{ in.}} = 1.026 \text{ k/in.}^2$$

Shear flow $q_v = 1.026$ k/in.$^2 \times 0.5$ in. $= 0.51$ k/in.

$$q_{max} = 0.35 \text{ k/in.} + 0.51 \text{ k/in.} = 0.86 \text{ k/in.}$$

For $\frac{5}{16}$-in. E70 weld,

$$W_c = 5 \times 0.928 \text{ k/in.} = 4.64 \text{ k/in.}$$

For a 1-ft length,

$$\frac{0.86 \text{ k/in.} \times 12 \text{ in./ft}}{4.64 \text{ k/in.}} = 2.22 \text{ in./1 ft}$$

Use $2\frac{1}{4}$ in. intermittent welds at 12 in. centers.

Beam 2

The maximum shear flow is produced in the web and is the superposition of the shear flow due to vertical loading and the shear flow due to the torsional moment.

$$\text{Shear } f_v = \frac{10 \text{ k}}{1 \text{ in.} \times 14 \text{ in.}} = 0.7 \text{ k/in.}^2$$

$$\frac{1}{2} \text{ shear flow } q_v = 0.7 \text{ k/in.}^2 \times 1.0 \text{ in.} \times \frac{1}{2} = 0.35 \text{ k/in.}$$

$$\text{Torsional } f_v = \frac{80 \text{ in.-k} \times 1.0 \text{ in.}}{10.67 \text{ in.}^4} = 7.5 \text{ k/in.}^2$$

Shear to be carried by weld on one side of web

$$f_v = 0.7 \text{ k/in.}^2 + 7.5 \text{ k/in.}^2 = 8.2 \text{ k/in.}^2$$

$$f_v = 0.7 \text{ k/in.}^2 - 7.5 \text{ k/in.}^2 = -6.8 \text{ k/in.}^2$$

$$q_{v \text{ max}} = \frac{8.2 \text{ k/in.}^2 \times 0.547 \text{ in.}}{2} = 2.24 \text{ k/in.}$$

For $\frac{5}{16}$-in. E70 weld,

$$W_c = 5 \times 0.928 \text{ k/in.} = 4.64 \text{ k/in.}$$

For a 1-ft length,

$$\frac{2.24 \text{ k/in.} \times 12 \text{ in./ft}}{4.64 \text{ k/in.}} = 5.79 \text{ in. per 1-ft length}$$

Use 6-in. intermittent welds at 1-ft centers.

8.7 MEMBRANE ANALOGY

Prandtl showed that the governing equation for the torsion problem is the same for that of a thin membrane stretched across the cross section and subjected to uniform pressure.

If we have a tube conforming to the outline of the twisted section, cover the tube with a thin membrane glued to the tube, and apply pressure to the inside, the membrane will deflect, and we will note the following: the shear stresses due to the torsion correspond to the slopes of the membrane, and the lines of equal shear correspond to the contour lines, the direction of the shear stress to the tangent at the contour lines, and the torque to the volume under the membrane.

In the case of a closed section, the boundaries of the tube will correspond to

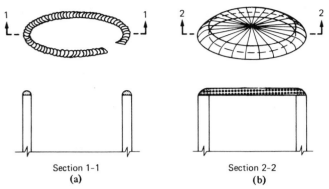

Section 1-1 | Section 2-2
(a) | (b)

Fig. 8.7. Membrane analogy of (a) open sections and (b) closed sections.

the limits of the material twisted, and the whole will be represented by a plate glued to the membrane (Fig. 8.7). From here, one can get a very good intuitive understanding of the stress conditions of a twisted member.

PROBLEMS TO BE SOLVED

8.1. For an applied torque of 15 in.-kips, determine the maximum shear stress for (a) a 2-in.-diameter steel bar and (b) a steel pipe having an outside diameter of $2\frac{1}{2}$ in. and an inside diameter of 2 in.

8.2. For the 2-in.-diameter steel bar and $2\frac{1}{2}$-in.-outside-diameter steel pipe mentioned in Problem 8.1, determine the maximum allowable torque, assuming A36 steel. Which member is more efficient, i.e., which carries more torque per unit weight?

8.3. For an applied torque of 18 in.-kips, determine the lightest circular bar necessary to resist the load, assuming F_y = 50-ksi steel.

8.4. For an applied torque of 18 in.-kips, determine the lightest standard weight pipe, assuming A36 steel. For the pipe selected, calculate the angle of twist if E_s = 29 × 10⁶ psi, μ = 0.25, and L = 10 ft.

8.5. For an applied torque of 15 in.-kips, determine the maximum shear stress for (a) a 2-in. by 2-in. square steel bar and (b) a 1-in. by 4-in. rectangular steel bar. Use A36 steel. Determine the angle of twist for both bars if L = 7 ft and G = 11.6 × 10⁶ psi.

8.6. For the two bars mentioned in Problem 8.5, determine the allowable torques and rotations corresponding to those torques if now the bars are made of aluminum with an allowable shear strength of 11.2 ksi and G = 3.86 × 10³ ksi. Assume linearly elastic behavior for aluminum.

8.7. For the bar shown, determine the maximum allowable load P if F_y = 36 ksi and G = 11.6 × 10³ ksi.

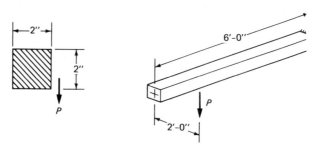

8.8. Design the most economical wide-flange section for an 8.0-ft cantilever on which a load of 5 k is applied at the free end, 12 in. to the right of the web centerline. Assume F_v = 13.0 ksi and F_b = 22 ksi. Neglect beam weight and combination of shear and bending stresses.

8.9. For the beam shown, determine the maximum allowable height for the concrete facia panel if lightweight concrete weighing 80 lb/ft³ is used. The beam is simply supported and restrained against torsion at the supports. It is laterally supported at ends only.

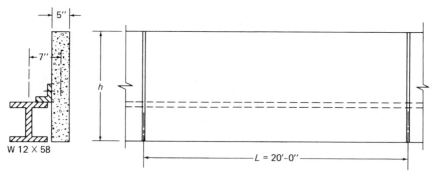

8.10. Design a square support column for the bus stop sign shown if wind pressure = 80 lb/ft², L = 10 ft, F_v = 14.5 ksi, G = 11.6 × 10³ ksi, and maximum allowable rotation = 4 degrees.

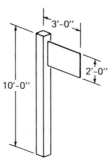

8.11. Compute the torsional constants J^* for the following open sections: (a) W 18 X 50, (b) W 16 X 31, and (c) C 12 X 20.7.

8.12. For a member subject to a torque of 3.5 ft-k, design the most economical W section and the most economical C section, if $F_v = 14.5$ ksi. For $L = 12$ ft, 0 in., which member has the greater resistance to twist (rotation) per unit weight?

8.13. For the fabricated edge beam shown, determine the maximum height of 8-in. brick wall that can be supported if the ends of the beam are fixed to very heavy columns. Assume A36 steel and that the brick weighs 80 pcf. $L = 20$ ft.

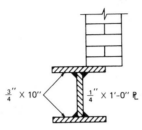

8.14. For the beam shown in Problem 8.13, design the welds, assuming E70 electrodes and the height of the brick wall to be 9 ft, 0 in.

8.15. Determine the maximum allowable load P that can be applied to a C 12 X 20.7 if the channel is simply supported over a span of 22 ft, 0 in. with the ends restrained against twisting. The load P is applied at the centerline of the lower flange.

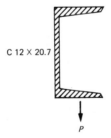

8.16. Determine the minimum thickness required for the angle shown if the wall weighs 400 lb/lin. ft and the angle is restrained from twisting at both ends of a 12-ft span.

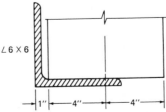

8.17. Determine the required weld for the section shown at a location where the shear force is 12 kips and the torque is 60 in.-kips. The member is constructed of 50-ksi steel. Use E70 weld.

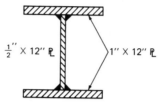

8.18. Locate the shear center for the following shapes.

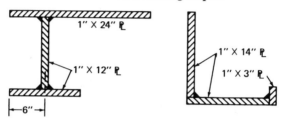

8.19. For the sections shown, determine the allowable torque if F_v = 14.5 ksi. Also, for each of the sections, calculate the rotation if L = 10 ft, T = 30 in.-kips, and G = 11.7 × 10³ ksi. Each section is made from four $\frac{1}{2}$-in. × 10-in. plates. For the left figure, assume the web to be one member, 1 in. thick.

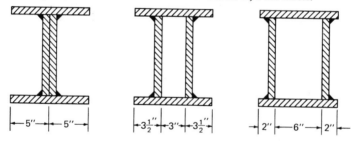

8.20. Calculate the required $\frac{1}{4}$-in. intermittent weld for the sections shown in the second and third sketches of Problem 8.19, if V = 30 kips, T = 300 in-kips, and F_v = 14.5 ksi. Use E70 weld.

9
Composite Design

9.1 COMPOSITE DESIGN

For many years, reinforced concrete slabs on steel beams were used without considering any composite effect. In recent years, it has been shown that concrete slabs and steel beams act as a unit when joined together to resist horizontal shear. This resistance is provided either by welding shear connectors to the beam or by encasing the beam in concrete (AISCS 1.11.1). Thus, a significant increase in strength is achieved. The specification recognizes this increase by allowing the bending stress to be $0.76\,F_y$ while disregarding the effect of the concrete (AISCS 1.11.2). Both materials are utilized to their fullest capacities, with the concrete slab acting in compression and the steel beam acting almost entirely in tension.

Composite sections have greater stiffness than the same slab and beam acting individually and, therefore, can carry larger loads or similar loads with smaller deflections on larger spans. When the upper portion of a beam is encased in the concrete slab to achieve composite action, the overall floor depth is reduced, and for high rise buildings the ultimate savings in volume, wiring, ductwork, walls, plumbing, etc. can be considerable.

If a part of the concrete is in tension, that portion is considered nonexisting and is neglected in calculations. Cover plates may be used on the bottom flange of the steel beams in composite sections to increase the efficiency of the assembly by lowering the neutral axis from within the concrete.[1]

Shores for composite constructions are temporary formworks that support the

---

[1] Sharp increases in labor costs have precluded most economic uses of cover plates in composite design, in spite of very large savings in steel poundage.

Fig. 9.1. A composite spandrel beam for a building. Note strands in conduits, perpendicular to the beam, for post-tensioning. By post-tensioning the slab, deflection and negative moment cracking are reduced in the direction perpendicular to the composite beam.

steel beams at sufficiently close intervals to prevent the beam from sagging before the concrete hardens. When the steel beams are not shored during the pouring and curing processes of the concrete, the section must be designed to support alone the wet concrete and all other construction loads. It is accepted practice to size the beam for dead plus live loads and full composite action, when this action is obtained by the use of shear connectors (AISCS Commentary 1.11.2, paragraph 3).

Fig. 9.2. An interior girder of the building whose spandrel beam is shown in Fig. 9.1. Note how reinforcing bars and post-tensioning strands drape over the girder toward the top of the slab to provide negative moment reinforcement.

9.2 EFFECTIVE WIDTH OF FLANGES

The AISC specifies the width of the slab that may participate in composite action (AISCS 1.11.1), as shown in Fig. 9.3. For slabs extending on both sides of the beam, the maximum effective flange width b may not exceed (1) one-fourth of the beam span L, nor (2) one-half the clear distances to the adjacent beams on both sides plus b_f, nor (3) 16 times the slab thickness t plus b_f. When the slab exists on only one side of the beam, the maximum effective width b may not exceed (1) one-twelfth of the beam span L, nor (2) one-half the clear distance to the adjacent beam plus b_f, nor (3) 6 times the slab thickness t plus b_f.

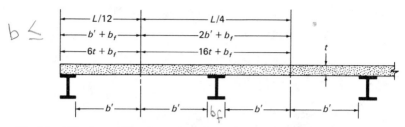

Fig. 9.3. Effective width of concrete flange in composite construction, as specified in AISCS 1.11.1.

[handwritten: → SOLVE FOR SMALLEST VALUE]

Example 9.1. Determine the effective width b for the interior and edge beams shown when span $L = 42$ ft, 0 in., slab thickness $t = 5\text{-}\frac{1}{2}$ in., and beam spacing is 8 ft, 0 in. on center for a W 18 × 35.

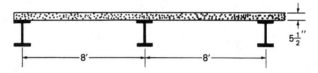

Solution. For a W 18 × 35, $b_f = 6$ in. *[handwritten: AISC]*

Clear distance between beams = (8 ft × 12 in./ft) – 6 in. = 90.0 in. *[handwritten: = 2b']*

For interior beams,

$$1)\ b \leqslant \frac{l}{4} = \frac{42.0\ \text{ft} \times 12\ \text{in./ft}}{4} = 126.0\ \text{in.}$$

$$2)\ b \leqslant \left(\frac{1}{2}\ \text{clear distance}\right) \times 2 + b_f = 2\left(\frac{90.0\ \text{in.}}{2}\right) + 6\ \text{in.} = 96.0\ \text{in.}$$

$$3)\ b \leqslant (8\ t) \times 2 + b_f = 2(8 \times 5.50\ \text{in.}) + 6\ \text{in.} = 94.0\ \text{in.}$$

Minimum $b = 94.0$ in. governs.

For edge beam,

$$1)\ b \leqslant \frac{l}{12} = \frac{42.0\ \text{ft} \times 12\ \text{in./ft}}{12} = 42.0\ \text{in.}$$

$$2)\ b \leqslant \left(\frac{1}{2}\ \text{clear distance}\right) + b_f = \frac{90.0\ \text{in.}}{2} + 6\ \text{in.} = 51.0\ \text{in.}$$

3) $b \leqslant (6\,t) + b_f = (6 \times 5.50\ \text{in.}) + 6\ \text{in.} = 39.0\ \text{in.}$

Minimum $b = 39.0$ governs.

Example 9.2. Rework Example 9.1 using a W 12 $\times$ 65 beam.

Solution. For a W 12 $\times$ 65, $b_f = 12$ in.

Clear distance between beams = 96 in. – 12 in. = 84.0 in.

For interior beams,

1) $b \leqslant \dfrac{l}{4} = \dfrac{42.0\ \text{ft} \times 12\ \text{in./ft}}{4} = 126.0\ \text{in.}$

2) $b \leqslant 2\left(\dfrac{1}{2}\ \text{clear distance}\right) + b_f = 2\left(\dfrac{84.0\ \text{in.}}{2}\right) + 12\ \text{in.} = 96.0\ \text{in.}$

3) $b \leqslant 2(8\,t) + b_f = 2(8 \times 5.50\ \text{in.}) + 12\ \text{in.} = 100.0\ \text{in.}$

Minimum $b = 96.0$ in. governs.

For edge beam,

1) $b \leqslant \dfrac{l}{12} = \dfrac{42.0\ \text{ft} \times 12\ \text{in./ft}}{12} = 42.0\ \text{in.}$

2) $b \leqslant \dfrac{1}{2}\ \text{clear distance} + b_f = \dfrac{84.0\ \text{in.}}{2} + 12\ \text{in.} = 54.0\ \text{in.}$

3) $b \leqslant 6\,t + b_f = 6 \times 5.50\ \text{in.} + 12\ \text{in.} = 45.0\ \text{in.}$

Minimum $b = 42.0$ in. governs.

9.3 STRESS CALCULATIONS

Stresses in composite sections are usually calculated by the transformed area method, in which one of the two materials is transformed into an equivalent area of the other. Usually the available effective area of concrete is transformed into an equivalent area of steel. Assuming the strain of both the materials to be the same at equal distances from the neutral axis, the unit stress in either material is then equal to its strain times its modulus of elasticity. The unit stress in the steel is therefore E_s/E_c times the unit stress in the concrete. Referring to the

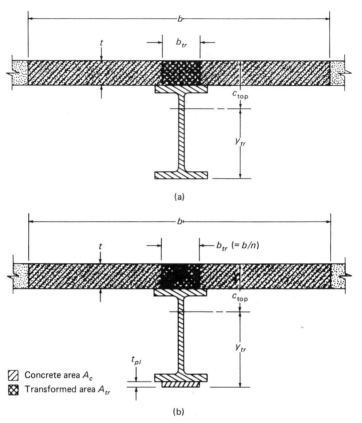

Fig. 9.4. Cross sections through composite construction: (a) noncover-plated section; (b) cover-plated section.

E_s/E_c ratio as the modular ratio n, the force resisted by 1 in.^{2} of concrete is equivalent to the force resisted by only $1/n$ in.^{2} of steel. Therefore, the effective area of concrete ($A_c = b_{eff} \times t$) is replaced by $A_{transformed}$ ($A_{tr} = A_c/n$).

After the neutral axis for the transformed section is located, its moment of inertia I_{tr} is then calculated. The maximum bending stress for the steel is expressed by

$$f_{b_s} = \frac{M\, y_{tr}}{I_{tr}} \tag{9.1}$$

where M is the bending moment, and y_{tr} is the distance from the neutral axis to the extreme steel fibers. The maximum bending stress for the concrete is expressed by

$$f_{b_c} = \frac{M\, c_{top}}{n\, I_{tr}}$$ (9.2)

where c_{top} is the distance from the neutral axis to the extreme concrete fibers, and n is the modular ratio. The value

$$S_{tr} = \frac{I_{tr}}{y_{tr}}$$ (9.3)

is sometimes referred to as the *transformed section modulus* of the beam. For unshored construction, because the steel alone must carry the dead loads, the bending stress for the steel is expressed by

$$f_{b_s} = \frac{M_D\, c}{I_s}$$ (9.4)

where M_D is the moment due to dead loads only, c is the distance from the neutral axis of the steel alone to the extreme steel fibers, and I_s is the total moment of inertia of the steel, including cover plates if used. The bending stress must be less than the allowable bending stress, as specified by AISCS 1.5.1.4.1.

Example 9.3. For Example 9.1, determine the transformed section of the concrete when $E_{steel} = 29 \times 10^6$ psi and $E_{concrete} = 2.9 \times 10^6$ psi.

Solution.

$$n = \frac{E_{steel}}{E_{conc}} = \frac{29 \times 10^6 \text{ psi}}{2.9 \times 10^6 \text{ psi}} = 10$$

a) For Example 9.1 interior beams, $t = 5\text{-}\frac{1}{2}$ in., $b = 94$ in.

Effective area of concrete

$$A_c = 5.50 \text{ in.} \times 94 \text{ in.} = 517 \text{ in.}^2$$

$$A_{trans} = \frac{A_c}{n} = \frac{517 \text{ in.}^2}{10} = 51.7 \text{ in.}^2$$

b) For Example 9.1 edge beam, $t = 5\text{-}\frac{1}{2}$ in., $b = 39$ in.

Effective area of concrete

$$A_c = 5.50 \text{ in.} \times 39 \text{ in.} = 214.5 \text{ in.}^2$$

$$A_{trans} = \frac{A_c}{n} = \frac{214.5 \text{ in.}^2}{10} = 21.5 \text{ in.}^2$$

Example 9.4. For the case shown, determine $y_{tr}, I_{tr}, f_{tr \, bot}, f_{tr \, top}$, and $f_{tr \, conc}$ when $M = 160$ ft-k. Use a W 18 × 35 with $b = 70$ in., $t = 4$ in., and $n = 10$.

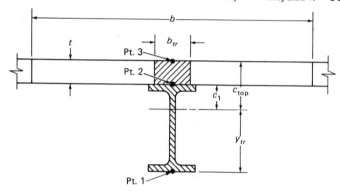

Solution. For W 18 × 35,

$$A = 10.3 \text{ in.}^2, \quad d = 17.70 \text{ in.}, \quad b_f = 6.00 \text{ in.}, \quad I = 510 \text{ in.}^4$$

Determine transformed dimensions

$$b_{tr} = \frac{b}{n} = \frac{70 \text{ in.}}{10} = 7 \text{ in.}$$

$$A_{tr} = 4 \text{ in.} \times 7 \text{ in.} = 28 \text{ in.}^2$$

$$y_{tr} = \frac{10.3 \text{ in.}^2 \times (17.70 \text{ in.}/2) + 28 \text{ in.}^2 \times (17.70 \text{ in.} + 4 \text{ in.}/2)}{10.3 \text{ in.}^2 + 28 \text{ in.}^2}$$

$$= 16.78 \text{ in.}$$

$$I_{tr} = I_0 + \Sigma (Ad^2)$$

$$= 510 + 10.3 \times \left(\frac{17.70}{2} - 16.78\right)^2 + \left(\frac{7 \times (4)^3}{12}\right)$$

$$+ 28.0 \times (17.70 + 2.0 - 16.78)^2$$

$$= 1434 \text{ in.}^4$$

Stresses at extreme fibers

Point 1

$$f_{tr\,bot} = \frac{My_{tr}}{I_{tr}} = \frac{160\text{ ft-k} \times 12\text{ in./ft} \times 16.78\text{ in.}}{1434\text{ in.}^4} = 22.47\text{ ksi}$$

Point 2

$$f_{tr\,top} = \frac{Mc_1}{I_{tr}} = \frac{160\text{ ft-k} \times 12\text{ in./ft} \times (17.70\text{ in.} - 16.78\text{ in.})}{1434\text{ in.}^4} = 1.23\text{ ksi}$$

Point 3

$$f_{tr\,conc} = \frac{Mc_{top}}{nI_{tr}} = \frac{160\text{ ft-k} \times 12\text{ in./ft} \times (21.70\text{ in.} - 16.78\text{ in.})}{10 \times 1434\text{ in.}^4} = 0.66\text{ ksi}$$

Example 9.5. For the situation shown, determine y_{tr}, I_{tr}, $f_{tr\,bot}$, $f_{tr\,conc}$, and $f_{tr\,top}$ when 560 ft-k, $A_{tr} = 44$ in.^{2}, $t = 5$ in., $b_{tr} = 8.8$, and $n = 10$.

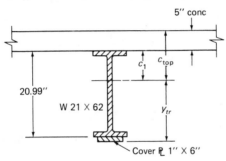

Solution. For a W 21 × 62,

$$A = 18.3\text{ in.}^2$$
$$d = 20.99\text{ in.}$$
$$b_f = 8.24\text{ in.}$$
$$I = 1330\text{ in.}^4$$

Determine centroid of transformed section

$$y_{tr} = \frac{18.3\text{ in.}^2 \times (1\text{ in.} + 20.99\text{ in.}/2) + 6.0\text{ in.}^2 \times (1\text{ in.}/2) + 44.0\text{ in.}^2 \times (21.99\text{ in.} + 5\text{ in.}/2)}{18.3\text{ in.}^2 + 6.0\text{ in.}^2 + 44.0\text{ in.}^2}$$

$$= 18.90\text{ in.}$$

$$I_{tr} = I_0 + \Sigma Ad^2 = 5831.5\text{ in.}^4$$

Stresses at extreme fibers

$$f_{tr\ bot} = \frac{My_{tr}}{I_{tr}} = \frac{560\ \text{ft-k} \times 12\ \text{in./ft} \times 18.90\ \text{in.}}{5831.5\ \text{in.}^4} = 21.78\ \text{ksi}$$

$$f_{tr\ conc} = \frac{Mc_{top}}{nI_{tr}} = \frac{560\ \text{ft-k} \times 12\ \text{in./ft} \times (26.99\ \text{in.} - 18.90\ \text{in.})}{10 \times 5831.5\ \text{in.}^4} = 0.93\ \text{ksi}$$

$$f_{tr\ top} = \frac{Mc_1}{I_{tr}} = \frac{560\ \text{ft-k} \times 12\ \text{in./ft} \times (21.99\ \text{in.} - 18.90\ \text{in.})}{5831.5\ \text{in.}^4} = 3.56\ \text{ksi}$$

As a general practice, the beam is designed to resist moments due to dead and live loads as a composite unit, regardless of the shoring condition. Experiments have shown that, at failure load, the effect of shoring or nonshoring is negligible. AISCS 1.11.2.2 furthermore requires that Formula 1.11-2 (Eq. 9.5) be satisfied to ensure that bending stresses will remain below yielding stress under service load for unshored beams.

$$S_{tr} = \left(1.35 + 0.35\ \frac{M_L}{M_D}\right) S_s \tag{9.5}$$

Here S_s is the section modulus of the steel alone. Bending stress in the steel is limited to $0.66\ F_y$, and in the concrete to $0.45\ f'_c$, where f'_c is the 28 days' resistance of the concrete.

Example 9.6. Determine the moment-resisting capacity of the composite section shown. Assume A36 steel, $f'_c = 3000$ psi, $n = 9$, $I_{tr} = 8636$ in.^{4}, $y_{tr} = 17.99$ in.

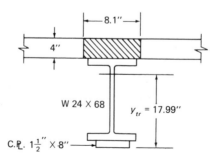

Solution.

$$F_y = 36 \text{ ksi} \quad F_b = 24 \text{ ksi} \quad \text{(AISCS 1.5.1.4.1)}$$

$$f_c = 0.45 f_c' = 0.45 \times 3000 \text{ psi} = 1.35 \text{ ksi} \quad \text{(AISCS 1.11.2.2)}$$

$$M_{max} = \frac{F_{bs}I_{tr}}{c} = \frac{24 \text{ ksi} \times 8636 \text{ in.}^4}{17.99 \text{ in.}} = 11520 \text{ in.-k} = 960.0 \text{ ft-k}$$

The maximum concrete stress can be transposed to determine the maximum moment capacity considering the concrete.

$$f_{bc} = \frac{Mc}{nI_{tr}}$$

$$M_{max} = \frac{nf_c I_{tr}}{c} = \frac{9 \times 1.35 \text{ ksi} \times 8636 \text{ in.}^4}{1.5 \text{ in.} + 23.73 \text{ in.} + 4 \text{ in.} - 17.99 \text{ in.}} = 9340 \text{ in.-k}$$

$$= 778.0 \text{ ft-k}$$

Since the moment capacity of the concrete is less than that of the steel, the concrete will fail first. Therefore,

$$M_{max} = 778 \text{ ft-k}$$

Example 9.7. Assuming that no shoring will be used, determine the dead load moment capacity of the composite section of Example 9.6.

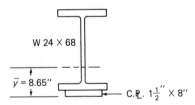

Solution. Since no shoring will be used, the steel alone will carry the dead load. For a W 24 × 68,

$$A = 20.1 \text{ in.}^2$$
$$d = 23.73 \text{ in.}$$
$$I = 1830 \text{ in.}^4$$

$$\bar{y}_{steel} = \frac{(1.5 \text{ in.} \times 8 \text{ in.})(1.5 \text{ in.}/2) + 20.1 \text{ in.}^2 (23.73 \text{ in.}/2 + 1.5 \text{ in.})}{(1.5 \text{ in.} \times 8 \text{ in.}) + 20.1 \text{ in.}^2}$$

$$= 8.65 \text{ in. from bottom}$$

$$I_{\text{steel}} = I_0 + \Sigma (Ad^2) = \frac{8 \text{ in.} (1.5 \text{ in.})^3}{12} + 12 \text{ in.}^2 (8.65 \text{ in.} - 0.75 \text{ in.})^2$$

$$+ 1830 \text{ in.}^4 + 20.1 \text{ in.}^2 \left(\frac{23.73 \text{ in.}}{2} + 1.5 \text{ in.} - 8.65 \text{ in.}\right)^2$$

$$= 3028 \text{ in.}^4$$

Note that the quantity $8 \times (1.5)^3/12$, which is the moment of inertia of a plate with respect to its weak axis, is usually neglected, being usually very small.

Maximum moment capacity of steel section

$$M_{\text{top}} = \frac{F_b I_s}{c_{\text{top}}} = \frac{24 \text{ ksi} \times 3028 \text{ in.}^4}{23.73 \text{ in.} + 1.5 \text{ in.} - 8.65 \text{ in.}} = 4383 \text{ in.-k} = 365 \text{ ft-k}$$

$$M_{\text{bot}} = \frac{F_b I_s}{c_{\text{bot}}} = \frac{24 \text{ ksi} \times 3028 \text{ in.}^4}{8.65 \text{ in.}} = 8401 \text{ in.-k} = 700 \text{ ft-k}$$

Moment capacity of the top governs

$$M_{\text{max}} = 365 \text{ ft-k}$$

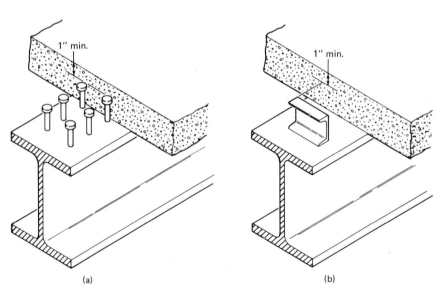

(a)　　　　　(b)

Fig. 9.5. Shear connectors for composite construction: (a) round-headed shear studs welded to the beam; (b) short length of a channel section welded to the beam.

9.4 SHEAR CONNECTORS

For a steel beam and concrete slab to act compositely, the two materials must be thoroughly connected to each other so that longitudinal shear may be transferred between them. When the steel beam is fully encased in the slab, mechanical connectors may not be necessary, since longitudinal shear could be fully transferred by the bond between the steel and concrete. Otherwise, mechanical shear connectors are necessary.

Round studs, channels, or spirals, welded to the top flange of beams, are the type of shear connectors most commonly used (see Fig. 9.5). The studs are round bars with one end upset to prevent vertical separation and the other welded to the beam. Because welded studs are often damaged when beams are shipped, field-welding of the studs by stud welding guns is preferable. The allowable horizontal shear loads for channels and round studs of various weights or diameters are tabulated in Table 1.11.4 of AISCS (reproduced as Table 9.1 here).

According to the AISC, the number of shear connectors required for full composite action is determined by dividing the total shear force V_h to be resisted, between the points of maximum positive moment and zero moment, by the shear capacity of one connector. This number is doubled to obtain the total number of connectors required for the entire span. The shear force V_h is the smaller of the two values as determined by the following formulas (AISCS 1.11.4)

$$V_h = \frac{0.85 f_c' A_c}{2} \tag{9.6}$$

Table 9.1. Allowable Horizontal Shear Load for One Connector (q), kips[a] (AISC Table 1.11.4, reprinted with permission).

Connector[b]	Specified Compressive Strength of Concrete (f_c'), ksi		
	3.0	3.5	$\geqslant 4.0$
$\frac{1}{2}$ in. diam. $\times$ 2 in. hooked or headed stud	5.1	5.5	5.9
$\frac{5}{8}$ in. diam. $\times$ $2\frac{1}{2}$ in. hooked or headed stud	8.0	8.6	9.2
$\frac{3}{4}$ in. diam. $\times$ 3 in. hooked or headed stud	11.5	12.5	13.3
$\frac{7}{8}$ in. diam. $\times$ $3\frac{1}{2}$ in. hooked or headed stud	15.6	16.8	18.0
Channel C3 $\times$ 4.1	$4.3\,w^c$	$4.7\,w^c$	$5.0\,w^c$
Channel C4 $\times$ 5.4	$4.6\,w^c$	$5.0\,w^c$	$5.3\,w^c$
Channel C5 $\times$ 6.7	$4.9\,w^c$	$5.3\,w^c$	$5.6\,w^c$

[a] Applicable only to concrete made with ASTM C33 aggregates.
[b] The allowable horizontal loads tabulated may also be used for studs longer than shown.
[c] w = length of channel, inches.

$$V_h = \frac{A_s F_y}{2} \qquad (9.7)$$

The AISC permits uniform spacing of the connectors between the points of maximum positive moment and the point of zero moment. For concentrated loads, the number of shear connectors required between any concentrated load and the nearest point of zero moment must be determined by AISCS Formula 1.11-7.

Example 9.8. Determine the number of $\frac{7}{8}$-in.-diameter studs required when $f_c' = 3000$ psi and $F_y = 36$ ksi are used for the interior beams in

a) Example 9.1.
b) Example 9.2.

Solution. Shear force V_h to be resisted by connectors

$$V_h \text{ for concrete} = \frac{0.85 f_c' A_c}{2}$$

$$V_h \text{ for steel} = \frac{A_s F_y}{2}$$

$$\text{Capacity of one } \frac{7}{8} \text{ in. stud} = 15.6 \text{ k} \qquad \text{(AISCS, Table 1.11.4)}$$

a) From Example 9.1, $b = 94$ in.

$$A_{conc} = b \times t = 94 \text{ in.} \times 5.50 \text{ in.} = 517 \text{ in.}^2$$

$$A_{stl} = 10.3 \text{ in.}^2$$

$$V_{h\ conc} = \frac{0.85 \times 3.0 \text{ ksi} \times 517 \text{ in.}^2}{2} = 659.2 \text{ k}$$

$$V_{h\ stl} = \frac{10.3 \text{ in.}^2 \times 36 \text{ ksi}}{2} = 185.4 \text{ k}$$

V_h for steel governs.

Number of shear connectors required

$$\frac{2 \times 185.4 \text{ k}}{15.6 \text{ k}} = 23.76$$

Use 24 $\frac{7}{8}$-in. studs

b) From Example 9.2, $b = 96$ in.

$$A_{conc} = b \times t = 96 \text{ in.} \times 5.50 \text{ in.} = 528 \text{ in.}^2$$

$$A_{stl} = 19.1 \text{ in.}^2$$

$$V_{h \, conc} = \frac{0.85 \times 3.0 \text{ ksi} \times 528 \text{ in.}^2}{2} = 673.2 \text{ k}$$

$$V_{h \, stl} = \frac{19.1 \text{ in.}^2 \times 36 \text{ ksi}}{2} = 343.8 \text{ k}$$

V_h for steel governs

Number of shear connectors required

$$\frac{2 \times 343.8 \text{ k}}{15.6 \text{ k}} = 44.08$$

Use 45 $\frac{7}{8}$-in. studs

Example 9.9. Determine the number of 4-in. channels needed for shear transfer when $f'_c = 3.0$ ksi and $F_y = 36$ ksi for the composite section shown. Assume the length of each channel to be 3 in.

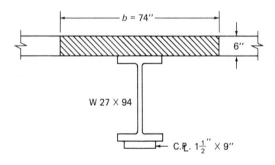

Solution.

$$A_c = tb = 6 \text{ in.} \times 74 \text{ in.} = 444 \text{ in.}^2$$

$$A_s = A_{WF} + A_{pl} = 27.7 \text{ in.}^2 + \left(1\frac{1}{2} \text{ in.} \times 9 \text{ in.}\right) = 41.2 \text{ in.}^2$$

$$V_{h \, conc} = \frac{0.85 f'_c A_c}{2} = \frac{0.85 \times 3.0 \text{ ksi} \times 444 \text{ in.}^2}{2} = 566.1 \text{ k}$$

$$V_{h \, steel} = \frac{A_s F_y}{2} = \frac{41.2 \text{ in.}^2 \times 36 \text{ ksi}}{2} = 741.6 \text{ k}$$

V_h for concrete governs

Number of shear connectors required

$$\frac{2 \times 566.1\,k}{15\,k} = 75.5 \quad \text{(capacity of shear connector = 15 k)}$$

Use 76 4-in. channels 3 in. long.

Example 9.10. Verify this composite beam and determine the number of $\frac{3}{4}$-in.-diameter studs required for full composite action between A and B and between B and C. Beams are as shown in the sketch and spaced 8 ft, 0 in. on center; $f_c' = 4000$ psi ($n = 9$). The beam is shored.

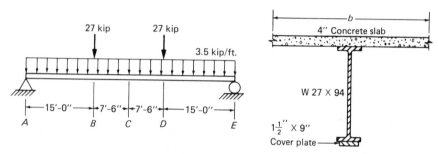

Solution.

$$R = \frac{wL}{2} + P = 105.75\,k$$

$$M_C = \frac{105.75\,k \times 45\,ft}{2} - \frac{27\,k \times 15\,ft}{2} - \frac{3.5\,k/ft \times (45\,ft)^2}{8}$$

$$= 1292\ \text{ft-k}$$

$$M_B = 105.75\,k \times 15\,ft - \frac{3.5\,k/ft \times (15)^2}{2}$$

$$= 1193\ \text{ft-k}$$

Determine transformed section properties.

$$b \leqslant \frac{L}{4} = \frac{45\,ft}{4} = 11.25\ \text{ft} = 135\ \text{in.}$$

$$b \leqslant 8.0\ \text{ft} \times 12\ \text{in./ft} = 96\ \text{in.}$$

$$b \leqslant 16\,t_f + b_f = (16 \times 4\ \text{in.}) + 10\ \text{in.} = 74\ \text{in. governs}$$

$$b_{tr} = \frac{b}{n} = \frac{74 \text{ in.}}{9} = 8.22 \text{ in.}$$

$$y_{tr} = \frac{(1.5 \text{ in.} \times 9 \text{ in.}) \times 0.75 \text{ in.} + 27.7 \text{ in.}^2 \times 14.96 \text{ in.} + [(4 \text{ in.} \times 8.22 \text{ in.}) \times 30.42 \text{ in.}]}{(1.5 \text{ in.} \times 9 \text{ in.}) + 27.7 \text{ in.}^2 + (4 \text{ in.} \times 8.22 \text{ in.})}$$

$$= 19.23 \text{ in.}$$

$$I_{tr} = \Sigma(I_0 + Ad^2) = (1.5 \times 9)(19.23 - 0.75)^2 + 3270$$

$$+ 27.7 \left(\frac{26.92}{2} + 1.5 - 19.23\right)^2 + \frac{8.22 \times (4.0)^3}{12} + (4 \times 8.22)(30.42 - 19.23)^2$$

$$= 12,550 \text{ in.}^4$$

$$S_{tr} = \frac{I_{tr}}{y_{tr}} = \frac{12,550 \text{ in.}^4}{19.23 \text{ in.}} = 652 \text{ in.}^3$$

$$M_{\text{all}} = 652 \text{ in.}^3 \times \frac{24 \text{ ksi}}{12 \text{ in./ft}} = 1304 \text{ ft-k} > 1292 \text{ ft-k} \quad \text{ok}$$

$$_{\text{conc}} = \frac{I_{tr}}{y_{\text{top}}} = 951.5 \text{ in.}^4$$

$$f_{bc} = 0.45 f_c' = 1.8 \text{ ksi}$$

$$M_{\text{all}} = n \times S_{\text{conc}} \times f_{bc} = 1285 \text{ ft-k} \approx 1292 \text{ ft-k}$$

$$y_s = \frac{(1.5 \text{ in.} \times 9 \text{ in.}) \times 0.75 \text{ in.} + 27.7 \text{ in.}^2 \times \left(\frac{26.92 \text{ in.}}{2} + 1.5 \text{ in.}\right)}{(1.5 \text{ in.} \times 9 \text{ in.}) + 27.7 \text{ in.}^2} = 10.3 \text{ in.}$$

$$I_s = \Sigma(I_0 + Ad^2) = (1.5 \times 9)(10.3 - 0.75)^2 + 3270$$

$$+ 27.7 \left(\frac{26.92}{2} + 1.5 - 10.3\right)^2$$

$$= 5103 \text{ in.}^4$$

$$S_{s \text{ bot}} = \frac{I_s}{y_s} = \frac{5103 \text{ in.}^4}{10.3 \text{ in.}} = 496 \text{ in.}^3$$

$$\beta = \frac{S_{tr}}{S_s} = \frac{652 \text{ in.}^3}{496 \text{ in.}^3} = 1.31$$

$$V_{h \text{ conc}} = \frac{0.85 f_c' A_c}{2} = \frac{0.85 \times 4.0 \text{ ksi} \times (74 \text{ in.} \times 4 \text{ in.})}{2} = 503.2 \text{ k}$$

$$V_{h \text{ steel}} = \frac{A_s F_y}{2} = \frac{41.2 \text{ in.}^2 \times 36 \text{ ksi}}{2} = 741.6 \text{ k}$$

Governing V_h is due to concrete, $V_h = 503.2$ k.

Shear capacity of $\frac{3}{4}$-in. stud = 12.5 k

Number of studs required between $M_{\max}$ and $M = 0$

$$N_1 = \frac{V_h}{q} = \frac{503.2 \text{ k}}{12.5 \text{ k}} = 40.3$$

Use 41 studs.

Minimum number of studs N_2 required between A and B

$$N_2 = \frac{N_1 (M\beta/M_{\max} - 1)}{\beta - 1} = \frac{40.3 \left(\dfrac{1193 \times 1.31}{1292}\right) - 1}{1.31 - 1} \qquad \text{(AISCS 1.11.4)}$$

$$= 27.3 \quad \text{say 28 studs}$$

Number of studs between A and B = 28

Number of studs between B and C = 41 - 28 = 13

When the required number of shear connectors for full composite action is not used, the AISC calls for an effective section modulus to be used as determined by

$$S_{\text{eff}} = S_s + \sqrt{\frac{V_h'}{V_h}} \, (S_{tr} - S_s) \qquad (9.8)$$

where S_s is the section modulus of the steel alone and V_h' is the shear capacity provided by the connectors used and is obtained by multiplying the number of connectors used and the shear capacity of one connector (AISCS 1.11.1). Conversely, when the required $S_{\text{transformed}}$ is less than the available, transposition of

Eq. 9.8 into

$$V_h' = V_h \left(\frac{S_{reqd} - S_s}{S_{avail} - S_s} \right)^2 \tag{9.9}$$

will give the horizontal force needed to be carried by shear connectors to accommodate the required section modulus. The effective moment of inertia, I_{eff} is determined by AISCS Formula 1.11-6 as

$$I_{eff} = I_s + \sqrt{\frac{V_h'}{V_h}} \, (I_{tr} - I_s) \tag{9.10}$$

where I_s is the moment of inertia of the steel beam, and I_{tr} is the moment of inertia of the transformed composite section. Also V_h' shall not be less than $\frac{1}{4} V_h$ (AISCS 1.11.4).

Example 9.11. Obtaining all the necessary values from the solution for the composite section in Example 9.10, determine S_{eff} if only 28 3-in. channels 3 in. long are used in half the beam.

Solution.

$$S_{eff} = S_s + \sqrt{\frac{V_h'}{V_h}} \, (S_{tr} - S_s)$$

$$S_{tr} = 652 \text{ in.}^3$$

$$S_s = 496 \text{ in.}^3$$

$$V_h = 503.2 \text{ k}$$

Shear capacity of each channel = 4.7 w (AISC, Table 1.11.4).

w = length of channel = 3 in.

4.7 k/in. $\times$ 3 in. = 14.1 k

$$\frac{503.2 \text{ k}}{14.1 \text{ k}} = 35.7 \quad \text{say 36 are needed for full capacity.}$$

Since only 28 channels are used,

$$V'_h = 28 \times 14.1 \text{ k} = 394.8 \text{ k}$$

$$S_{eff} = 496 + \sqrt{\frac{394.8}{503.2}} \ (652 - 496) = 634.2 \text{ in.}^3$$

9.5 FORMED STEEL DECK

When formed steel deck (FSD) is used as permanent formwork for composite construction, AISCS 1.11.5 should be consulted. Certain points to note are that decking rib height shall not exceed 3 in. and that rib width shall not be less than 2 in. Welded stud shear connectors shall be $\frac{3}{4}$ in. or less in diameter and shall extend at least $1\frac{1}{2}$ in. over the top of the steel deck after installation, and the slab thickness above the steel deck shall be not less than 2 in. (see Fig. 9.7).

When the deck ribs are oriented perpendicular to the steel beams or girders, the main requirements of the AISCS are:

1. Concrete below the top of steel decking shall be neglected in computations of section properties.

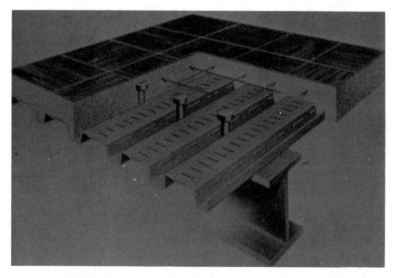

Fig. 9.6.   Cutaway view of a typical composite floor system using formed steel deck (FSD).

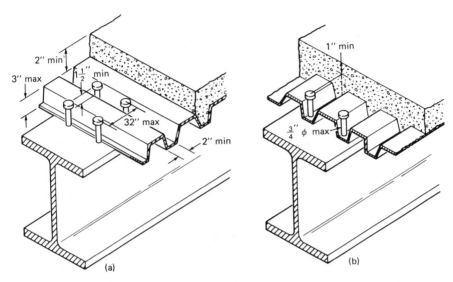

Fig. 9.7. Formed steel deck (FSD) used as permanent formwork for composite construction: (a) deck ribs oriented perpendicular to the steel beam; (b) deck ribs oriented parallel to the steel beam. Dimension and clearance restrictions shown in either (a) or (b) apply to both.

2. The spacing of stud shear connectors shall not exceed 32 in. along the length of the beam.
3. The steel deck shall be anchored to the beam or girder by welding or anchoring devices whose spacing shall not exceed 16 in.
4. The allowable horizontal shear capacity of stud connectors shall be reduced from the values given in AISCS Tables 1.11.4 and 1.11.4A by the reduction factor given in AISC Formula 1.11-8.

For composite sections with the metal deck ribs oriented parallel to the steel beam or girder, reference is made to AISCS 1.11.5.3. The major difference to note between perpendicular and parallel orientation of deck ribs is that, for the latter, the concrete below the top of decking may be included in the calculations for section properties and must be included when calculating A_c for AISC Formula 1.11-3.

9.6 COVER PLATES

When cover plates are used in composite construction, it is not generally required to extend the cover plate to the ends of the beam where the moments are small. To determine the theoretical cutoff point for cover plates, the easiest method is to draw the moment diagram of the span and locate the parts of the beam that will require the extra moment of capacity provided by the cover plate.

It is generally a good practice to use cover plates that are at least 1 in. narrower than the bottom flange of the beam, to leave sufficient space along the edge for welding. The AISC requires that partial-length cover plates be extended beyond the theoretical cutoff point and welded or high-strength-friction-bolted to the beam flange to transfer the flexural stresses that the cover plate would have received to the beam (AISCS 1.10.4) (see Section 3.1). The cover plate force to be developed by the fasteners in the extension is equal to MQ/I, where M is the moment at the cutoff point, Q is the statical moment of the cover plate about the neutral axis of the cover-plated section, and I is the moment of inertia of the cover-plated section (AISCS Commentary 1.10.4). Refer to Chapter 3 of this book.

Example 9.12. Determine the theoretical cutoff point and the connection of the cover plates for loading as shown for the beam in Example 9.10. Use f_c' of 3500 psi, and neglect the weight of the beam.

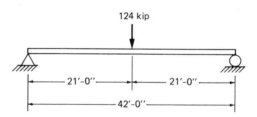

Solution.

$$M_{\max} = \frac{PL}{4} = \frac{124 \text{ k} \times 42 \text{ ft}}{4} = 1302 \text{ ft-k}$$

S_{tr} with cover plates = 652 in.^{3} (see Example 9.10)

S_{tr} without cover plates = 316 in.^{3} (AISCM, Composite Beam Selection Table)

Moment capacity of beam without cover plate

$$M = S_{tr} \times F_b = 316 \text{ in.}^3 \times 24.0 \text{ ksi} = 7584 \text{ in.-k}$$

$$= 632 \text{ ft-k}$$

Moment capacity of beam with cover plate

$$M_{\max} = 652 \text{ ft-k} \times 24 \text{ ksi} = 15650 \text{ in.-k} = 1304 \text{ ft-k}$$

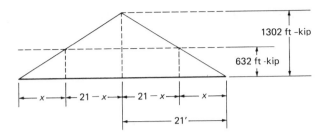

To determine x,

$$\frac{x}{632} = \frac{21}{1302}, \quad x = 10.2 \text{ ft}$$

$$21.0 - x = 21.0 - 10.2 = 10.8 \text{ ft}$$

$1\frac{1}{2}$-in. $\times$ 9-in. cover plate (2×10.8 ft) = 21.6 ft long theoretically required 10.2 ft from either end. See Chapter 3 for development length necessary according to AISC requirements (AISCS 1.10.4).

Extension length

$$L = 1.5 \times 9 \text{ in.} = 13.5 \text{ in.}$$

Shear at cutoff point

$$V = 62.0 \text{ k}$$

$$Q = A_{pl} \times y = 13.5 \text{ in.}^2 \times (19.23 \text{ in.} - 0.75 \text{ in.}) = 249.5 \text{ in.}^3$$

Shear flow

$$q = \frac{VQ}{I} = \frac{62.0 \text{ k} \times 249.5 \text{ in.}^3}{12550 \text{ in.}^4} = 1.23 \text{ k/in.}$$

Using minimum E70 intermittent fillet welds at 12 in. on center,

$$\text{Weld size} = \frac{5}{16} \text{ in.} \quad (\text{AISCS } 1.17.2)$$

$$W_c = 5 \times 0.928 \text{ k/in./sixteenth} = 4.64 \text{ k/in.}$$

$$l_w = \frac{12 \text{ in./ft} \times 1.23 \text{ k/in.}}{4.64 \text{ k/in.} \times 2 \text{ sides}} = 1.64 \text{ in./ft length}$$

Say $l_w = 1\frac{3}{4}$ in. per 1 ft length E70 $\frac{5}{16}$ in. intermittent fillet weld

Termination welds (AISCS 1.10.4.2)

$$H \text{ (force needed)} = \frac{632 \text{ ft-k} \times 12 \text{ in./ft} \times 249.5 \text{ in.}^3}{12550 \text{ in.}^4} = 151 \text{ k}$$

Total weld length

$$l_w = 1.5 \times 2 \times 9 \text{ in.} + 9 \text{ in. (end)} = 36 \text{ in.}$$

Force developed

$$36 \text{ in.} \times 4.64 \text{ k/in.} = 167.0 \text{ k} > 151.0 \text{ k} \quad \text{ok}$$

Example 9.13. Rework Example 9.12 with distributed loading of 5.8 kips per foot.

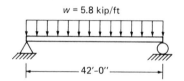

w = 5.8 kip/ft

42'-0"

Solution.

$$M_{\max} = \frac{wL^2}{8} = \frac{5.8 \text{ k/ft} \times (42 \text{ ft})^2}{8}$$

$$= 1279 \text{ ft-k}$$

From Example 9.12 moment capacity of the section without cover plate is 632 ft-k

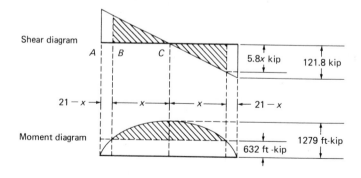

Shear diagram

Moment diagram

From statics, the area under the shear diagram is equal to the change in moment.

Change in moment between B and C = 1279 - 632 = 647 ft-k

Area under shear diagram between B and C = 647 ft-k

Area under shear diagram between B and $C = (x) (5.8 x) \times \frac{1}{2}$

$$= 2.90 x^2$$

$2.90 x^2 = 647$ ft-k

$x^2 = 223.1; \quad x = 14.94$ ft

$21 - x = 6.06$ ft

The theoretical length of cover plate is $(2 \times 15 \text{ ft}) = 30$ ft long, 6 ft from either end.

Shear at termination point

$$V_B = 5.8 \text{ k/ft} \times 14.94 \text{ ft} = 86.65 \text{ k}$$

$$Q = 249.5 \text{ in.}^3 \quad \text{(Example 9.12)}$$

$$q = \frac{VQ}{I} = \frac{86.65 \text{ k} \times 249.5 \text{ in.}^3}{12550 \text{ in.}^4} = 1.72 \text{ k/in.}$$

Use $\frac{5}{16}$-in. E70 intermittent fillet weld at 12 in. on center.

$$l_w = \frac{12 \text{ in./ft} \times 1.72 \text{ k/in.}}{4.64 \text{ k/in.} \times 2} = 2.2 \text{ in.}$$

Use $l_w = 2\frac{1}{2}$ in E70 $\frac{5}{16}$ in. intermittent fillet weld.

Termination welds—same as in Example 9.12.

9.7 DEFLECTION COMPUTATIONS

To simplify deflection computations of simply supported beams under uniformly distributed loading, the following formulas may be used. They are derived from the expression $\Delta = 5 \, wl^4 / 384 \, EI$. For unshored construction

$$\Delta_D = \frac{M_D L^2}{160 \, I_s} \tag{9.11}$$

$$\Delta_L = \frac{M_L L^2}{160 \, I_{tr}} \qquad (9.12)$$

and for shored construction

$$\Delta_D = \frac{M_D L^2}{160 \, I_{tr}} \qquad (9.13)$$

$$\Delta_L = \frac{M_L L^2}{160 \, I_{tr}} \qquad (9.14)$$

where M is the moment in ft-kips, L is the span in feet, and I is the moment of inertia in inch^4, and where subscript D is for dead load, subscript L is for live load, subscript s is for steel alone, and subscript tr is for transformed.

Example 9.14. Show that the deflection formula commonly used for composite construction for uniformly distributed loading is derived from simple beam deflection.

Solution. The equation for deflection of simple supported beams is

$$\Delta = \frac{5 \, wl^4}{384 \, EI}, \quad \text{in inches}$$

When the moment due to uniform loading is introduced,

$$\Delta = \frac{5 \, l^2}{48 \, EI} \times \frac{wl^2}{8} = \frac{5 \, Ml^2}{48 \, EI}$$

Inserting the value for E_{steel} yields

$$\Delta = \frac{5 \, Ml^2}{48 \times 29,000 \times I} = \frac{Ml^2}{278400 \, I}$$

Converting units,

$$\Delta \text{ in.} = \frac{M \text{ ft-k } L \text{ ft}^2}{278400 \text{ ksi } I \text{ in.}^4} \, (12 \text{ in./ft})^3$$

$$\Delta = \frac{ML^2}{161\,I}, \quad \text{approximated as } \frac{ML^2}{160\,I}, \text{ where } \Delta \text{ is in inches, } M \text{ is in}$$

foot-kips, I is in inches^4, and L is in feet.

Example 9.15. For the composite section in Example 9.10, determine dead load deflection, live load deflection, and total deflection when the dead load moment is 748 ft-k and live load moment is 554 ft-k. Assume unshored construction.

Solution. From Example 9.10,

$$I_{tr} = 12,549 \text{ in.}^4$$

$$I_s = 5105 \text{ in.}^4$$

$$\Delta_D = \frac{M_D L^2}{160\,I_s} = \frac{554 \text{ ft-k} \times (42 \text{ ft})^2}{160 \times 5105 \text{ in.}^4} = 1.20 \text{ ft}$$

$$\Delta_L = \frac{M_L L^2}{160\,I_{tr}} = \frac{748 \text{ ft-k} \times (42 \text{ ft})^2}{160 \times 12550 \text{ in.}^4} = 0.66 \text{ in.}$$

$$\Delta_T = \Delta_D + \Delta_L = 1.20 \text{ in.} + 0.66 \text{ in.} = 1.86 \text{ in.}$$

Example 9.16. Determine the dead load deflection, live load deflection, and total deflection of Example 9.15 if shored construction is used.

Solution.

$$\Delta_D = \frac{M_D L^2}{160\,I_{tr}} = \frac{554 \text{ ft-k} \times (42 \text{ ft})^2}{160 \times 12550 \text{ in.}^4} = 0.49 \text{ in.}$$

$$\Delta_L = \frac{M_L L^2}{160\,I_{tr}} = \frac{748 \text{ ft-k} \times (42 \text{ ft})^2}{160 \times 12550 \text{ in.}^4} = 0.66 \text{ in.}$$

$$\Delta_T = \Delta_D + \Delta_L = 0.49 \text{ in.} + 0.66 \text{ in.} = 1.15 \text{ in.}$$

9.8 AISCM COMPOSITE DESIGN TABLES

Because the steel beam in composite construction is continuously laterally supported, $F_b = 0.66\,F_y$ (AISCS 1.5.1.4.1), and therefore $S_{tr} = M/0.66\,F_y$. For the design of a composite section without cover plates, the tables on AISCM, pp. 2-108 and 2-109 are extremely helpful. By preselecting a slab thickness, the lightest steel beam may be obtained from the tables based on the required S_{tr}. The tables are intended to assist the designer in selecting a trial section, after which he must calculate all necessary section properties and complete the design. It should be noted that the tables are for 3000-psi concrete ($n = 9$) with an effective flange width of $16t + b_f$ and no formed steel deck. The tables are arranged in descending order of transformed section modulus (S_{tr}) for the 4-in. slab thickness, and slight inconsistencies may be noted in the descending order for other slab thicknesses. The tables are applicable to all grades of steel.

Example 9.17. Design a composite section for dead load of 90 lb/ft² (including beam weight) and live load of 100 lb/ft² when W 24 beams are placed 10 ft on center and no cover plates are to be used. The beam length is 45 ft, concrete thickness 5 in., $f_c' = 3000$ psi and $n = 9$. Check stresses top and bottom for unshored construction.

Solution. Total load per square foot

$$q_D + q_L = 90 \text{ lb/ft}^2 + 100 \text{ lb/ft}^2 = 190 \text{ lb/ft}^2$$

$$w_T = 190 \text{ lb/ft}^2 \times 10 \text{ ft} = 1.90 \text{ k/ft}$$

$$M_{\max} = \frac{wL^2}{8} = \frac{1.9 \text{ k/ft} \times (45 \text{ ft})^2}{8} = 481 \text{ ft-k}$$

$$S_{tr \, req} = \frac{M}{F_b} = \frac{481 \text{ ft-k} \times 12 \text{ in./ft}}{24 \text{ ksi}} = 240.5 \text{ in.}^3$$

From AISC Composite Design (Manual, Part 2),

$$\text{Use W 24} \times 76, \quad S_{tr} = 248 \text{ in.}^4$$

The values of I_{tr} and $\bar{y}$ are

$$I_{tr} = 5420 \text{ in.}^4$$

$$\bar{y} = 21.90 \text{ in.}$$

Dead load moment

$$M_D = \frac{(90 \text{ lb/ft}^2 \times 10 \text{ ft})(45 \text{ ft})^2}{8} = 228 \text{ ft-k}$$

Because of unshored construction,

$$f_{bs} = \frac{M_D c}{I_s} = \frac{228 \text{ ft-k} \times 12 \text{ in./ft} \times \dfrac{23.92 \text{ in.}}{2}}{2100 \text{ in.}^4} = 15.58 \text{ ksi}$$

15.58 ksi $<$ 24.0 ksi   steel stress ok

Stress on concrete

$$f_{bc} = \frac{M_T c_{\text{top}}}{n I_{tr}} = \frac{481 \text{ ft-k} \times 12 \text{ in./ft} \times (23.92 \text{ in.} + 5 \text{ in.} - 21.9 \text{ in.})}{9 \times 5320 \text{ in.}^4}$$

$$= 0.85 \text{ ksi}$$

$$F_{c \text{ max}} = 0.45 f'_c = 1.35 \text{ ksi}$$

0.85 ksi $<$ 1.35 ksi   concrete stress ok

Check transformed section modulus for unshored construction.

$$M_L = .10 \text{ ksf} \times 10 \text{ ft} \times \frac{(45 \text{ ft})^2}{8} = 253 \text{ ft-k}$$

$$M_D = 228 \text{ ft-k}$$

$$S_{tr} \leqslant \left(1.35 + 0.35 \frac{M_L}{M_D}\right) S_s$$

$$\leqslant \left(1.35 + 0.35 \times \frac{253}{228}\right) \times 176 \text{ in.}^3 = 306 \text{ in.}^3$$

248 in.^{3} $<$ 306 in.^{3} ok

Example 9.18. For the same span, loading, and specifications as in Example 9.17, but for concrete thickness of 4 in., select a composite section without cover plate for

a) the lightest W 18 beam.
b) the lightest composite section.

Solution.

$$M_{\text{max}} = 481 \text{ ft-k}$$

$$S_{tr \text{ req}} = 240.5 \text{ in.}^3$$

a) For $t = 4$ in., select W 18 X 97

$$S_{tr} = 249 \text{ in.}^3$$

b) For $t = 4$ in., select W 24 X 84 or W 27 X 84

$$S_{tr} = 259 \text{ in.}^3 \quad \text{for} \quad \text{W24} \times 84$$
$$S_{tr} = 282 \text{ in.}^3 \quad \text{for} \quad \text{W27} \times 84$$

Example 9.19. For a span of 50 ft with beams spaced at 10 ft on center and a distributed loading of 4.1 k/ft (including beam weight), design a cover-plated W 27 X 94 beam. Determine the required width and length of the cover plate, its extension for development, welding, and the number of $\frac{3}{4}$-in. studs. Use $f'_c = 3000$ psi, concrete slab thickness of $4\text{-}\frac{1}{2}$ in., E70 weld, and 9 in. X $1\text{-}\frac{1}{2}$-in.-thick cover plate.

Solution.

$$M_{\text{max}} = \frac{wL^2}{8} = \frac{4.1 \text{ k/ft} \times (50 \text{ ft})^2}{8} = 1281 \text{ ft-k}$$

$$F_{bs} = 0.66 \times 36 \text{ ksi} = 24.0 \text{ ksi}$$

$$S_{tr \text{ req}} = \frac{M}{F_b} = \frac{1281 \text{ ft-k} \times 12 \text{ in./ft}}{24.0 \text{ ksi}} = 640.5 \text{ in.}^3$$

Width of plate is satisfactory as it allows $\frac{1}{2}$ in. on either side for welding to flange.

$$b = \frac{1}{4} = \frac{50 \text{ ft} \times 12 \text{ in./ft}}{4} = 150 \text{ in.}$$

$$b = 2\,b' + b_f = 2\left(\frac{10\text{ ft} \times 12\text{ in./ft} - 10\text{ in.}}{2}\right) + 10\text{ in.} = 10\text{ ft} \times 12\text{ in./ft} = 120\text{ in.}$$

$$b = 16\,t + b_f = 16 \times 4.50\text{ in.} + 10\text{ in.} = 82\text{ in. governs}$$

$$b_{tr} = b/n = 82\text{ in./}9 = 9.11\text{ in.}$$

$$y_{tr} = \frac{(9\text{ in.} \times 1.5\text{ in.}) \times 0.75\text{ in.} + 27.7\text{ in.}^2 \times \left(\dfrac{26.92}{2} + 1.50\right) + (4.5\text{ in.} \times 9.11\text{ in.}) \times \left(1.50 + 26.92 + \dfrac{2.5}{2}\right)}{(9\text{ in.} \times 1.5\text{ in.}) + 27.7\text{ in.}^2 + (4.5\text{ in.} \times 9.11\text{ in.})}$$

$$= 20.46\text{ in.}$$

$$I_{tr} = (9 \times 1.5)\,(20.46 - 0.75)^2 + 3270 + 27.7\left(20.46 - \frac{26.92}{2} - 1.5\right)^2$$

$$+ \frac{9.11 \times (4.5)^3}{12} + (9.11 \times 4.5)\left(\frac{4.5}{2} + 26.92 + 1.5 - 20.46\right)^2$$

$$= 13{,}700\text{ in.}^4$$

$$S_{tr} = \frac{13700\text{ in.}^4}{20.46\text{ in.}} = 669.5\text{ in.}^3 > 640.5\text{ in.}^3\text{ required}\quad\text{ok}$$

Find theoretical length of cover plate

Transformed section modulus of W 27 × 94 only

$$S_{tr} = 325\text{ in.}^3 \quad \text{(AISC Composite Beam Selection Table)}$$

Moment capacity of section without cover plate

$$M = S_{tr}F_b = \frac{325\text{ in.}^3 \times 24\text{ ksi}}{12\text{ in./ft}} = 650\text{ ft-k}$$

Referring to procedure outlined in Example 9.14,

Change in moment between maximum and theoretical cutoff point

$$1281\text{ ft-k} - 650\text{ ft-k} = 631\text{ ft-k}$$

Area under the shear diagram between the same two points

$$631\text{ ft-k} = (X)\,(4.1\,X) \times \tfrac{1}{2} = 2.05\,X^2$$

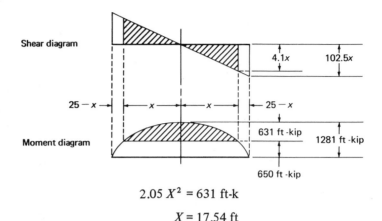

$$2.05\,X^2 = 631 \text{ ft-k}$$

$$X = 17.54 \text{ ft}$$

Theoretical length of cover plate

$$2 \times 17.54 \text{ ft} = 35 \text{ ft-0 in.}$$

Theoretical end point of cover plate

$$\frac{50 \text{ ft} - 35 \text{ ft}}{2} = 7.5 \text{ ft from beam ends}$$

For W 27 × 94,

$$t_f = 0.745 \text{ in.}$$

$$t_{cp} = 1.5 \text{ in.}$$

Minimum size of weld = $\frac{5}{16}$ in.　(AISCS 1.17.2)

Capacity of $\frac{5}{16}$-in. weld for both sides of cover plate

$$2 \times (5 \times 0.928) = 9.28 \text{ k/in.}$$

Shear at cutoff points

$$V = \frac{35.0 \text{ ft} \times 4.1 \text{ k/ft}}{2} = 71.8 \text{ k}$$

Shear flow

$$q = \frac{VQ}{I}$$

$$Q = (1.5 \text{ in.} \times 9 \text{ in.}) \times \left(20.46 \text{ in.} - \frac{1.5 \text{ in.}}{2}\right) = 266.1 \text{ in.}^3$$

$$q = \frac{71.8 \text{ k} \times 266.1 \text{ in.}^3}{13700 \text{ in.}^4} = 1.39 \text{ k/in.}$$

Assume $1\text{-}\frac{1}{2}$-in. weld length will be used (minimum length),

$$\frac{W_{cap}}{q} = \frac{1.5 \text{ in.} \times 9.28 \text{ k/in.}}{1.39 \text{ k/in.}} = 10.0 \text{ in.}$$

Maximum spacing allowed = 12 in. or 24 t (thinner member)

$$24 \times 0.745 \text{ in.} = 17.9 \text{ in.}$$

Use $1\text{-}\frac{1}{2}$-in. long $\frac{5}{16}$ in. weld 10 in. on center.

Moment at cutoff (7.5 ft from end)

$$R = \frac{wL}{2} = \frac{4.1 \text{ k/ft} \times 50 \text{ ft}}{2} = 102.5 \text{ k}$$

$$102.5 \text{ k} \times 7.5 \text{ ft} - \left(7.5 \text{ ft} \times 4.1 \text{ k/ft} \times \frac{7.5 \text{ ft}}{2}\right) = 653 \text{ ft-k}$$

Force to be resisted by end weld

$$\frac{MQ}{I} = \frac{653 \text{ ft-k} \times 12 \text{ in./ft} \times 266.1 \text{ k/in.}}{13,700 \text{ in.}^4} = 152.2 \text{ k}$$

Using AISCS 1.10.4.2, extend plate $13\frac{1}{2}$ in. beyond theoretical cutoff point with continuous fillet weld.

Force available

$$L = 1.5 \times 2 \times 9 \text{ in.} + 9 \text{ in.} = 36 \text{ in.}$$

$$H = 36 \text{ in.} \times 4.64 \text{ k/in.} = 167.0 \text{ k} > 152.2 \text{ k} \qquad \text{ok}$$

Number of studs required for steel for $\frac{1}{2}$ span

$$V_{h \text{ steel}} = \frac{A_s F_y}{2} = \frac{(27.7 + 13.5) \times 36}{2} = 741.6 \text{ k}$$

$$A_c = t \times b = 4.5 \text{ in.} \times 82 \text{ in.} = 369 \text{ in.}^2$$

Number of studs required for concrete

$$V_{h \text{ conc}} = \frac{0.85 \, f_c' A_c}{2} = \frac{0.85 \times 3 \times 369}{2} = 470.5 \text{ k governs}$$

Capacity of $\frac{3}{4}$-in. studs = 11.5 k   (AISC, Table 1.11.4)

$$N = \frac{470.5 \text{ k}}{11.5 \text{ k}} = 40.9 \quad \text{say 41 studs for } \frac{1}{2} \text{ span}$$

Use total of 82 $\frac{3}{4}$-in. ϕ studs.

NOTE: This is the number of studs needed for full moment capacity. As the required moment is less than the available, the number of shear connectors may be reduced by the use of formula 1.11-1 transposed (AISCS 1.11.2.2).

$$V_h' = V_h \left(\frac{S_{\text{eff}} - S_s}{S_{tr} - S_s} \right)^2$$

Calculating S_s (section modulus of steel beam referred to its bottom flange),

$$y_s = \frac{9 \times 1.5 \times 0.75 + 27.7 \times \left(\dfrac{26.92}{2} + 1.50 \right)}{9 \times 1.5 + 27.7} = 10.30 \text{ in.}$$

$$I_s = 9 \times 1.5 \times (10.30 - 0.75)^2 + 3270 + 27.7 \times \left(10.30 - \frac{26.92}{2} - 1.5 \right)^2$$

$$= 5103 \text{ in.}^4$$

$$S_s = \frac{5103 \text{ in.}^4}{10.30 \text{ in.}} = 495.4 \text{ in.}^3$$

$$V_h' = 470.5 \times \left(\frac{640.5 - 495.4}{669.5 - 495.4} \right)^2 = 326.8 \text{ k}$$

$$N_{\text{tot}} = \frac{326.8 \text{ k}}{470.5 \text{ k}} \times 40.9 \times 2 = 56.8 \quad \text{say 57 } \frac{3}{4}\text{-in. } \phi \text{ studs}$$

Check shear

$$V = 102.5 \text{ k}$$

$$f_v = \frac{102.5 \text{ k}}{26.92 \text{ in.} \times 0.490 \text{ in.}} = 7.77 \text{ ksi} < 14.5 \text{ ksi}$$

As the beam is not to be coped, there is no need to verify safety in tearing of web (block shear).

Example 9.20. Design beams A and B for the situation shown. Because of clearance limitations, beam A cannot exceed 17 in. in depth, and beam B cannot exceed 30 in. Beam B will be cover-plated (W 27 $\times$ 102 + 1-$\frac{1}{2}$ in. $\times$ 9 in. cover plate suggested), and beam A will be coped at supports on beam B to have both top flanges at the same level.

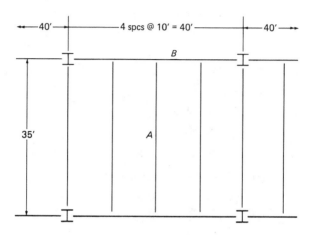

Solution.

Beam A

DL	$\dfrac{4.5 \text{ in.}}{12 \text{ in. /ft}} \times 0.15 \text{ kcf} \times 10 \text{ ft}$	$= 0.563 \text{ k/ft}$
Beam (estimated)		$= \underline{0.060 \text{ k/ft}}$
		$w_D = 0.623 \text{ k/ft}$
Ceiling, flooring, partition	$0.030 \text{ ksf} \times 10 \text{ ft}$	$= 0.30 \text{ k/ft}$
LL	$0.100 \text{ ksf} \times 10 \text{ ft}$	$= \underline{1.00 \text{ k/ft}}$
		$w_L = 1.30 \text{ k/ft}$

$$M_D = \frac{w_D L^2}{8} = \frac{0.623 \text{ k/ft} \times (35 \text{ ft})^2}{8} = 95.4 \text{ ft-k}$$

$$M_L = \frac{w_L L^2}{8} = \frac{1.30 \text{ k/ft} \times (35 \text{ ft})^2}{8} = 199.1 \text{ ft-k}$$

$$\frac{M_L}{M_D} = \frac{1.30}{0.623} = 2.09$$

$$M_{\text{tot}} = M_D + M_L = 294.5 \text{ ft-k}$$

$$S_{\text{req}} = \frac{12 \text{ in./ft} \times M_{\text{tot}}}{F_b} = \frac{12 \text{ in./ft} \times 294.5 \text{ ft-k}}{24 \text{ ksi}} = 147.3 \text{ in.}^3$$

From Composite Beam Selection Table, select W 16 × 67

$$S_{tr} = 162 \text{ in.}^3$$

$$d = 16.33 \text{ in.}$$

$$A = 19.70 \text{ in.}^2$$

$$I_s = 954 \text{ in.}^4$$

$$S_s = 117 \text{ in.}^3$$

$$b \leqslant \frac{35 \text{ ft} \times 12 \text{ in./ft}}{4} = 105 \text{ in.}$$

$$b \leqslant 10 \text{ ft} \times 12 \text{ in./ft} = 120 \text{ in.}$$

$$b \leqslant 4.5 \text{ in.} \times 16 + 10.0 \text{ in.} = 82.0 \text{ in. governs}$$

$$b_{tr} = \frac{82.0 \text{ in.}}{9} = 9.11 \text{ in.}$$

Calculate properties of beam A

$$y_{tr} = \frac{19.70 \times \dfrac{16.33}{2} + 9.11 \times 4.5 \times \left(16.33 + \dfrac{4.5}{2}\right)}{19.70 + 9.11 \times 4.5} = 15.20 \text{ in.}$$

$$I = 954 + 19.70 \times \left(\frac{16.33}{2} - 15.20\right)^2 + 9.11 \times \frac{(4.5)^3}{12}$$

$$+ 9.11 \times 4.5 \times \left(16.33 + \frac{4.5}{2} - 15.20\right)^2 = 2466 \text{ in.}^4$$

$$S_{tr} = \frac{2466 \text{ in.}^4}{15.20 \text{ in.}} = 162.2 \text{ in.}^3 \quad \text{ok}$$

Verify section modulus for no shoring (AISCS 1.11.2.2)

$$S_{tr} < \left(1.35 + 0.35 \frac{M_L}{M_D}\right) S_s$$

$$(1.35 + 0.35 \times 2.09) \times 117 \text{ in.}^4 = 244 \text{ in.}^4 > 162.2 \text{ in.}^4 \quad \text{ok}$$

Dead load stress

$$f_b = \frac{95.4 \text{ ft-k} \times 12 \text{ in./ft}}{117 \text{ in.}^3} = 9.78 \text{ ksi} < 24 \text{ ksi} \quad \text{ok}$$

Dead load plus live load stress

$$f_b = \frac{294.5 \text{ ft-k} \times 12 \text{ in./ft}}{162 \text{ in.}} = 21.81 \text{ ksi} < 24 \text{ ksi} \quad \text{ok}$$

Use $\frac{3}{4}$-in.-diameter shear connectors, $r_v = 11.5$ k (AISCS 1.11.4)

$$V_h \leqslant 0.85 f_c' A_c/2 = \frac{0.85 \times 3.0 \text{ ksi} \times 82.0 \text{ in.} \times 4.5 \text{ in.}}{2} = 470.5 \text{ k}$$

$$V_h \leqslant A_s F_y/2 = \frac{19.70 \text{ in.}^2 \times 36 \text{ ksi}}{2} = 354.5 \text{ k}$$

Use $V_h = 354.5$ k

Considering that $S_{\text{req}} < S_{\text{avail}}$, reduce V_h to V_h'.

$$V_h' = V_h \times \left[\frac{S_{\text{eff}} - S_s}{S_{tr} - S_s}\right]^2 = 354.5 \times \left(\frac{147.3 - 117.0}{162.2 - 117.0}\right)^2 = 159.3 \text{ k}$$

$$N = \frac{159.3 \text{ k}}{11.5 \text{ k}} \times 2 = 27.7$$

Use 28 connectors total

Beam end shear

$$V = 1.923 \text{ k/ft} \times \frac{35 \text{ ft}}{2} = 33.7 \text{ k}$$

$$f_v = \frac{33.7 \text{ k}}{16.33 \text{ in.} \times 0.395 \text{ in.}} = 5.22 \text{ ksi} < 14.5 \text{ ksi} \quad \text{ok}$$

Framed connection

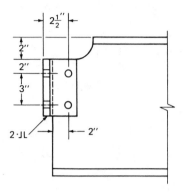

W 16 × 67 beam will be coped 2 in. deep with reaction of 33.7 k. Bolting will be with two $\frac{3}{4}$-in. ϕ A325-N high-strength bolts.

From Table I-D (AISCM, p. 4–5),

Shear capacity = 2 × 18.6 k = 37.2 k > 33.7 k

From Table I-E (AISCM, p. 4-6),

Bearing capacity = 65.3 k × 0.395 × 2 N 51.6 k

$$51.6 \text{ k} > 33.7 \text{ k}$$

Use two angles 4 × 4 × $\frac{5}{16}$ in. × 7 in. long

From Table I-F, for $l_v = l_h = 2''$,

$$58.0 \text{ k} × 0.395 × 2 = 45.8 \text{ k} > 33.7 \text{ k} \quad \text{ok}$$

From Table I-G,

$$(1.60 + 0.33) × 58 × 0.395 = 44.2 \text{ k} > 33.7 \text{ k} \quad \text{ok}$$

Deflection

$$I_{\text{eff}} = I_s + \sqrt{\frac{V_h'}{V_h}} (I_{tr} - I_s)$$

$$= 954 + \sqrt{\frac{159.3}{354.5}} (2466 - 954) = 1968 \text{ in.}^4$$

$$\Delta_D = \frac{95.4 × (35)^2}{160 × 954} = 0.77 \text{ in.}$$

$$\Delta_L = \frac{199.1 \times (35)^2}{160 \times 1968} = 0.77 \text{ in.}$$

$$\frac{l}{360} = \frac{35 \text{ ft} \times 12 \text{ in./ft}}{360} = 1.17$$

Beam B

Assume 150 lb/ft weight.

Allowance for concentrated load

$$\frac{0.15 \text{ k/ft} \times 40 \text{ ft}}{3} = 2 \text{ k}$$

Concentrated DL 0.623 k/ft $\times$ 35 ft/2 $\times$ 2 + 2.0 k = 21.8 k

Concentrated LL 1.30 k/ft $\times$ 35 ft/2 $\times$ 2 $\qquad$ = 45.5 k

Total load $\qquad\qquad\qquad\qquad$ = 67.3 k

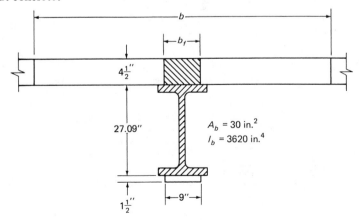

$P_D = 21.8$ k $\qquad R_D = 32.7$ k

$P_L = 45.5$ k $\qquad R_L = 68.3$ k

$M_D = 32.7$ k $\times$ 20 ft $- 21.8$ k $\times$ 10 ft = 436 ft-k

$M_L = 68.3$ k $\times$ 20 ft $- 45.5$ k $\times$ 10 ft = 911 ft-k

Using W 27 $\times$ 102 + $1\frac{1}{2}$-in. $\times$ 9-in. cover plate, calculate properties with and without concrete.

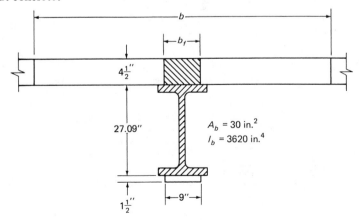

Steel properties

$b = 82$ in., as for beam A

$$y_s = \frac{9 \text{ in.} \times 1.5 \text{ in.} \times 0.75 \text{ in.} + 30 \text{ in.}^2 \times \left(\dfrac{27.09 \text{ in.}}{2} + 1.5 \text{ in.}\right)}{9 \text{ in.} \times 1.5 \text{ in.} + 30 \text{ in.}^2} = 10.61 \text{ in.}$$

$I_s = 13.5 \text{ in.}^2 \times (.75 \text{ in.} - 10.61 \text{ in.})^2 + 30.0$

$$\times \left(\frac{27.09 \text{ in.}}{2} + 1.5 \text{ in.} - 10.61 \text{ in.}\right)^2 + 3620 \text{ in.}^4 = 5522 \text{ in.}^4$$

$$S_{s \text{ top}} = \frac{5522 \text{ in.}^4}{27.09 \text{ in.} + 1.5 \text{ in.} - 10.61 \text{ in.}} = 307 \text{ in.}^3$$

$$S_{s \text{ bot}} = \frac{5522 \text{ in.}^4}{10.61 \text{ in.}} = 520 \text{ in.}^3$$

$$M_{\text{all}} = \frac{307 \text{ in.}^3 \times 24 \text{ ksi}}{12 \text{ in./ft}} = 614 \text{ ft-k} > 436 \text{ ft-k} \quad \text{ok}$$

Transformed section properties

$$b' = \frac{82 \text{ in.}}{9} = 9.11 \text{ in.}$$

$$y_{tr} = \frac{\begin{array}{c}9 \text{ in.} \times 1.5 \text{ in.} \times 0.75 \text{ in.} + 30 \text{ in.}^2 \times (27.09 \text{ in.}/2 + 1.5 \text{ in.}) + \\ 4.5 \text{ in.} \times 9.11 \text{ in.} \times (27.09 \text{ in.} + 1.5 \text{ in.} + 2.25 \text{ in.})\end{array}}{9 \text{ in.} \times 1.5 \text{ in.} + 30 \text{ in.}^2 + 9.11 \text{ in.} \times 4.5 \text{ in.}}$$

$$= 20.42 \text{ in.}$$

$I_{tr} = 13.5 \text{ in.}^2 \times (0.75 \text{ in.} - 20.42 \text{ in.})^2 + 30 \text{ in.}^2$

$$\times \left(\frac{27.09 \text{ in.}}{2} + 1.5 \text{ in.} - 20.42 \text{ in.}\right)^2 + 3620 \text{ in.}^4 + 9.11 \text{ in.} \times \frac{(4.5 \text{ in.})^3}{12 \text{ in./ft}}$$

$$+ 4.5 \text{ in.} \times 9.11 \text{ in.} \times (27.09 \text{ in.} + 1.5 \text{ in.} + 2.25 \text{ in.} - 20.42 \text{ in.})^2$$

$$= 14230 \text{ in.}^4$$

$$S_{tr \text{ top}} = \frac{14230 \text{ in.}^4}{27.09 \text{ in.} + 1.5 \text{ in.} + 4.5 \text{ in.} - 20.42 \text{ in.}} = 1123 \text{ in.}^3$$

$$S_{tr \text{ bot}} = \frac{14230 \text{ in.}^4}{20.42 \text{ in.}} = 697 \text{ in.}^3$$

$$M_{max} = \frac{697 \text{ in.}^3 \times 24 \text{ ksi}}{12 \text{ in./ft}} = 1394 \text{ ft-k} > 1347 \text{ ft-k} \quad \text{ok}$$

$$S_{req} = 1347 \text{ ft-k} \times \frac{12 \text{ in./ft}}{24 \text{ ksi}} = 673.5 \text{ in.}^3$$

$$f_{s \, bot} = \frac{1347 \text{ ft-k} \times 12 \text{ in./ft}}{697 \text{ in.}^3} = 23.19 \text{ ksi}$$

$$f_{c \, top} = \frac{911 \text{ ft-k} \times 12 \text{ in./ft}}{1123 \text{ in.}^3 \times 9} = 1.08 \text{ ksi} < 1.35 \text{ ksi} \quad \text{ok}$$

Verify section modulus for no shoring

$$\left(1.35 + 0.35 \times \frac{911}{436}\right) \times 520 \text{ in.}^3 = 1082 \text{ in.}^3 > 703 \text{ in.}^3 \quad \text{ok}$$

Shear connectors

$$V_h \leqslant (30 \text{ in.}^2 + 13.5 \text{ in.}^2) \times \frac{36 \text{ ksi}}{2} = 783 \text{ k}$$

$$V_h \leqslant 0.85 \times 3.0 \text{ ksi} \times \frac{82 \text{ in.} \times 4.5 \text{ in.}}{2} = 470 \text{ k governs}$$

$$V_h' = 470 \text{ k} \times \left(\frac{673.5 - 520}{697 - 520}\right)^2 = 353.5 \text{ k}$$

$$N = 2 \times \frac{353.5 \text{ k}}{11.5 \text{ k}} = 62 \quad \tfrac{3}{4}\text{-in. studs}$$

Use 31 shear connectors in each half of beam B

Connectors needed between load at quarter points and support

$$N_2 = \frac{N_1\left(\dfrac{M_\beta}{M_{max}} - 1\right)}{\beta - 1} \quad \text{(AISCS 1.11.4)}$$

$$\beta = \frac{673.5 \text{ in.}^3}{520 \text{ in.}^3} = 1.295$$

$$M = (32.7 \text{ k} + 68.3 \text{ k}) \times 10 \text{ ft} = 1010 \text{ ft-k}$$

$$M_{max} = 1347 \text{ ft-k}$$

$$N_2 = \frac{29 \times \left(\frac{1010}{1347} \times 1.295 - 1\right)}{1.295 - 1} < 0 \text{ (negative value)}$$

Studs may be uniformly spaced.

Cover plate cutoff

Calculating transformed properties for W 27 × 102 without cover plate

$$y_{tr} = 22.67 \text{ in.}$$

$$I_{tr} = 8009 \text{ in.}^4$$

$$S_{bot} = \frac{8009 \text{ in.}^4}{22.67 \text{ in.}} = 353 \text{ in.}^3$$

$$S_{top} = \frac{8009 \text{ in.}^4}{27.09 \text{ in.} + 4.5 \text{ in.} - 22.67 \text{ in.}} = 898 \text{ in.}^3$$

Allowable moment without cover plate

$$M_{all} = 353 \text{ in.}^3 \times \frac{24 \text{ ksi}}{12 \text{ in./ft}} = 706 \text{ ft-k}$$

Distance of cutoff point to end beam

$$101 \text{ k} \times X = 706 \text{ ft-k} \qquad X = 6.99 \text{ ft}$$

Welding

$$Q = 13.5 \text{ in.}^2 \times (20.42 \text{ in.} - 0.75 \text{ in.}) = 265.5 \text{ in.}^3$$

Shear flow between concentrated load at quarter point and cutoff

$$V = 101 \text{ k}$$

$$q = \frac{101 \text{ k} \times 265.5 \text{ in.}^3}{14230 \text{ in.}^4} = 1.88 \text{ k/in.}$$

Shear flow between two concentrated loads

$$V = 101 \text{ k} - 67.3 \text{ k} = 33.7 \text{ k}$$

$$q = \frac{33.7 \text{ k} \times 265.5 \text{ in.}^3}{14230 \text{ in.}^4} = 0.63 \text{ k/in.}$$

Use $\frac{5}{16}$-in. E70 intermittent fillet welds

$$W_c = 5 \times 0.928 \text{ k/in.} = 4.64 \text{ k/in.}$$

Minimum length = 1.5 in.

Spacing of welds between concentrated load to cutoff

$$l_w = \frac{4.64 \text{ k/in.} \times 1.5 \text{ in.} \times 2}{1.88 \text{ k/in.}} = 7.40 \text{ in.}$$

Use $\frac{5}{16}$-in. weld $1\text{-}\frac{1}{2}$-in.-long at 7-in. centers

If 12-in. spacing is desired,

$$l_w = \frac{1.87 \text{ k/in.} \times 12 \text{ in./ft}}{4.64 \text{ k/in.} \times 2} = 2.42 \text{ in.}$$

Use $\frac{5}{16}$-in. weld $2\text{-}\frac{1}{2}$-in.-long at 12-in. centers

Spacing of welds between two concentrated loads

$$\frac{4.64 \text{ k/in.} \times 1.5 \text{ in.} \times 2}{0.63 \text{ k/in.}} = 22.09 \text{ in.}$$

Use $\frac{5}{16}$-in. weld $1\text{-}\frac{1}{2}$-in.-long at 12-in. centers

Termination welds (AISCS 1.10.4.2)

$$l_w = 9 \text{ in.} \times 1.5 \text{ in.} \times 2 + 9 \text{ in.} = 36 \text{ in.}$$

Weld capacity

$$W_c = 36 \times 4.64 \text{ k/in.} = 167 \text{ k}$$

Moment at cutoff point

$$M = 706 \text{ ft-k}$$

$$H = \frac{MQ}{I} = \frac{706 \text{ ft-k} \times 12 \text{ in./ft} \times 265.5 \text{ in.}^3}{14230 \text{ in.}^4} = 158 \text{ k} < 167 \text{ k} \quad \text{ok}$$

PROBLEMS TO BE SOLVED

9.1. Determine the effective width b for the interior and edge beams for W 16 × 40 beams with 5-in. concrete slab. Assume the beams are spaced 8 ft, 6 in. on center and span 35 ft.

9.2. Rework Problem 9.1, but for W 18 × 76 beams and 4-in. slab.

9.3. Assuming $E_{steel} = 29 \times 10^6$ psi and $E_{concrete} = 2.9 \times 10^6$ psi, determine the transformed area A_{tr} of the sections described in Problems 9.1 and 9.2.

9.4. For the composite beam shown, determine y_{tr}, I_{tr}, S_{top}, S_{tr}, f_{top}, and f_s when $M = 420$ ft-kips, $b = 82$ in.^{2}, $t = 5$ in., and $n = 10$.

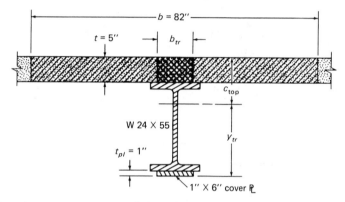

9.5. For the composite beam (interior) described in Problem 9.1, determine y_{tr}, I_{tr}, S_{top}, S_{tr}, f_c, and f_s if $M = 165$ ft-kips and $n = 9$.

9.6. For the composite beam shown in Problem 9.4, determine the dead load and live load deflections if $w_{DL} = 400$ lb/lin. ft, $w_{LL} = 1700$ lb/lin. ft, and $L = 40$ ft.

9.7. Determine the properties and the allowable moment, considering both maximum steel and concrete stresses for a composite section made from a W 18 × 35 beam, $1\frac{1}{2}$-in. × 5-in. cover plate and 5-in. slab. Assume $b = 86$ in., $f_c' = 3000$ psi, and $n = 9$.

9.8. Rework Problem 9.7, but for a W 16 × 40 beam and 1 × 5-in. cover plate of 50-ksi steel.

9.9. Determine the number of $\frac{3}{4}$-in.-diameter studs required for the composite section of Problem 9.4 to develop full composite action. Assume $f_c' = 3000$ psi.

9.10. Rework Problem 9.9, but for C3 × 4.1 channel shear connectors, each 4 in. long. Use minimum number of connectors to develop the moment resistance of 220 ft-kips only.

9.11. Determine the number of $\frac{7}{8}$-in.-diameter stud shear connectors required to develop the full composite capacity of the section described in Problem 9.7.

9.12. Show that a W 24 X 76 beam with 2 X 7-in. cover plate is satisfactory for the given loads. Assume A36 steel, $\frac{5}{16}$-in. E70 welds, 5-in. slab, $f_c' = 3500$ psi, $n = 9$, and $b_{tr} = 7$ in. Also, calculate the theoretical cutoff point of cover plate, connections of cover plate to the flange, termination welds, and the required number of $\frac{3}{4}$-in.-diameter stud shear connectors.

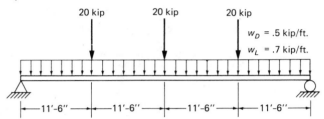

9.13., 9.14., and 9.15. For the following problems, assume A36 steel, E70 welds, $\frac{3}{4}$-in.-diameter stud shear connectors, 5-in. slab, $f_c' = 4000$ psi, $n = 8$, dead load = 75 lb/ft^2 (including beam weight), and live load = 125 lb/ft^2. In addition, assume that spandrel girder 3 supports a facia panel weighing 400 lb/in. ft.

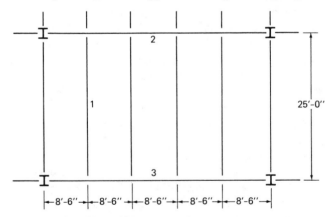

9.13. Design beam 1. Due to clearance limitations, depth cannot exceed 18 in. (including slab). Do not use cover-plated beam.

9.14. Design girder 2. Due to clearance limitations, depth, including slab, cannot exceed 36 in. Use 2-in.-thick cover plate.

9.15. Design girder 3. There are no clearance limitations.

10
Plastic Design of Steel Beams

10.1 PLASTIC DESIGN OF STEEL BEAMS

Plastic design is a method by which structural elements are selected considering the system's overall ultimate capacity. For safety, the applied loads are increased by load factors dictated by appropriate codes. This design is based on the yield property of the steel. Because not all steels have properties suitable for plastic design, only those appropriate are specified in AISC, Section 2.2.

Consider the stress–strain diagram of a steel appropriate for plastic design, (Fig. 10.2). It can be seen that until the steel reaches point 1, corresponding to a stress F_y, stress and strain are proportional, and the material is linearly elastic in the range 0-1. Beyond point 1, the material begins to exhibit some yield, so that if the stress is released at a point between 1 and 3, the material will experience some permanent deformation. For example, if the stress is released at point A, the member will undergo a permanent deformation ϵ_A. We note that between points 2 and 3 the material seems to flow, i.e., to increase its strain without any appreciable increase in stress. The stress corresponding to point 2 is called the *yield stress* and is indicated by F_y. Point 3 is the beginning of the strain-hardening state, where the material needs additional stress to increase its strain. Finally, the material's ultimate capacity is reached at point 4 with the corresponding stress F_u.

In working stress (elastic) design, the beam is considered to fail when the stress in the steel reaches F_y. In view of the safety factor imposed on the stresses, the working stress is approximately one-third below the yield stress. Plastic design is based on the behavior of steel in the range 2-3. In Fig. 10.2b, the stress is idealized, showing one stress range below yield (linearly elastic) and one range at yield. In practice, as a result of the introduction of load factors, the stress F_y is very seldom reached in plastic design; only the additional load-carrying capacity of the structure is considered.

372

Fig. 10.1. Bank of Villa Park, Villa Park, Illinois.

To clarify the behavior of a beam, consider a rectangular member under flexure. The relationship between stress and moment is given by

$$f = \frac{M}{S} \tag{10.1}$$

where $S = bh^2/6$. When the moment increases, the stress f also increases and reaches F_y, the yield stress. We can recover the expression (10.1) by considering the equilibrium of moments in Fig. 10.3c

$$M_y = \frac{bh}{2} \frac{1}{2} \frac{2h}{3} F_y = \frac{bh^2}{6} F_y \tag{10.2}$$

From this point on, fibers stressed at F_y cannot increase their stress and hence begin to yield. If the moment M is increased, fibers closer to the neutral axis start to yield, and a stress distribution, illustrated in Fig. 10.3d, occurs. When the stress distribution in Fig. 10.3e is reached, the member will have exhausted its carrying capacity, and all the fibers will have yielded, i.e., have plastified. The moment corresponding to this stage is called the *plastic moment* and is represented by M_p.

Obtaining the relationship between the plastic moment and the yield stress, we have

$$M_p = b \frac{h}{2} \frac{h}{2} F_y = \frac{bh^2}{4} F_y \tag{10.3}$$

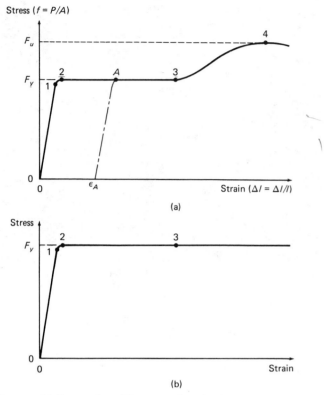

Fig. 10.2. Stress-strain diagram for mild structural steel: (a) normal diagram; (b) idealized diagram for plastic design.

The quantity $bh^2/4$ is the plastic section modulus and is represented by Z. The ratio $M_p/M_y = Z/S$ is called the *shape factor*. This shape factor, which is 1.5 for a rectangle, varies considerably with the shape of the section. Common values for wide flanges range from 1.08 to 1.20.

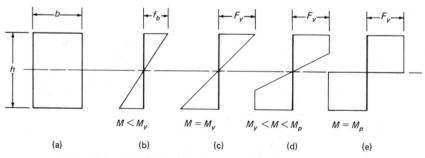

Fig. 10.3. Plastic hinge development in a rectangular section.

The location where there is no stress, and consequently no strain, is the *plastic neutral axis*. In plastic design, for tension to equal compression, the tension area must be equal to the compression area, as both the tensile and compressive stresses are equal to F_y. Therefore, the plastic neutral axis must divide the cross section into two parts, both having equal areas. In a section that is symmetrical with respect to the horizontal axis, the elastic neutral axis (center of gravity) and the plastic neutral axis coincide.

Example 10.1. Calculate the values of S, Z, and the shape factor about the X axis for the figure shown.

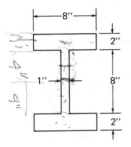

Solution. By inspection, the neutral axis (NA) is at midheight, 6 in. from the bottom.

$$I = \Sigma \left(\frac{bh^3}{12} + Ad^2 \right) = 2 \times \left[\frac{8 \text{ in. } (2 \text{ in.})^3}{12} + (2 \text{ in.} \times 8 \text{ in.}) (5 \text{ in.})^2 \right]$$

$$+ \frac{1 \text{ in. } (8 \text{ in.})^3}{12}$$

$$I = 853 \text{ in.}^4$$

$$S = \frac{I}{c} = \frac{853 \text{ in.}^4}{6 \text{ in.}} = 142.2 \text{ in.}^3$$

The plastic section modulus can be determined by summing the area of the stress blocks times their distance to the neutral axis.

$$Z = \Sigma Ad = 2 \times [(2 \text{ in.} \times 8 \text{ in.}) \times 5 \text{ in.} + (1 \text{ in.} \times 4 \text{ in.}) \times 2 \text{ in.}] = 176 \text{ in.}^3$$

Shape factor

$$\frac{Z}{S} = \frac{176 \text{ in.}^3}{142.2 \text{ in.}^3} = 1.24$$

Example 10.2. Calculate the values of S, Z, and the shape factor about the X axis for the figure shown.

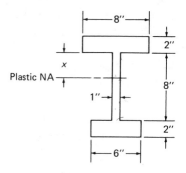

Solution. The elastic properties are:

$$\bar{y} = \frac{(6 \text{ in.} \times 2 \text{ in.}) \times 1 \text{ in.} + (1 \text{ in.} \times 8 \text{ in.}) \times 6 \text{ in.} + (8 \text{ in.} \times 2 \text{ in.}) \times 11 \text{ in.}}{(6 \text{ in.} \times 2 \text{ in.}) + (8 \text{ in.} \times 1 \text{ in.}) + (8 \text{ in.} \times 2 \text{ in.})}$$

$$= 6.56 \text{ in.}$$

$$I = \frac{8 \text{ in.} (2 \text{ in.})^3}{12} + 16 \text{ in.}^2 (11 \text{ in.} - 6.56 \text{ in.})^2 + \frac{1 \text{ in.} (8 \text{ in.})^3}{12}$$

$$+ 8 \text{ in.}^2 (6 \text{ in.} - 6.56 \text{ in.})^2 + \frac{6 \text{ in.} (2 \text{ in.})^3}{12} + 12 \text{ in.}^2 (1 \text{ in.} - 6.56 \text{ in.})^2$$

$$= 740.9 \text{ in.}^4$$

$$S = \frac{I}{c} = \frac{740.9 \text{ in.}^4}{6.56 \text{ in.}} = 112.9 \text{ in.}^3$$

To locate the plastic NA, find the distance X, such that the area above the neutral axis is equal to the area below.

$$(8 \text{ in.} \times 2 \text{ in.}) + (1 \text{ in.} \times (X)) = (1 \text{ in.} \times (8 \text{ in.} - X) + (6 \text{ in.} \times 2 \text{ in.})$$

$$X = 2 \text{ in.}$$

$$Z = (8 \text{ in.} \times 2 \text{ in.}) \times 3 \text{ in.} + (2 \text{ in.} \times 1 \text{ in.}) \times 1 \text{ in.} + (1 \text{ in.} \times 6 \text{ in.})$$

$$X 3 \text{ in.} + (6 \text{ in.} \times 2 \text{ in.}) \times 7 \text{ in.} = 152 \text{ in.}^3$$

$$\text{Shape factor} = \frac{Z}{S} = \frac{152 \text{ in.}^3}{112.9 \text{ in.}^3} = 1.35$$

Examples 10.3 and 10.4. Find the values of S, Z, and the shape factor for the shapes shown.

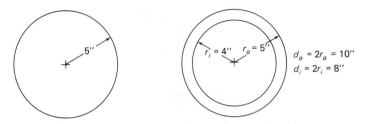

Solution. For a solid circular section, the elastic and plastic section moduli are found to be

$$S = \frac{\pi r^3}{4} = \frac{\pi d^3}{32} \qquad \text{(AISCM p. 6-32 Properties of Geometric Sections)}$$

$$Z = \frac{d^3}{6}$$

$$S = \frac{\pi (10 \text{ in.})^3}{32} = 98.2 \text{ in.}^3$$

$$Z = \frac{(10 \text{ in.})^3}{6} = 166.7 \text{ in.}^3$$

$$\text{Shape factor } \frac{Z}{S} = \frac{166.7 \text{ in.}^3}{98.2 \text{ in.}^3} = 1.70$$

For a hollow circular section, the elastic and plastic section moduli are found to be

$$S = \frac{\pi (d_o^4 - d_i^4)}{32 d_o} \qquad \text{(AISCM p. 6-32)}$$

$$Z = \frac{d_o^3 - d_i^3}{6}$$

$$S = \frac{\pi \times [(10 \text{ in.})^4 - (8 \text{ in.})^4]}{32 \times 10 \text{ in.}} = 57.96 \text{ in.}^3$$

$$Z = \frac{(10 \text{ in.})^3 - (8 \text{ in.})^3}{6} = 81.33 \text{ in.}^3$$

$$\text{Shape factor } \frac{Z}{S} = \frac{81.33 \text{ in.}^3}{57.96 \text{ in.}^3} = 1.40$$

Example 10.5. Calculate the S, Z, and shape factor for a W 24 X 104, and compare with values given in the AISC tables.

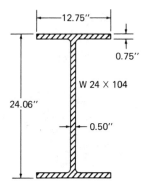

Solution.

$d = 24.06$ in.; $t_w = 0.50$ in.

$b_f = 12.75$ in.; $t_f = 0.75$ in.

$$I = \frac{0.5 \text{ in.} \times (24.06 \text{ in.} - 2(0.75 \text{ in.}))^3}{12} + 2 \times \left(\frac{12.75 \text{ in.} \times (0.75 \text{ in.})^3}{12}\right.$$

$$+ (0.75 \text{ in.} \times 12.75 \text{ in.}) \times \left(\frac{24.06 \text{ in.}}{2} - 0.375 \text{ in.}\right)^2\right) = 3077 \text{ in.}^4$$

$$S = \frac{I}{c} = \frac{3077 \text{ in.}^4}{12.03 \text{ in.}} = 255.8 \text{ in.}^3$$

$$Z = 2 \times \left((0.75 \text{ in.} \times 12.75 \text{ in.}) \times \frac{24.06 \text{ in.} - 0.75 \text{ in.}}{2} + \frac{24.06 \text{ in.} - 1.5 \text{ in.}}{2}\right.$$

$$+ \frac{24.06 \text{ in.} - 1.5 \text{ in.}}{2} \times 0.50 \text{ in.} \times \frac{24.06 \text{ in.} - 1.5 \text{ in.}}{4}\right) = 286.5 \text{ in.}^3$$

The AISCM lists a value of 258 in.^{3} for the member section modulus, and 289 in.^{3} for the member plastic modulus.[1]

$$\text{Shape factor } \frac{Z}{S} = \frac{286.5 \text{ in.}^3}{255.8 \text{ in.}^3} = 1.12 \quad \text{(shape factor same by AISC)}$$

10.2 PLASTIC HINGES

When a flexural member is subjected to loading, so as to cause certain parts of the member to acquire its plastic moment, the member will begin to yield at

---

[1] The differences are due to the fillets at the web/flange corners.

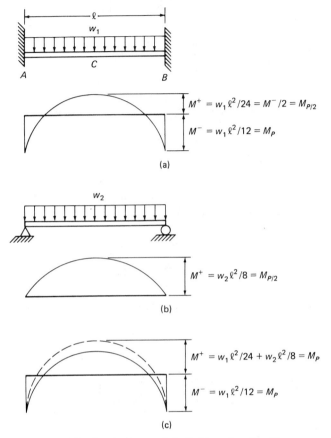

$$M^+ = w_1 \ell^2/24 = M^-/2 = M_{P/2}$$

$$M^- = w_1 \ell^2/12 = M_P$$

(a)

$$M^+ = w_2 \ell^2/8 = M_{P/2}$$

(b)

$$M^+ = w_1 \ell^2/24 + w_2 \ell^2/8 = M_P$$

$$M^- = w_1 \ell^2/12 = M_P$$

(c)

Fig. 10.4. Development of plastic hinges in a fixed beam.

those points and transfer any moment caused by further loading, which could have been resisted at that location had it not yielded, to other parts of the beam still below the plastic moment. Where the moment reaches the plastic moment, the steel section yields and the structure, for any additional load, behaves as if a hinge were located there.

Consider a beam with fixed supports uniformly loaded with a load w (Fig. 10.4a). If the load w is increased until the plastic moment is reached at the supports, plastic hinges will form at A and B. The load corresponding to this stage is $w_1 = 12 M_p/L^2$. If the load is further increased, no more moment will be carried at points A and B because they have reached the plastic hinge conditions at M_p. For increased loads, the beam will behave as a simply supported beam (Fig. 10.4b). The collapse will occur when the third hinge forms at the center of the beam when the moment M_p is attained there. The total load applied to the beam is then $w = w_1 + w_2$. At collapse, the moment M at the center is

$$\frac{w_1 L^2}{24} + \frac{w_2 L^2}{8} = M_p \qquad (10.4)$$

which yields

$$w_2 = \frac{4 M_p}{L^2} \qquad (10.5)$$

The total capacity of the beam at collapse is $12 M_p/L^2 + 4 M_p/L^2 = 16 M_p/L^2$ instead of only $12 M_p/L^2$ for the elastic approach. This behavior is due to redistribution of moments.[2] (Here safety and load factors have not been considered.)

Example 10.6. Calculate P, considering elastic and plastic design approaches. Omit load and safety factors.

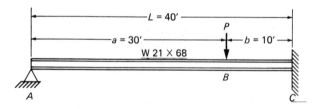

Solution. Elastic Design Approach

Formulas are from AISCM beam diagram and Formulas No. 14

$$M_C = \frac{Pab}{2L^2} (a + L)$$

$$a = 30 \text{ ft}; \quad b = 10 \text{ ft}; \quad L = 40 \text{ ft}$$

$$M_C = \frac{30 \times 10}{2 (40)^2} (30 + 40) P = 6.56 P$$

$$M_B = R_A a$$

---

[2] This fact is recognized by the AISC code in the working stress design approach in 1.5.1.4.1.

$$R_A = \frac{Pb^2}{2L^3}(a + 2L) = 0.086 \, P$$

$$M_B = 0.086 \, P \times 30 = 2.58 \, P$$

$$F_y = \frac{M_{max}}{S}; \quad M_{max} = F_y S$$

$$F_y = 36.0 \, \text{ksi}$$

$$S = 140 \, \text{in.}^3$$

$$M_{max} = 6.56 \, P = \frac{36.0 \, \text{ksi} \times 140 \, \text{in.}^3}{12 \, \text{in./ft}} = 420 \, \text{ft-k}$$

$$P = 64.0 \, \text{k}$$

Elastic moment diagram

Plastic Design Approach

$$F_y = \frac{M_p}{Z}; \quad M_p = Z F_y$$

$$F_y = 36 \, \text{ksi}$$

$$Z = 160 \, \text{in.}^3$$

$$M_p = 160 \, \text{in.}^3 \times \frac{36 \, \text{ksi}}{12 \, \text{in./ft}} = 480 \, \text{ft-k}$$

The load P_1 is the load that will cause a plastic hinge at C.

$$M_p = 6.56 \, P_1$$

$$P_1 = \frac{480}{6.56} = 73.17 \, \text{k}$$

To plastify B (i.e., to obtain the third hinge, the second being already at A), we need an additional moment at B of

$$M = 480 - (2.58 \times 73.17) = 291.22 \, \text{ft-k}$$

This moment will be obtained by an additional load P_2 acting on a simply supported beam. As A is a normal hinge and C a plastic one,

$$P_2 \times \frac{10 \times 30}{40} = 291.22$$

$$P_2 = 38.83 \text{ k}$$

$$P = P_1 + P_2 = 73.17 \text{ k} + 38.83 \text{ k} = 112.0 \text{ k}$$

480 ft-kip

188.78 ft-kip

480 ft-kip

Plastic moment diagram

10.3 BEAM ANALYSIS BY VIRTUAL WORK

In mechanics, energy is defined as the capacity to do work. Work is the product of a force and the distance traveled by the force in its own direction or the product of a moment and the angle of its rotation. From conservation of energy, internal work U_i is equal to the external work U_o. By introducing a small fictitious (virtual) displacement in a structure, and thereby causing the forces and moments to do work, the plastic moment in the structure may be determined.

In plastic design, because hinges form at the locations of plastic moments, the internal virtual work is the sum of the products of the rotations of the member at the hinge locations and the plastic moment of the member.

External virtual work of a flexural member is the summation of the externally applied loads and the respective deformations of the member at the locations of the loads.

In the process of analysis by virtual work for a one-span beam, hinges occur at the supports and under the concentrated load. In case of more than one concentrated load, the location of the hinge in the span cannot be readily determined.

The location of the hinge in the span is assumed, the plastic moment M_p is calculated, and the corresponding moment diagram is drawn. If at no point there is a moment larger in absolute value than M_p, then the assumption is correct, and M_p is the plastic moment. If at some point the moment is larger than

M_p, then the hinge is erroneously located and has to be moved to that point. It can be proved that the plastic moment corresponding to the proper hinge location is the maximum M_p.

A quick way to determine the plastic moment of a beam section involves the superposition of moment diagrams due to hinged supports and fixed supports. The three necessary plastic hinges are formed at the two ends and the point of maximum positive bending (Fig. 10.5). Using these concepts, the plastic moment can be expressed as the moment diagram of a statically determinate span superimposed on the moment diagram due to given end conditions. Examples 10.7a through 10.10a have been reworked to exemplify this method.

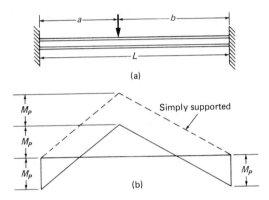

Fig. 10.5. a) Load diagram of fixed end span with single concentrated force. b) Moment diagram of simply supported beam, fixed end beam, and superimposed diagram.

Example 10.7. Calculate the plastic moment M_p for the beam shown.

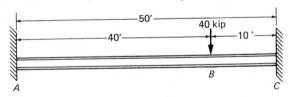

Solution. We assume that plastic hinges form at points $A, B,$ and C.

Consider a virtual displacement given to the system, so that AB rotates by θ. The virtual displacement at B becomes $40\,\theta$ ft, and the rotation at C, $40\,\theta/10 = 4\,\theta$.

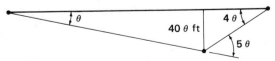

The angle at B is the sum of the angles at A and C, which equals 5θ. The virtual work of the external force is 40θ ft $\times$ 40 k = 1600θ ft-k The virtual work of the internal effects is

$$(M_p \times \theta) + (M_p \times 5\theta) + (M_p \times 4\theta) = 10\theta M_p$$

Internal work is equal to the external work

$$10 M_p \theta = 1600\theta \text{ (ft-k)}$$

$$M_p = 160 \text{ ft-k}$$

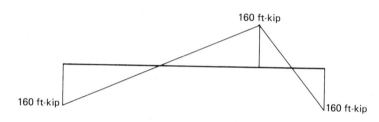

Example 10.7a.

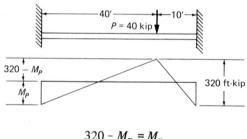

$$320 - M_p = M_p$$

$$M_p = 160 \text{ ft-k}$$

Example 10.8. Calculate the plastic moment for the uniformly loaded beam shown, using the principle of virtual work.

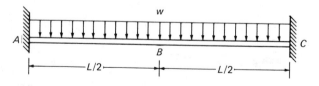

Solution. Hinges will form at A, B, and C.

Virtual work of internal effects

$$(M_p \times \theta) + (M_p \times 2\theta) + (M_p \times \theta) = 4 M_p \theta$$

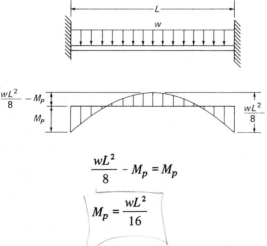

Virtual work of external forces

We can consider that the virtual work in member AB is equal to the load in that member times the displacement of its center. Hence, it is equal to

$$\frac{wL}{2} \times \frac{\theta L}{2} \times \frac{1}{2} \times 2 \text{ sections}$$

$$4 M_p \theta = \frac{wL^2 \theta}{4}$$

$$M_p = \frac{wL^2}{16}$$

Example 10.8a.

$$\frac{wL^2}{8} - M_p = M_p$$

$$M_p = \frac{wL^2}{16}$$

Example 10.9. Determine the plastic moment for a beam that has loads applied as shown.

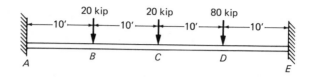

Solution. Assume plastic hinges at A, C, and E.

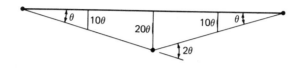

$$M_p \times 4\,\theta = (20 \times 10\,\theta) + (20 \times 20\,\theta) + (80 \times 10\,\theta)$$

$$M_p = \frac{(200 + 400 + 800)\,\theta}{4\,\theta} = 350 \text{ ft-kip}$$

$$R_A = \frac{20\text{ k} \times 30\text{ ft}}{40\text{ ft}} + \frac{20\text{ k} \times 20\text{ ft}}{40\text{ ft}} + \frac{80\text{ k} \times 10\text{ ft}}{40\text{ ft}} = 45\text{ k}$$

$$R_B = 120\text{ k} - 45\text{ k} = 75\text{ k}$$

$$M_A = -350 \text{ ft-k}$$

$$M_B = -350 \text{ ft-k} + (45\text{ k} \times 10\text{ ft}) = 100 \text{ ft-k}$$

$$M_C = -350 \text{ ft-k} + (45\text{ k} \times 20\text{ ft}) - (20\text{ k} \times 10\text{ ft}) = 350 \text{ ft-k}$$

$$M_D = (75\text{ k} \times 10\text{ ft}) - 350 \text{ ft-k} = 400 \text{ ft-k} > M_p = 350 \text{ ft-k}$$

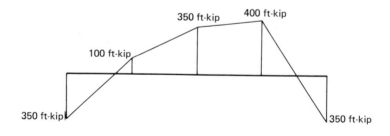

We see that the hinge selection is wrong and that the hinge located at C must be moved to D because $M_D > M_p$.

Selecting the new location for hinges A, D, E,

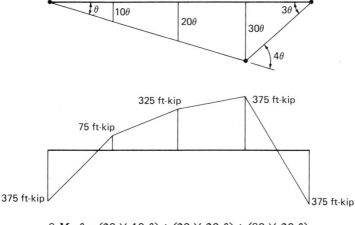

$$8 M_p \theta = (20 \times 10 \theta) + (20 \times 20 \theta) + (80 \times 30 \theta)$$

$$M_p = \frac{(200 + 400 + 2400) \theta}{8 \theta} = 375 \text{ ft-k}$$

Recalculating,

$$R_A = 45 \text{ k} \quad R_B = 75 \text{ k}$$

$$M_A = -375 \text{ ft-k}$$

$$M_B = -375 \text{ ft-k} + (45 \text{ k} \times 10 \text{ ft}) = 75 \text{ ft-k}$$

$$M_C = -375 \text{ ft-k} + (45 \text{ k} \times 20 \text{ ft}) - (20 \text{ k} \times 10 \text{ ft}) = 325 \text{ ft-k}$$

$$M_D = (75 \text{ k} \times 10 \text{ ft}) - 375 \text{ ft-k} = 375 \text{ ft-k}$$

We see then that the location selection is correct, for no other moment is larger than M_p.

Example 10.9a.

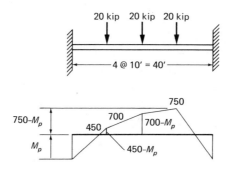

$$M_p = 750 - M_p$$

$$M_p = 375 \text{ ft-k}$$

Example 10.10. Find the plastic moment for the given situation.

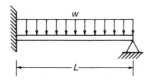

Solution. Since the hinge location is not known, the location must be found by the method of sections and assuming the location of zero shear and maximum moment equal to M_p.

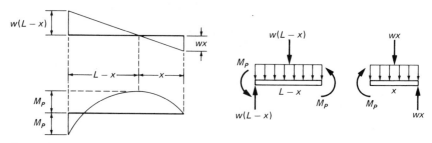

From the right section,

$$M_p + (wx)\left(\frac{x}{2}\right) - (wx)(x) = 0$$

$$M_p = \frac{wx^2}{2}$$

From the left section,

$$-2 M_p + w(L - x)(L - x) - w(L - x)\left(\frac{L - x}{2}\right) = 0$$

$$M_p = \frac{w}{4}(L^2 - (2 Lx) + x^2)$$

Equating the two equations,

$$\frac{wx^2}{2} = \frac{wL^2}{4} - \frac{wLx}{2} + \frac{wx^2}{4}$$

$$wx^2 + (2 wLx) - wL^2 = 0$$

$$x = 0.414\,L$$

$$L - x = 0.586\,L$$

$$M_p = \frac{wx^2}{2} = \frac{w(0.414\,L)^2}{2} = \frac{wL^2}{11.66}$$

$$M_p = \frac{wL^2}{11.66}$$

Example 10.10a.

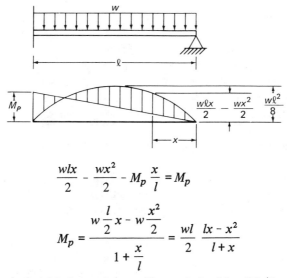

$$\frac{wlx}{2} - \frac{wx^2}{2} - M_p \frac{x}{l} = M_p$$

$$M_p = \frac{w\dfrac{l}{2}x - w\dfrac{x^2}{2}}{1 + \dfrac{x}{l}} = \frac{wl}{2}\,\frac{lx - x^2}{l + x}$$

x must be such that M_p is maximum. To maximize M_p, dM_p/dx must be equal to zero

$$(l - 2x)(l + x) - (lx - x^2) = 0$$

$$l^2 - 2lx + lx - 2x^2 - lx + x^2 = 0$$

$$x^2 + 2lx - l^2 = 0 \qquad x = -l \pm \sqrt{l^2 + l^2}$$

$$x = 0.414\,l$$

$$M_p = \frac{wl^2}{2}\,\frac{0.414 - 0.171}{1.414} = 0.0858\,wl^2 = \frac{wl^2}{11.67}$$

10.4 PLASTIC DESIGN OF BEAMS

Plastic analysis enables the designer to calculate the maximum load that the structure can support just before it collapses; thus, the working loads must be multiplied by load factors to compensate for uncertainties in the loading and to provide the structure safety against collapse. The AISC calls for a load factor of 1.70, for dead and live loads only, or a load factor of 1.30, for dead, live, and wind or earthquake loads combined (AISCS 2.1). It may be noted that the load factor 1.70 is approximately equal to the average shape factor 1.12 (for sections having a compression and tension flange) times the safety factor of 1.50 for a simply supported beam, and the load factor 1.30 is approximately equal to the load factor 1.70 reduced by 25%, which is equivalent to the 25% reduction in moment allowed in elastic design when wind or earthquake forces are taken into account.

To determine the required plastic moment capacity of any span of a structure:

1. Multiply loads by the appropriate load factors.
2. Assume the locations of the plastic hinges.
3. Introducing a virtual displacement in the structure, determine the rotations of the member at all plastic hinge locations and the displacements under all applied concentrated loads or at the interior hinge location for distributed loads.
4. Equating the internal virtual work to the external, calculate the required plastic moment.
5. Determine proper locations of hinges, and determine M_p.
6. Once the largest required plastic moment is obtained, determine Z required and select a satisfactory member from the AISCM.

Up to now, only single-span beams were designed. The failure of a continuous beam is essentially the failure of one of its spans. Hence, in the case of a continuous beam, each of the spans is analyzed individually, and the largest moment M_p obtained becomes the design moment of the entire beam.

Example 10.11. Select the lightest beams according to elastic and plastic theories. Assume full lateral support and that the loads include the weight of the beam. Consider also all appropriate safety and load factors.

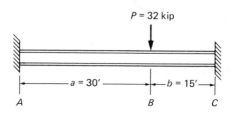

Solution. Elastic Approach

$$M_A = \frac{Pab^2}{L^2} = -106.7 \text{ ft-k}$$

$$M_B = \frac{2\,Pa^2\,b^2}{L^3} = 142.2 \text{ ft-k}$$

$$M_C = \frac{Pa^2\,b}{L^2} = -213.3 \text{ ft-k}$$

The moments can be redistributed in accordance with AISCS 1.5.1.4.

$$M_A = 0.9 \times -106.7 \text{ ft-k} = -96.0 \text{ ft-k}$$

$$M_C = 0.9 \times -213.3 \text{ ft-k} = -192.0 \text{ ft-k}$$

$$M_B = 142.2 \text{ ft-k} + \left(\frac{10.7 + 21.3}{2}\right) \text{ ft-k} = 158.2 \text{ ft-k}$$

$$S_{req} = \frac{M_{max}}{F_b} = \frac{192.0 \text{ ft-k} \times 12 \text{ in./ft}}{0.66 \times 36.0 \text{ ksi}} = 97.0 \text{ in.}^3$$

Use W 18 × 55 $S_x = 98.3 \text{ in.}^3$

Plastic Approach

$$\bar{P} = 1.7\,P = 1.7 \times 32.0 \text{ k} = 54.4 \text{ k}$$

Assume plastic hinges at A, B, and C

$$6\,\phi\,M_p = 54.4 \text{ k} \times 30\,\phi$$

$$M_p = \frac{1632\,\phi}{6\,\phi} = 272 \text{ ft-k}$$

$$Z = M_p/F_y = \frac{272 \text{ ft-k} \times 12 \text{ in./ft}}{36.0 \text{ ksi}} = 90.7 \text{ in.}^3$$

The section must be chosen from the plastic design selection table.

Use W 21 × 44 $Z_x = 95.4 \text{ in.}^3$

Example 10.12. Select the lightest beams according to elastic and plastic theories. Assume full lateral support and that the loads include the weight of the beam.

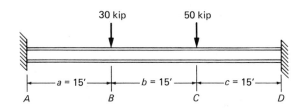

Solution. Elastic Approach

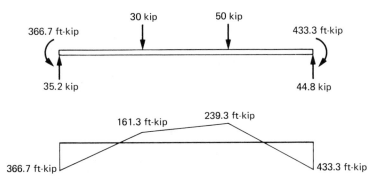

Redistributing moments in accordance with AISCS 1.5.1.4,

$$M_A = 0.9 \times -366.7 \text{ ft-k} = 330.0 \text{ ft-k}$$

$$M_D = 0.9 \times -433.3 \text{ ft-k} = 390.0 \text{ ft-k}$$

$$M_C = 239.3 \text{ ft-k} + \left(\frac{36.7 + 43.3}{2}\right) = 279.3 \text{ ft-k}$$

$$S_{\text{req}} = \frac{M_{\text{max}}}{F_b} = \frac{390.0 \text{ ft-k} \times 12 \text{ in./ft}}{0.66 \times 36.0 \text{ ksi}} = 195.0 \text{ in.}^3$$

Use W 24 × 84 $S_x = 196.0$ in.^{3}

Plastic Approach

$$\bar{P}_1 = 1.7 \times 30.0 \text{ k} = 51.0 \text{ k}$$

$$\bar{P}_2 = 1.7 \times 50.0 \text{ k} = 85.0 \text{ k}$$

Assume hinges at A, B, and D

$$6 \, \phi M_p = (51.0 \text{ k} \times 30 \, \phi) + (85.0 \text{ k} \times 15 \, \phi)$$

$$M_p = 467.5 \text{ ft-k}$$

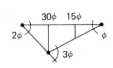

Assume hinges at A, C, and D

$$6 \phi M_p = (51.0 \text{ k} \times 15 \phi) + (85.0 \text{ k} \times 30 \phi)$$

$$M_p = 552.5 \text{ ft-k governs}$$

$$Z = M_p/F_y = \frac{552.5 \text{ ft-k} \times 12 \text{ in./ft}}{36.0 \text{ ksi}} = 184.0 \text{ in.}^3$$

Use W 24 × 76 $Z = 200 \text{ in.}^3$

Example 10.13. Select beams according to elastic and plastic theories. Assume full lateral support and that the loads include the weight of the beam.

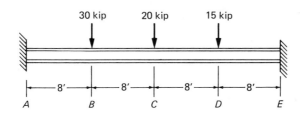

Solution. Elastic Approach

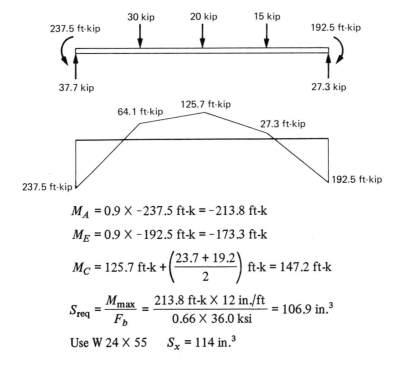

$$M_A = 0.9 \times -237.5 \text{ ft-k} = -213.8 \text{ ft-k}$$

$$M_E = 0.9 \times -192.5 \text{ ft-k} = -173.3 \text{ ft-k}$$

$$M_C = 125.7 \text{ ft-k} + \left(\frac{23.7 + 19.2}{2}\right) \text{ ft-k} = 147.2 \text{ ft-k}$$

$$S_{req} = \frac{M_{max}}{F_b} = \frac{213.8 \text{ ft-k} \times 12 \text{ in./ft}}{0.66 \times 36.0 \text{ ksi}} = 106.9 \text{ in.}^3$$

Use W 24 × 55 $S_x = 114 \text{ in.}^3$

Plastic Approach

$\bar{P}_1 = 1.7\,P = 1.7 \times 30\ \text{k} = 51\ \text{k}$

$\bar{P}_2 = 1.7 \times 20\ \text{k} = 34\ \text{k}$

$\bar{P}_3 = 1.7 \times 15\ \text{k} = 25.5\ \text{k}$

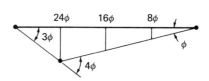

Assume hinges at A, B, and E

$6\,\phi\,M_p = (51\ \text{k} \times 24\,\phi) + (34\ \text{k} \times 16\,\phi) + (25.5 \times 8\,\phi)$

$M_p = 246.5\ \text{ft-k}$

Assume hinges at A, C, and E

$4\,\phi\,M_p = (51\ \text{k} \times 8\,\phi) + (34\ \text{k} \times 16\,\phi) + (25.5 \times 8\,\phi)$

$M_p = 289\ \text{ft-k}$

Assume hinges at A, D, and E

$8\,\phi\,M_p = (51 \times 8\,\phi) + (34 \times 16\,\phi) + (25.5 \times 24\,\phi)$

$M_p = 195.5\ \text{ft-k}$

Hinges at A, C, and E govern

$M_p = 289\ \text{ft-k}$

$Z = M_p/F_y = 289\ \text{ft-k} \times 12\ \text{in./ft}\ 36\ \text{ksi} = 96.3\ \text{in.}^3$

Use W 21×50 $Z = 110\ \text{in.}^3$
or W 18×50 ($Z = 101.0\ \text{in.}^3$) which has same weight but less depth.

Example 10.14. For the situation shown, design the lightest beam, using the plastic approach.

Solution.

$\bar{P}_1 = 1.7\,P = 25 \times 1.7 = 42.5\ \text{k}$

$\bar{P}_2 = 1.7 \times 15 = 25.5\ \text{k}$

Assume plastic hinges at A and B, with an actual hinge at D.

Since plastic hinges are only at A and B,

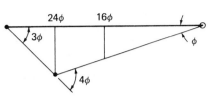

$$3\,\phi M_p + 4\,\phi M_p = (42.5 \text{ k} \times 24\,\phi)$$
$$+ (25.5 \text{ k} \times 16\,\phi)$$

$M_p = 204$ ft-k

Assume plastic hinge at A and C, with an actual hinge at D.

Since plastic hinges are only at A and C,

$$1\,\phi M_p + 2\,\phi M_p = (42.5 \text{ k} \times 8\,\phi)$$
$$+ (25.5 \text{ k} \times 16\,\phi)$$

$M_p = 249$ ft-k   governs

$Z = M_p/F_y = (249 \text{ ft-k} \times 12 \text{ in./ft})/36 \text{ ksi} = 83 \text{ in.}^3$

Use W 21 $\times$ 44 $Z = 95.4$ in.^{3}

Example 10.15. Select, by plastic design, a continuous rolled section that is satisfactory for the entire structure.

Solution.

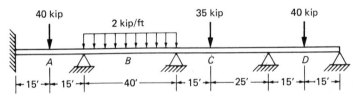

Because plastic hinges will form at the supports where negative moment is greatest, each span will be analyzed separately and the critical moment determined.

Span A $4\,\phi M_p = 40 \text{ k} (1.7)\, 15\,\phi$

$M_p = 255$ ft-k

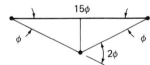

Span B $M_p = w(L)^2/16 = \dfrac{2\,(1.7)\,(40)^2}{16}$

$M_p = 340$ ft-k   governs

Span C $(80/15)\,\phi M_p = 35 \text{ k} (1.7)\, 25\,\phi$

$M_p = 279$ ft-k

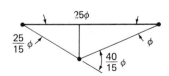

Span D $3 \phi M_p = 40$ k (1.7) 15ϕ

$M_p = 340$ ft-k   governs

$Z = 340$ ft-k $\times$ 12 in./ft/36 ksi = 113 in.^{3}

Use W 24 $\times$ 55 $Z = 134$ in.^{3}

Example 10.16. Select, by plastic design, a continuous rolled section that is satisfactory for the entire structure.

Solution.

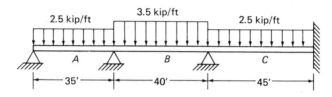

Span A $M_p = w(L)^2/11.66 = \dfrac{2.5\,(1.7)\,(35)^2}{11.66}$

$M_p = 446.5$ ft-k

Span B $M_p = \dfrac{w(L)^2}{16} = \dfrac{3.5\,(1.7)\,(40)^2}{16}$

$M_p = 595.0$ ft-k   governs

Span C $M_p = \dfrac{w(L)^2}{16} = \dfrac{2.5\,(1.7)\,(45)^2}{16}$

$M_p = 537.9$ ft-k

Maximum $M_p = 595$ ft-k

$Z = 595$ ft-k $\times$ 12 in./ft/36 ksi = 198 in.^{3}

Use W 24 $\times$ 76 $Z = 200$ in.^{3}

10.5 COVER PLATE DESIGN

It is generally uneconomical to use the same rolled section for all spans in a continuous multispan structure; the rolled section must be satisfactory for the span with the largest plastic moment, and, therefore, it will quite possibly be overdesigned for some if not all of the other spans. Usually in such cases, the rolled section is selected to satisfy the moment requirement of the span with the least moment, and cover plates are used at all other locations reducing them. Initially, the shear and moment diagrams of the structure are drawn, and the lengths and locations of the cover plates are determined for all portions of the bam exceeding in plastic moment requirement from that provided by the rolled section. Assuming a width b for the cover plate, the thickness t is then calculated.

A cost analysis has to be performed to determine if a single rolled section covering all spans is cheaper than the lighter section with cover plates (including the cost of welds).

Example 10.17. Using the plastic design method,

a) Design a single rolled section satisfactory for the entire structure.
b) Select a section for the span with the least moment, and design cover plates as necessary for the other spans.

Assume full lateral bracing.

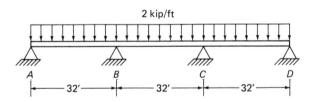

Solution. The plastic moments of span AB and CD are equal.

$$(M_p)_{AB} = (M_p)_{CD} = \frac{wL^2}{11.66} = \frac{(2 \times 1.7)(32)^2}{11.66}$$

$$(M_p)_{AB} = (M_p)_{CD} = 298.6 \text{ ft-k}$$

$$(M_p)_{BC} = \frac{wL^2}{16} = \frac{(2 \times 1.7)(32)^2}{16}$$

$$(M_p)_{BC} = 217.6 \text{ ft-k}$$

a) Governing $(M_p)_{AB \& CD} = 298.6$ ft-k

$$Z = M_p/F_y = \frac{298.6 \text{ ft-k} \times 12 \text{ in./ft}}{36.0 \text{ ksi}} = 99.5 \text{ in.}^3$$

Use W 18 × 50 $Z_x = 101$ in.^{3}

b) Least $(M_p)_{BC} = 217.6$ ft-k

$$Z = M_p/F_y = \frac{217.6 \text{ ft-k} \times 12 \text{ in./ft}}{36.0 \text{ ksi}} = 72.53 \text{ in.}^3$$

Use W 16 × 40 $Z_x = 72.9$ in.^{3}

Section will carry plastic moment of span BC only. Cover plates need to be designed for plastic moments in other spans.

The plastic moment capacity of a W 16 × 40 beam is

$$M_p = \frac{Z_x F_y}{12 \text{ in./ft}} = \frac{72.9 \text{ in.}^3 \times 36 \text{ ksi}}{12 \text{ in./ft}} = 218.7 \text{ ft-k}$$

If W 16 × 40 is used for the entire structure, a plastic hinge of 218.7 ft-k moment will form at B, C, and at the midspan BC.

Summing moments about point A, the free body diagram of span AB becomes

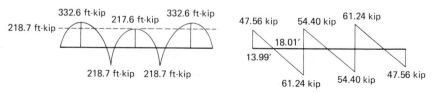

$$M = \left(32 \times 2 \times 1.7 \times \frac{32}{2}\right) + 219 - (R_B \times 32) = 0$$

$$R_B = 61.24 \text{ k}$$

$$R_A = 47.56 \text{ k}$$

$$M_{AB} = \frac{(47.56)^2}{2 \times 2 \times 1.7} = 332.6 \text{ ft-k}$$

$$M_{CD} = M_{AB} = 332.6 \text{ ft-k}$$

Moment and shear diagrams for the entire structure are

Cover plates must compensate for the deficient moment capacity of the W 16 X 40, as shown by the crosshatched area.

Since the change in moment equals the area under the shear diagram, R and S are found by letting $R = S$,

$$R \times \left(\frac{R}{13.99} \times 47.56 \right) \times \frac{1}{2} = 113.9 \text{ ft-k}$$

$R = S = 8.19$ ft

$Q = 13.99 - 8.19 = 5.80$ ft

$T = 18.01 - 8.19 = 9.82$ ft

Theoretical length of cover plates

$$8.19 \text{ ft} \times 2 = 16.38 \text{ ft}$$

To design cover-plates for the beam, assume the width of the cover plate to be 4 in. to leave space along the sides for welding.

$$d = 16.01 \text{ in.}; \quad b_f = 6.995 \text{ in.}$$

Summing area moments about the neutral axis for the cover plates and equating them to the required moment capacity of the cover plates, we get

$$2 \times \left(\frac{16.01}{2} + \frac{t}{2} \right) \times 4 \times t = \frac{113.9 \times 12}{36}$$

$$t^2 + 16.01 \, t - 9.49 = 0$$

$t = 0.572$ in.   Use $\frac{5}{8}$ in. $\times$ 4 in. cover plates

Connection of cover plates to flanges (AISCS 2.8 paragraph 4 & 1.10.4).

Shear force at the theoretical end

$$8.19 \text{ ft} \times 2 \text{ k/ft} \times 1.7 = 27.85 \text{ k}$$

Shear flow

$$q = \frac{VQ}{I}$$

$$Q = \left(\frac{16.01}{2} + \frac{1}{2} \times \frac{5}{8}\right) \times 4 \times \frac{5}{8} = 20.79 \text{ in.}^3$$

$$I = 518 + 2 \times \left(\frac{16.01}{2} + \frac{1}{2} \times \frac{5}{8}\right)^2 \times 4 \times \frac{5}{8} = 863.9 \text{ in.}^4$$

$$q = \frac{27.85 \text{ k} \times 20.79 \text{ in.}^3}{863.9 \text{ in.}^4} = 0.67 \text{ k/in.}$$

Minimum size of E70 weld that can be used is $\frac{1}{4}$ in. (AISCS 1.17.2)

Capacity of weld (AISC, Table 1.5.3)

0.30 × nominal tensile strength on throat =

$$1.7 \times 0.30 \times 70 \times \frac{\sqrt{2}}{2} = 25.24 \text{ k/in. long/in. leg}$$

0.40 × yield stress of base metal =

$$1.7 \times 0.4 \times 36 = 24.48 \text{ k/in. long/in. leg} \qquad \text{governs}$$

For $\frac{1}{4}$-in. weld, two-sided,

$$W_c = 2 \times 0.25 \times 24.48 = 12.24 \text{ k/in.}$$

If intermittent fillet weld is to be used,

$$l_{\min} = 1.5 \text{ in. corresponds to } 1.5 \times 12.24 = 18.36 \text{ k/weld}$$

Spacing between welds

$$s = \frac{18.36 \text{ k/weld}}{0.67} = 27.40 \text{ in.}$$

Use at 12 in. on center.

Termination of welds (AISCS 1.10.4.3)

Force to be developed is MQ/I

$$\frac{218.7 \times 12 \times 20.79}{863.9} = 63.16 \text{ k}$$

Length of termination fillets is 8 in.

$$W_c \times l = 12.24 \text{ k/in.} \times 8 \text{ in.} = 97.92 \text{ k} > 63.16 \text{ k} \qquad \text{ok}$$

$$\text{width/thickness ratio} = \frac{6.995}{2 \times 0.505} = 6.93 < 8.5 \qquad \text{ok} \qquad \text{(AISCS 2.7)}$$

$$\text{depth/thickness ratio} = \frac{16.01}{.305} = 52.49 < \frac{412}{\sqrt{F_y}} = 68.67 \qquad \text{ok} \qquad \text{(AISCS 2.7)}$$

Example 10.18. Using the plastic design method,

a) Design a single rolled section satisfactory for the entire structure.
b) Select a section for the span with the least moment, and design cover plates as necessary for the other spans.

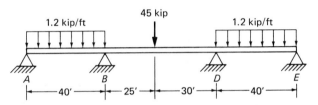

Solution. The plastic moments of spans AB and DE are equal.

$$(M_p)_{AB} = (M_p)_{DE} = \frac{wL^2}{11.66} = \frac{(1.2 \times 1.7)\,(40)^2}{11.66} = 279.9 \text{ ft-k}$$

For M_p of span BD,

$$4.4\,\phi\,M_p = 45 \text{ k} \times 1.7 \times 30\,\phi$$

$$M_p = 521.6 \text{ ft-k for span } BD$$

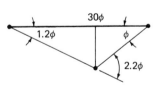

a) governing $(M_p)_{BD} = 521.6$ ft-k

$$Z = M_p/F_y = \frac{521.6 \text{ ft-k} \times 12 \text{ in./ft}}{36.0 \text{ ksi}} = 173.9 \text{ in.}^3$$

Use W 24 × 68

b) least $(M_p)_{AB\,\&\,DE} = 279.9$ ft-k

$$Z = M_p/F_y = \frac{279.9 \text{ ft-k} \times 12 \text{ in./ft}}{36.0 \text{ ksi}} = 93.3 \text{ in.}^3$$

Use W 21 X 44

$$M_p \text{ of W 21 X 44 beam is 286 ft-k} \quad (Z = 95.4 \text{ in.}^3)$$

The free body diagram of span BD gives

$\Sigma M_D = -45 \times 1.7 \times 30 + 55 R_B = 0$

$R_B = 41.7 \text{ k}$

$R_D = 34.8 \text{ k}$

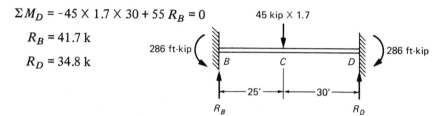

Moment and shear diagrams for the entire structure are

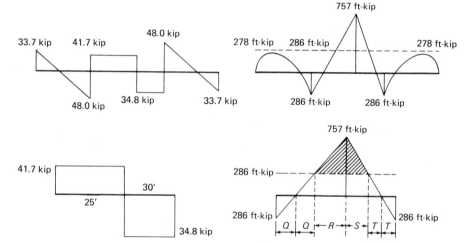

Cover plates must compensate for the deficient moment capcity of the W 21 X 44, as shown by the crosshatched area.

Since the change in moment equals the area under the shear diagram,

$Q \times 41.7 = 286 \text{ ft-k}$

$Q = 6.85 \text{ ft}$

$R = 25 - 2(6.85 \text{ ft}) = 11.28 \text{ ft}$

$T = 286/34.8 = 8.22 \text{ ft}$

$S = 30 \text{ ft} - 2(8.22 \text{ ft}) = 13.56 \text{ ft}$

Theoretical length of cover plate

$$R + S = 11.28 + 13.56 = 24.84 \text{ ft}$$

To design cover plates for the beam, assume the width of the cover plate to be 4 in. to leave space along the sides for welding.

$$d = 20.66 \text{ in.}, \qquad b_f = 6.50 \text{ in.}$$

$$2 \times \left(\frac{20.66}{2} + \frac{t}{2}\right) \times (4\ t) \times F_y = M_{req} \times 12 \text{ in./ft}$$

$$(20.66 + t) \times (4\ t) \times 36 = (757 - 286) \times 12$$

$$t^2 + 20.66\ t - 39.25 = 0$$

$$t = 1.75 \text{ in.} \qquad 1\frac{3}{4} \text{ in.}$$

Checking the section,

$$Z = 95.4 \text{ in.}^3 + 4 \text{ in.} \times 1.75 \text{ in.} \times 2 \times \left(\frac{20.66}{2} + \frac{1.75}{2}\right) = 252.3 \text{ in.}^3$$

$$M_p = F_y \times Z = \frac{36 \text{ ksi} \times 252.3 \text{ in.}^3}{12 \text{ in./ft}} = 757 \text{ ft-k} \qquad \text{ok}$$

Connection of plate to flange (AISCS 2.8)

Shear force

$$V_{max} = 41.7 \text{ k}$$

$$I = 843 + 4 \times 1.75 \times \left(\frac{20.66}{2} + \frac{1.75}{2}\right)^2 \times 2 = 2601 \text{ in.}^4$$

$$Q = 4 \times 1.75 \times \left(\frac{20.66}{2} + \frac{1.75}{2}\right) = 78.44 \text{ in.}^3$$

$$q = \frac{VQ}{I} = \frac{41.7 \text{ k} \times 78.44 \text{ in.}^3}{2601 \text{ in.}^4} = 1.26 \text{ k/in.}$$

Design weld using $\frac{5}{16}$-in. E70 weld.

Capacity of weld

$$W_c = 1.7 \times 0.30 \times 70 \text{ ksi} \times \frac{\sqrt{2}}{2} \times \frac{5}{16} \text{ in.} = 7.89 \text{ k/in.}$$

$$W_c = 1.7 \times 0.40 \times 36 \text{ ksi} \times \frac{5}{16} \text{ in.} = 7.65 \text{ k/in.} \quad \text{governs}$$

Capacity of weld for two sides = 2×7.65 k/in. = 15.30 k/in.

Use intermittent fillet welds spaced at 12 in. on center.

$$q = 1.26 \text{ k/in.} \times 12 \text{ in./ft} = 15.12 \text{ k/ft}$$

$$l_{req} = \frac{15.12 \text{ k/ft}}{15.30 \text{ k/in.}} \cong 1 \text{ in. weld per foot of plate}$$

Use minimum weld length of $1\frac{1}{2}$ in.

Termination of welds

$$M = 286 \text{ ft-k}$$

$$H = \frac{MQ}{I} = \frac{286 \text{ ft-k} \times 12 \text{ in./ft} \times 78.44 \text{ in.}^3}{2601 \text{ in.}^4} = 103.5 \text{ k}$$

Try length of 8 in.

$$W_{c \text{ term}} = 8 \text{ in.} \times 15.30 = 122.4 \text{ k} > 103.5 \text{ k} \quad \text{ok}$$

Check width/thickness and depth/thickness ratios (AISCS 2.7).

$$\frac{6.50 \text{ in.}}{2 \times 0.45 \text{ in.}} = 7.22 < 8.5 \qquad \text{ok}$$

$$\frac{20.66 \text{ in.}}{0.35 \text{ in.}} = 59.0 < \frac{412}{\sqrt{F_y}} = 68.67 \qquad \text{ok}$$

10.6 ADDITIONAL CONSIDERATIONS

SHEAR

The AISC Code requires, in Section 2.5, that the ultimate shear be governed by

$$V_u \leqslant 0.55 \, F_y \, td \qquad (10.6)$$

unless the web is reinforced. Nevertheless, the design of additional reinforcement for shear is quite involved, and it is suggested that the design of rolled sections meet the requirements of Eq. 10.6.

MINIMUM THICKNESS RATIO

Similar to the requirements for width-thickness ratios in Part 1 of the AISCS, some rules have been established in Section 2.7 for beams. For most commonly used A36 steel, the ratio

$$\frac{b_f}{2 \, t_f} \leqslant 8.5 \qquad (10.7)$$

Also, the depth-thickness ratio of beam webs shall not exceed

$$\frac{d}{t} = \frac{412}{F_y} \qquad (10.8)$$

CONNECTIONS

In AISCM, Section 2.8, some rules for connections are established. The most important rule is that the connection elements shall resist the factored loads with a capacity equal to 1.7 times their respective stresses, given in Part 1.

PROBLEMS TO BE SOLVED

10.1 through 10.5. Calculate the values S, Z, and shape factor for the following shapes with respect to a centroidal horizontal axis.

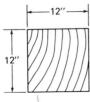
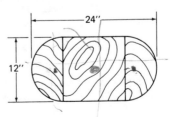

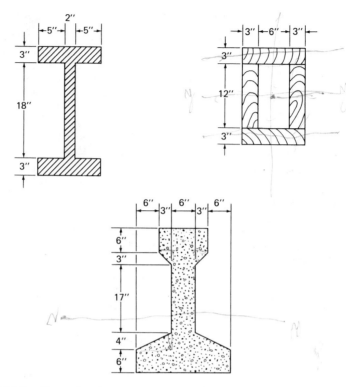

10.6 and 10.7. Draw the elastic and plastic moment diagrams for the following beams loaded as shown. Disregard load factors.

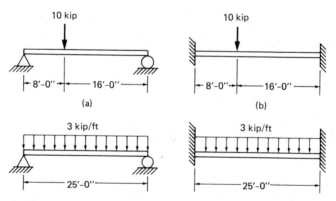

10.8. Calculate the plastic moment M_p for the W 18 × 60 beam loaded as shown assuming (a) simple supports at both ends, (b) the left support hinged and the other fixed, and (c) both ends fixed.

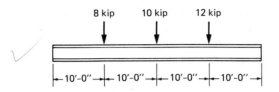

10.9. Calculate the plastic moment for the beam shown.

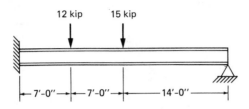

10.10. Calculate the plastic moment for the beam shown, using the principles of virtual work.

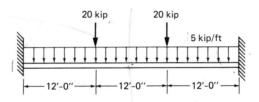

10.11 and 10.12. For the loadings shown, select the lightest beams according to elastic and plastic theories. Assume full lateral support, and disregard the weight of the beam.

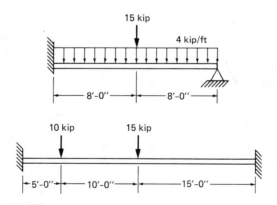

10.13 through 10.15. Select a continuous wide-flange section, using plastic design that is satisfactory for the entire structure.

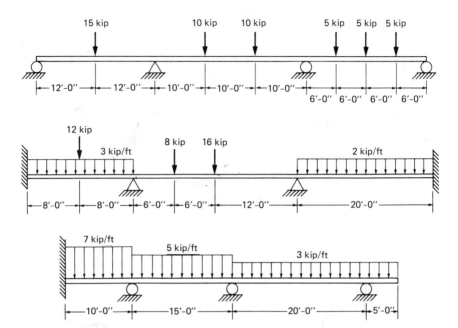

10.16 and 10.17. For the loadings shown determine:

a) A single rolled section satisfactory for this beam using elastic analysis and the 0.9 rule

b) A single rolled section satisfactory for this beam using plastic analysis, and

c) A section for the span with the least moment, and design cover plates as necessary.

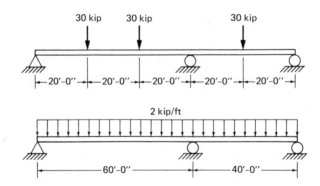

Index